T0224476

Communications
in Computer and Information Science　　1150

Commenced Publication in 2007
Founding and Former Series Editors:
Phoebe Chen, Alfredo Cuzzocrea, Xiaoyong Du, Orhun Kara, Ting Liu,
Krishna M. Sivalingam, Dominik Ślęzak, Takashi Washio, Xiaokang Yang,
and Junsong Yuan

More information about this series at http://www.springer.com/series/7899

Hien Nguyen (Ed.)

Statistics and Data Science

Research School on Statistics and Data Science, RSSDS 2019
Melbourne, VIC, Australia, July 24–26, 2019
Proceedings

 Springer

Editor
Hien Nguyen 🆔
La Trobe University
Bundoora, VIC, Australia

ISSN 1865-0929   ISSN 1865-0937 (electronic)
Communications in Computer and Information Science
ISBN 978-981-15-1959-8   ISBN 978-981-15-1960-4 (eBook)
https://doi.org/10.1007/978-981-15-1960-4

This Springer imprint is published by the registered company Springer Nature Singapore Pte Ltd.
The registered company address is: 152 Beach Road, #21-01/04 Gateway East, Singapore 189721, Singapore

Preface

Welcome to the proceedings of the RSSDS: Research School in Statistics and Data Science, held during July 26–28, 2019, in Melbourne, Australia. This was the third edition in a series of workshops; the first and second of which were called the S4D: International Research Summer School in Statistics and Data Science, held in 2017 and 2018, at the University of Caen, in Normandy, France.

The workshop was organized as a collaboration between the French and Australian statistical research communities, and was sponsored by AFRAN, ANR, Inria, La Trobe University, the region of Normandy, and the University of Caen. The workshop brought together academics, researchers, and industry practitioners of statistics and data science, to discuss numerous advances in the disciplines and their impact on the sciences and society. Attendees and presenters at the workshop covered numerous topics, including data analysis, data science, data mining, data visualization, bioinformatics, machine learning, neural networks, statistics, and probability.

This year, RSSDS received 23 submissions. After a thorough peer-review process, 11 English papers were selected for inclusion in these proceedings, which implies an acceptance rate of 47.83%. These 11 papers were presented at the workshop via poster presentations. In addition to these 11 papers, 7 invited English papers were solicited from the invited speakers of the workshop. These 7 invited papers were subjected only to a technical editing process. In total, the proceedings contain 18 high-quality English papers on various topics from data science to statistics.

The high-quality program would not have been possible without the authors who chose RSSDS 2019 as their preferred venue for their publications. Furthermore, the workshop would not have been successful if not for the work of the Program Committee members and Organizing Committee members, who put a tremendous amount of effort into organizing the event, and soliciting and reviewing the research papers that make up the program.

We hope that you enjoy reading and benefit from the proceedings of RSSDS 2019.

October 2019   Hien Nguyen

Organization

RSSDS 2019 was organized via a collaborative effort between staff and students from La Trobe University, Melbourne, Australia, along with fellow Australian academics, and French and international collaborators.

Organizing Committees

Workshop Co-chairs

Hien Nguyen	La Trobe University, Australia
Natalie Karavarsamis	La Trobe University, Australia

Program Co-chairs

Faicel Chamroukhi	University of Caen, France
Florence Forbes	Inria Grenoble Rhône-Alpes, France

Program Committee

Daniel Ahfock	The University of Queensland, Australia
Julyan Arbel	Inria Grenoble Rhône-Alpes, France
Daniel Fryer	The University of Queensland, Australia
Charles Gray	La Trobe University, Australia
Paul Kabaila	La Trobe University, Australia
Sharon Lee	University of Adelaide, Australia
Steffen Maeland	University of Bergen, Norway
Geoffrey McLachlan	The University of Queensland, Australia
Shu-Kay Ng	Griffith University, Australia
Tin Nguyen	University of Caen, France
Andriy Olenko	La Trobe University, Australia
Kok-Leong Ong	La Trobe University, Australia
Luke Prendergast	La Trobe University, Australia
Kai Qin	Swinburne University, Australia
Inga Strumke	University of Oslo, Norway
Emi Tanaka	The University of Sydney, Australia
Dianhui Wang	La Trobe University, Australia
Benjamin Wong	Monash University, Australia

Organizers

Organized and hosted by

La Trobe University, Melbourne, Australia
In Cooperation with

The University of Caen, Inria, and Springer

Sponsors

Sponsored by

AFRAN, ANR, and the region of Normandy

Contents

Invited Papers

Symbolic Formulae for Linear Mixed Models

Emi Tanaka[1,2(✉)] and Francis K. C. Hui[3]

[1] The University of Sydney, Camperdown, NSW 2008, Australia
[2] Monash University, Clayton, VIC 3800, Australia
emi.tanaka@monash.edu
[3] Australian National University, Acton, ACT 2601, Australia
francis.hui@anu.edu.au

Abstract. A statistical model is a mathematical representation of an often simplified or idealised data-generating process. In this paper, we focus on a particular type of statistical model, called linear mixed models (LMMs), that is widely used in many disciplines e.g. agriculture, ecology, econometrics, psychology. Mixed models, also commonly known as multi-level, nested, hierarchical or panel data models, incorporate a combination of fixed and random effects, with LMMs being a special case. The inclusion of random effects in particular gives LMMs considerable flexibility in accounting for many types of complex correlated structures often found in data. This flexibility, however, has given rise to a number of ways by which an end-user can specify the precise form of the LMM that they wish to fit in statistical software. In this paper, we review the software design for specification of the LMM (and its special case, the linear model), focusing in particular on the use of high-level symbolic model formulae and two popular but contrasting R-packages in lme4 and asreml.

Keywords: Multi-level model · Hierarchical model · Model specification · Model formulae · Model API · Fixed effects · Random effects

1 Introduction

Statistical models are mathematical formulation of often simplified real world phenomena, the use of which is ubiquitous in many data analyses. These models are fitted or trained computationally, often with practitioners using some readily available application software package. In practice, statistical models in its mathematical (or descriptive) representation would require translation to the right input argument to fit using an application software. The design of these input arguments (called application programming interface, API) can help ease

---

Supported by R Consortium.

H. Nguyen (Ed.): RSSDS 2019, CCIS 1150, pp. 3–21, 2019.
https://doi.org/10.1007/978-981-15-1960-4_1

the friction in fitting the user's desired model and allow focus on important tasks, e.g. interpreting or using the fitted model for purposes downstream.

While there are an abundance of application software for fitting a variety of statistical models, the API is often inconsistent and restrictive in some fashion. For example, in linear models, the intercept may or may not be included by default; and the random error typically assumed to be identical and independently distributed (i.i.d) with no option to modify these assumptions straightforwardly. Some efforts have been made in this front such as by the `parsnip` package (Kuhn 2018) in the R language (R Core Team 2018) to implement a tidy unified interface to many predictive modelling functions (e.g. random forest, logistic regression, linear regression etc) and the `scikit-learn` library (Pedregosa et al. 2011) for machine learning in the `Python` language (Van Rossum and Drake Jr 1995) that provides consistent API across its modules (Buitinck et al. 2013). There is, however, little effort on consistency or discussion for the software specification of many other types of statistical models, including the class of linear mixed models (LMMs), which is the focus of this article.

LMMs (a special case of mixed models in general, which are also sometimes referred to as hierarchical, panel data, nested or multi-level models) are widely used across many disciplines (e.g. ecology, psychology, agriculture, finance etc) due to their flexibility to model complex, correlated structures in the data. This flexibility is primarily achieved via the inclusion of random effects and their corresponding covariance structures. It is this flexibility, however, that results in major differences in model specification between software for LMMs. In R, arguably the most popular general purpose package to fit LMMs is `lme4` (Bates et al. 2015) – total downloads from RStudio Comprehensive R Archive Network (CRAN) mirror from `cranlogs` (Csárdi 2019) indicate there were over two million downloads for `lme4` in the whole of 2018, while other popular mixed model packages e.g. `nlme`, `rstan`, and `brms` (Bürkner 2017; Pinheiro et al. 2019; Stan Development Team 2019) in the same year have less than half a million downloads, albeit `rstan` and `brms` are younger packages. Another general purpose LMM package is `asreml` (Butler et al. 2009), which wraps the proprietary software ASreml (Gilmour 2009) into the R framework. As this package is not available on CRAN, there are no comparable download logs, although, citations of its technical document indicates popular usage particularly in the agricultural sciences. In this paper, we discuss only `lme4` and `asreml` due to their active maintenance, maturity and contrasting approaches to LMM specification.

The functions to fit LMM in `lme4` and `asreml` are `lmer` and `asreml`, respectively. Both of these functions employ high-level symbolic formulae as part of their API to specify the model. In brief, symbolic model formulae define the structural component of a statistical model in an easier and often more accessible terms for practitioners. The earlier instance of symbolic formulae for linear models was applied in Genstat (VSN International 2017) and GLIM (Aitkin et al. 1989), with a detailed description by Wilkinson and Rogers (1973). Later on, Chamber and Hastie (1992) describe the symbolic model formulae implementation for linear models in the S language, which remains much the same in the R language. While the symbolic formula of linear models generally have a consistent representation and evaluation rule as implemented in stats::formula, this is not the case for LMMs (and mixed models more generally) – the inconsistency of symbolic formulae arises primarily in the representation of the random effects, with the additional need to specify the covariance structure of the random effects as well as structure of the associated model matrix that governs how the random effects are mapped to (groups of) the observational units.

In Sect. 2, we briefly describe the symbolic formula in linear models. We then describe the symbolic model formula employed in the LMM functions lmer and asreml in Sect. 3. We follow by illustrating a number of statistical models motivated by the analysis of publicly available agricultural datasets, with corresponding API for lmer and asreml in Sect. 4. We limit the discussion of symbolic model formulae to mostly those implemented in R, however, it is important to note that the conceptual framework is not limited to this language. We conclude with a discussion and some recommendations for future research in Sect. 5.

2   Symbolic Formulae for Linear Models

A special case of LMMs is linear models, which comprises of only fixed effects and a single random term (i.e. the error or noise), given in a matrix notation as

$$y = \mathbf{X}\beta + e, \tag{1}$$

where y is a n-vector of responses, $\mathbf{X}$ is the $n \times p$ design matrix with an associated p-vector of fixed effects coefficients β, and e is the n-vector of random errors. Typically we assume $e \sim N(\mathbf{0}, \sigma^2 \mathbf{I}_n)$.

The software specification of linear model is largely divided into two approaches: (1) input of arrays for the response y and design matrix for fixed effects $\mathbf{X}$, and (2) input of a symbolic model formula along with a data frame that define the variables in the formula. The input of the data frame may be optional if the variables are defined in the parental environment, although such approach is not recommended due to larger potential for error (e.g. one variable is sorted while others are not).

Symbolic model formulae have been heavily used to specify linear models in R since its first public release in 1993, inheriting most of its characteristic from S. In R, formulae have a special class formula, and can be used for other purposes other than model specification, such as case_when function in dplyr R-package (Wickham et al. 2019), which uses the left hand side (LHS) to denote cases to substitute with given value on the right hand side (RHS) - these type of use is not within the scope of this paper. The history of the formula class in R (and S) is considerably longer than other popular languages, e.g. the patsy Python library (Smith et al. 2018), which imitates R's formula, was introduced in 2011 and used in Statsmodels library (Seabold and Perktold 2010) to fit a statistical model.

Symbolic model formulae makes use of the variable names defined in the environment (usually through the data frame) for specifying the precise model formulation. With linear models, the LHS indicate the response; the RHS resembles its mathematical form; and the LHS and RHS are separated by ~ which can be read as "modelled by". For example, the symbolic model formula y ~ 1 + x can be thought of as the vector y is modelled by a linear predictor consisting of an overall intercept term and the variable x.

When the variables are numerical then the connection between the formula to its regression equation is obvious – the LHS is y, while the RHS corresponds to the columns of the design matrix $\mathbf{X}$ in the linear model (1). One advantage of this symbolic model formula approach is that any transformation to the variable can be parsed in the model formula and may be used later in the pipeline (e.g. prediction in its original scale). This contrasts to when the input arguments are the design matrix and the corresponding response vector – there is now an additional step required by the user to transform the data before model fitting and subsequently afterwards for extrapolation. Such manual transformation also likely results in manual back-transformation later in the analysis pipeline for interpretation reasons. This no doubt creates extra layer of friction for the practitioner in their day-to-day analysis. Figure 1 illustrates this connection using the trees dataset.

The specification of the intercept by 1 in the formula, as done in Fig. 1, is unnecessary in R since this is included by default. In turn, the removal of the intercept can be done by including -1 or +0 on the RHS. In this paper, the intercept is explicitly included as the resemblance to its model equation form is lost without it. While the omission of 1 is long ingrained within R, we recommend to explicitly include 1 and do not recommend designing software to require explicit specification to remove intercept as currently required in R; see Sect. 2.3 for further discussion on this.

Categorical or factor variables are typically converted to a set of dummy variables consisting of 0s and 1s indicating whether the corresponding observation belongs to the respective level. For parameter identifiability, a constraint needs to be applied, e.g. the treatment constraint will estimate effects in comparison with a particular reference group (the default behaviour in R). Note that in the presence of categorical variables, the direct mapping of the symbolic formula to the regression equation is lost. However, the mapping is clear in converting the model equation to the so-called Analysis of Variance (ANOVA) model specification as illustrated in Fig. 2, which represents the fit of a two-way factorial ANOVA model to the herbicide data.

Interaction effects are specified easily with symbolic model formula by use of the : operator as seen in Fig. 2. More specifically, the formula in Fig. 2 can also be written more compactly as sqrt(Weight) ~1 + Block + Population * Herbicide where the * operator is a shorthand for including both main effects and the interaction effects. Further shorthand exists for higher order interactions, e.g. y ~1 + (x1 + x2 + x3)^3 is equivalent to y ~1 + x1 + x2 + x3 + x1:x2 + x1:x3 + x2:x3 + x1:x2:x3, a model that contains main effects as well as two-way and three-way interaction effects. The 1 can be included in the bracket as y ~(1 + x1 + x2 + x3)^3 to yield the same result.

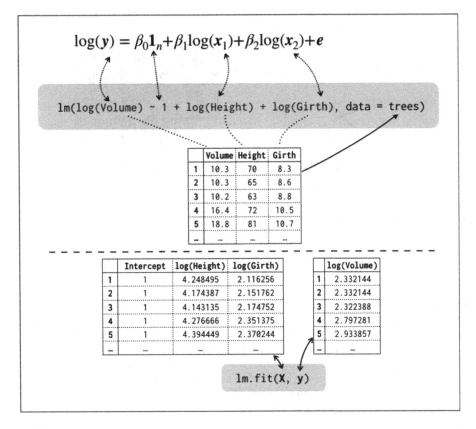

Fig. 1. There are two main approaches to fitting a linear model illustrated above with the fit of a linear model to the `trees` dataset: (1) the top half uses the `lm` function with the input argument as a symbolic model formulae (in blue); (2) the bottom half uses the `lm.fit` function which requires input of design matrix and the response. The latter approach is not commonly used in R, however, it is the common approach in other languages; see Sect. 2.1 about the data and the model.

Perhaps surprisingly, `y ~(0 + x1 + x2 + x3)^3` does not include the intercept in the fitted model, since 0 is converted to −1 and carried outside the bracket and power operator. The formula simplification rule, say for `y ~(0 + x1 + x2 + x3)^3`, in R can be found by

```
formula(terms(y ~ (0 + x1 + x2 + x3)^3, simplify = TRUE))
```

```
## y ~ x1 + x2 + x3 + x1:x2 + x1:x3 + x2:x3 + x1:x2:x3 - 1
```

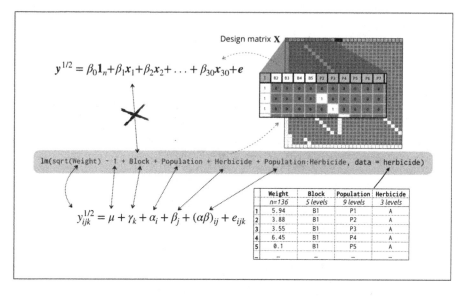

Fig. 2. In the presence of categorical variables, the resemblance of the symbolic model formulae to its regression model form is not immediately obvious. In this case, categorical variables are transformed to a set of dummy variables with constraint applied for parameter identifiability. As such, a single categorical variable span a number of columns in the design matrix. On the other hand, if the model equation is written using the ANOVA model specification (with index notation), then the categorical variables have an immediate connection to the fixed effects in the model; see Sect. 2.2 for more information about the data and the model.

2.1 Trees Volume: Linear Model

The `trees` data set (original data source from Ryan et al. 1976 built-in data in R) contain 31 observations with 3 numerical variables. The model shown in Fig. 1 is a linear model in (2) with the 31×3 design matrix $\mathbf{X} = \begin{bmatrix} \mathbf{1}_{31} & \log(\boldsymbol{x}_1) & \log(\boldsymbol{x}_2) \end{bmatrix}$, where $\boldsymbol{x}_1$ is the tree height and $\boldsymbol{x}_2$ is the tree diameter (named `Girth` in the data). Finally, $\boldsymbol{y}$ is the log of the volume of the tree.

In Fig. 1, the connection of the data column names to symbolic model formula and its resemblance to the model equation is immediately obvious. As discussed before, transformations may be saved for later analysis using the symbolic model formulae (e.g. prediction in original scale), however, this likely requires manual recovery when the API requires design matrix as input.

2.2 Herbicide: Categorical Variable

The `herbicide` data set (original source from R. Hull, Rothamsted Research, data sourced from Welham et al. 2015) contains 135 observations with 1 numerical variable (weight response) and 3 categorical variables: block, herbicide, and population of black-grass with 5, 3 and 9 levels respectively. The experiment employed has a factorial treatment structure (i.e. 27 treatments which are combinations of herbicide and population), with the complete set of treatment randomised within each of the five blocks (i.e. it employs a randomised complete block design).

The model in Fig. 2 is a linear model to the square root of the weight of the black-grass with the design matrix $\mathbf{X} = \begin{bmatrix} \mathbf{1}_{135} \ \boldsymbol{x}_1 \cdots \boldsymbol{x}_{30} \end{bmatrix}$, where $\boldsymbol{x}_1, ..., \boldsymbol{x}_4$ are dummy variables for Block B2, B3 and B4, $\boldsymbol{x}_5, ..., \boldsymbol{x}_{12}$ are dummy variables for Population P2 to P9, $\boldsymbol{x}_{13}$ and $\boldsymbol{x}_{14}$ are dummy variables for Herbicide B and C, and $\boldsymbol{x}_{15}, ..., \boldsymbol{x}_{30}$ are dummy variables for the corresponding interaction effects. Alternatively, the model can be written via the ANOVA model specification,

$$y_{ijk} = \mu + \gamma_k + \alpha_i + \beta_j + (\alpha\beta)_{ij} + e_{ijk},$$

where index i denotes for level of population, index j for level of herbicide and index k for the replicate block. With dummy variables, the relevant constraints are $\alpha_1 = \beta_1 = \gamma_1 = (\alpha\beta)_{1j} = (\alpha\beta)_{i1} = 0$. This form is equivalent to the linear regression model given in Eq. (1) with the fixed effects vector

$$\boldsymbol{\beta} = (\mu, \gamma_2, ..., \gamma_5, \alpha_2, \alpha_3, ..., \alpha_9, \beta_2, \beta_3, (\alpha\beta)_{22}, (\alpha\beta)_{23}, ..., (\alpha\beta)_{93})^{\top}.$$

2.3   Specification of Intercept

Wilkinson and Rogers (1973) described many of the operators and evaluation rules associated with symbolic model formulae, that to this day remain a mainstay of R as well as other languages. These include simplification rules such as y ~ x + x and y ~ x:x to y ~ x. Their description however did not include any discussion about the intercept. The symbolic evaluation rules governing the intercept are classified as special cases in the current implementation of R, although they may not be as overly intuitive on first glance, e.g.

- y ~ 1:x simplifies to y ~ 1, although one may expect y ~ x;
- y ~ 1*x simplifies to y ~ 1, which may be surprising in light of the proceeding point;
- y ~ x*1 simplifies to y ~ x, which makes the cross operator unsymmetric for this special case.

Further ambiguity arises when we consider cases where we wish to explicitly remove the intercept, e.g.

- y ~ -1:x simplifies to the nonsensical y ~ 1 - 1, which is equivalent to y ~ 0,
- y ~ 1 + (-1 + x) simplifies to y ~ x - 1.

The last point was raised by Smith et al. (2018), and subsequently the formula evaluation differs in the patsy Python library on this particular aspect. These complications arise due to the explicit specification for removing the intercept. Furthermore, the symbolic model formulae that includes -1 or 0 removes the resemblance to the model equation, detracting from the aim of symbolic model formula to make model formulation straightforward and accessible for practitioners. It should be noted, however, that these cases are all somewhat contrived and would rarely be used in practice.

3   Linear Mixed Models

Consider a n-vector of response $\boldsymbol{y}$, which is modelled as

$$\boldsymbol{y} = \mathbf{X}\boldsymbol{\beta} + \mathbf{Z}\boldsymbol{b} + \boldsymbol{e}, \tag{2}$$

where the $\mathbf{X}$ is the design matrix for the fixed effects coefficients $\boldsymbol{\beta}$; $\mathbf{Z}$ is the design matrix of the random effects coefficients $\boldsymbol{b}$, and $\boldsymbol{e}$ is the vector of random errors. We typically assume that the random effects and errors are independent of each other and both multivariate normally distributed,

$$\begin{bmatrix} \boldsymbol{b} \\ \boldsymbol{e} \end{bmatrix} \sim N \left(\begin{bmatrix} \mathbf{0} \\ \mathbf{0} \end{bmatrix}, \begin{bmatrix} \mathbf{G} & \mathbf{0} \\ \mathbf{0} & \mathbf{R} \end{bmatrix} \right)$$

where $\mathbf{G}$ and $\mathbf{R}$ are the covariance matrices of $\boldsymbol{b}$ and $\boldsymbol{e}$, respectively.

In Sect. 4, we present examples with different variables and structures for model (2). In the next sections, we briefly describe and contrast the fitting functions `lmer` and `asreml` from the `lme4`, `asreml` R-packages, respectively.

3.1   lme4

The `lme4` R package fits a LMM with the function `lmer`. The API consists of a *single* formula that extends the linear model formula as follows – the random effects are added by surrounding the term in round brackets with grouping structure specified on the right of the vertical bar, and the random terms within each group on the left of the vertical bar, e.g. `(formula | group)`. The `formula` is evaluated under the same mechanism for symbolic model formula as linear models in Sect. 2, with `group` specific effects from `formula`. These `group` specific effects are assumed to be normally distributed with zero mean and unstructured variance, as given above in (2). Examples of its use are provided in Sect. 4.

3.2   asreml

In `asreml`, the random effects are specified as another formula to the argument `random`. One of the main strength of LMM specification in `asreml`, in contrast to `lme4` in wide array of flexible covariance structures. The full list of covariance structures available in `asreml` Version 3 are given in Butler et al. (2009); `asreml` version 4 has some slight differences as outlined in Butler et al. (2018), although the main concept is similar: variance structures are specified with function-like terms in the model formulae, e.g. `us(factor)` will fit a `factor` effect with unstructured covariance matrix; `diag(factor)` will fit a `factor` effect with diagonal covariance matrix, i.e. zero off-diagonal and different parameterisation in the diagonal elements. Note `factor` corresponds to a categorical variable in the data; see Sect. 4 for examples of its usage.

4   Motivating Examples for LMMs

This section presents motivating examples with model specification by `lmer` or `asreml`. It should be noted that the models are not advocated to be "correct", but rather a plausible model that a practitioner may consider in light of the data and design. For succinctness, we omit all `data` argument to model fit functions. Also, this paper uses `lme4` version 1.1.21; `pedigreemm` version 0.3.3 and `asreml` version 3.

4.1   Chicken Weight: Longitudinal Analysis

The chicken weight data is originally sourced from Crowder and Hand (1990) and found as a built-in data set in R. It consists of the weights of 50 chickens tracked over regular time intervals (not all weights at each time points are observed). Each chicken are fed one of 4 possible diets.

In this experiment, we are interested in the influence of different diets on chicken weight. We can model the weight of each chicken over time that includes diet effect, overall intercept and slope for time. Fitting these effects as fixed and assuming that the error is i.i.d. means that the observations from same chicken are uncorrelated and there is no variation for the intercept and slope between chickens. This motivates the inclusion of random intercept and random slope for each chicken. More explicitly, and using an ANOVA model specification, the weight may be modelled as

$$y_{ij} = \beta_0 + \beta_1 x_{ij} + \alpha_{T(i)} + b_{0i} + b_{1i}x_{ij} + e_{ij}, \tag{3}$$

where y_{ij} is the weight of the i-th chicken at time index j, x_{ij} is the days since birth at time index j for the i-th chicken, b_{0i} and b_{1i} are the random intercept and random time slope effects for the i-th chicken, β_0 and β_1 are the overall fixed intercept and fixed time slope, and e_{ij} is the random error.

The above model is incomplete without distributional assumptions for the random effects. As intercept and slope clearly measure different units, the variance will be on different scales. Furthermore, we make an assumption that the random intercept and random slope are correlated within the same chicken, but independent across chickens. With the typical assumption of mutual independence of random effects and random error, and normally and identically distributed (NID) effects, we thus have the distribution assumptions,

$$\begin{bmatrix} b_{0i} \\ b_{1i} \end{bmatrix} \sim NID \left(\begin{bmatrix} 0 \\ 0 \end{bmatrix}, \begin{bmatrix} \sigma_0^2 & \sigma_{01} \\ \sigma_{01} & \sigma_1^2 \end{bmatrix} \right) \quad \text{and} \quad e_{ij} \sim NID(0, \sigma^2). \tag{4}$$

If the effects in model (3) are vectorised as in model (2) with

$$b = (b_{01}, b_{02}, ..., b_{0,50}, b_{11}, b_{12}, ..., b_{1,50})^\top \quad \text{and} \quad e = (e_{ij}),$$

then the model assumption (4) can also be written as

$$b \sim N\left(0, \begin{bmatrix} \sigma_0^2 & \sigma_{01} \\ \sigma_{01} & \sigma_1^2 \end{bmatrix} \otimes \mathbf{I}_m\right) \quad \text{and} \quad e \sim N(0, \sigma^2 \mathbf{I}_n)$$

where $\otimes$ is the Kronecker product. In `asreml`, a separable covariance structure, $\Sigma_1 \otimes \Sigma_2$, is specified by the use of an interaction operator where the dimensions and structures of Σ_1 and Σ_2 are specified by the `factor` input or its number of levels and the function that wraps the `factor`, e.g. `us(2):id(50)` is equivalent to $\Sigma_{2\times 2} \otimes \mathbf{I}_{50}$ where $\Sigma_{2\times 2}$ is a 2×2 unstructured covariance matrix.

The symbolic model formulae that encompasses the model (3) coupled with assumption in (4) for `lmer` and `asreml` are shown in Fig. 3. The two symbolic model formulae share the same syntax for fixed effects, however, in this case the random effects syntax is more verbose for `asreml`.

One may wish to modify their assumption such that now we assume

$$b \sim N\left(0, \begin{bmatrix} \sigma_0^2 & 0 \\ 0 & \sigma_1^2 \end{bmatrix} \otimes \mathbf{I}_m\right),$$

That is, the random slope and random intercept are assumed to be uncorrelated. This uncorrelated model may be specified in `lme4` by replacing | with || as below.

```
lmer(weight ~ 1 + Time + Diet + (1 + Time || Chick))
```

It should be noted that the effects specified on the LHS of the || are uncorrelated if the variables are numerical only; we refer to the example in Table 1 for a case where this does not work when the variable is a factor.

```
lmer(weight ~ 1 + Time + Diet + (1 | Chick) + (0 + Time | Chick))
```

The same model is specified as below for `asreml` where now `us(2)` is replaced with `diag(2)`. The correspondence to the covariance structure is more explicit, but again involves the random effects being (implicitly) vectorised as show in Fig. 3 and care is needed with orders of separable structure.

```
asreml(weight ~ 1 + Time + Diet,
random=~ str(~Chick + Chick:Time, ~diag(2):id(50)))
```

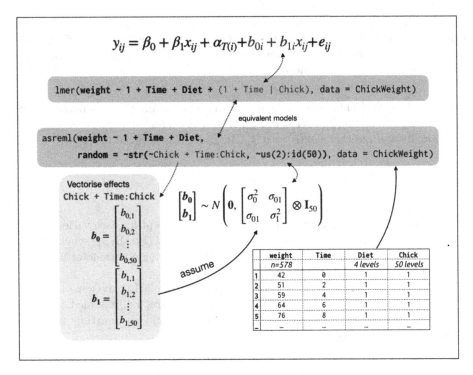

Fig. 3. This figure shows a longitudinal analysis of the chicken data (see Sect. 4.1). The index form of the model equation shows direct resemblance for symbolic model formula in `lmer` for the fixed and random effects, however, its covariance form is not as easily inferred. In contrast, the symbolic model formula in `asreml` show resemblance of the covariance structure specified in the second argument of `~str`, however, the corresponding random effects specified in the first argument of `~str` must be vectorised as show in the above figure and requires implicit knowledge of the Kronecker product of relevant matrices.

4.2   Field Trial: Covariance Structure

In this example, we consider wheat yield data sourced from the **agridat** R-package (Wright 2018), which originally appeared in Gilmoure et al. (1997). This data set consists of $n = 330$ observations from a near randomised complete block experiment with $m = 107$ varieties, of which 3 varieties have 6 replicates while the rest have 3 replicates. The field trial that the yield data was collected from was laid out in a rectangular array with $r = 22$ rows and $c = 15$ columns. Each of the variety replicates are spread uniformly to $b = 3$ blocks. The columns 1–5, columns 6–10 and columns 11–15 form three equal blocks of contiguous area within the field trial. The data frame `gilmour.serpentine` contains the columns for `yield`, `gen` (variety), `rep` (block), `col` (column) and `row`. Further columns `colf` and `rowf`, which are factor versions of `col` and `row`, have also been added.

We may model the yield observations $\boldsymbol{y}$, ordered by the rows within columns, using the model (2) where here $\boldsymbol{\beta}$ is the b-vector of replicate block effects and $\boldsymbol{b}$ is the m-vector of variety random effects. We consider next a few potential covariance structures for $\boldsymbol{b}$ and $\boldsymbol{e}$.

Scaled Identity Structure. One of the simplest assumptions to make would be to assume that $var(\boldsymbol{b}) = \mathbf{G} = \sigma_g^2 \mathbf{I}_m$ i.e., a scaled identity structure. We may additionally assume that $var(\boldsymbol{e}) = \sigma^2 \mathbf{I}_n$. In lmer, this is fitted as below.

```
lmer(yield ~ 0 + rep + (1 | gen))
```

To elaborate further, lmer specifies a random intercept for each variety. This variety intercept will each be assumed to arise from $NID(0, \boldsymbol{\Sigma}_{1 \times 1})$ where $\boldsymbol{\Sigma}_{1 \times 1}$ is a 1×1 unstructured variance matrix (essentially a single parameter variance component).

The same model is fitted in asreml as below. Particularly, idv(gen) signifies a vector of variety effects with idv variance structure, i.e. a scaled identity structure. This is the default structure in asreml, and so omitting variance structure, random = ~gen, results in the same fit.

```
asreml(yield ~ 0 + rep, random = ~idv(gen))
```

Crossed Random Effects. Field trials often employ rows and/or columns as blocking factors in the experimental design. Furthermore, it is common practice that the management practices of field experiments follow some systematic routine, e.g., harvesting may occur in a serpentine fashion from the first to the last row. These occasionally introduce obvious unwanted noise in the data that are often removed by including random row or column effects assuming that they are i.i.d. for simplicity. These so-called crossed random effects are fitted as below for lmer and asreml.

```
lmer(yield ~ 0 + rep + (1 | gen) + (1 | rowf) + (1 | colf))
```

```
asreml(yield ~ 0 + rep, random = ~idv(gen) + idv(rowf) + idv(colf))
```

Error Covariance Structure. A field trial is often laid out in a rectangular array and observations from each plot indexed by row and column within this array. Consequently, the assumption that $var(\boldsymbol{e}) = \sigma^2 \mathbf{I}_n$ may be restrictive when there is likely to be some sort of spatial correlation, i.e. plots that are geographically closer would be similar than plots further apart. A range of models may be considered for this potential correlation. In practice, a separable autoregressive process of order one, denoted AR1×AR1, has worked well as a compromise between parsimony and flexibility as a structure (Gilmour et al. 1997). More specifically, we assume $var(e) = \sigma^2 \Sigma_c \otimes \Sigma_r$ where Σ_c is a $c \times c$ matrix with (i, j)-th entry of Σ_c given as $\rho_c^{|i-j|}$ with autocorrelation parameter ρ_c, and a similar definition holds for $r \times r$ matrix Σ_r except the autocorrelation parameter is denoted bvy ρ_r. This model is fitted in asreml by supplying a symbolic formula, ar1(colf):ar1(rowf), to the argument rcov as below.

```
asreml(yield ~ 0 + rep, random = ~gen, rcov = ~ar1(colf):ar1(rowf))
```

Here, the ar1 specifies an autoregressive process of order one with dimension given by number of levels in rowf and colf. It is important to note that ar1 denotes a correlation matrix and a covariance matrix may be specified by ar1v. Care needs to be taken in covariance specification for separable models, as clearly there is a lack of variance parameter(s) where Σ_1 and Σ_2 are both correlation structures only, while if both are covariance structure then the model is over-parameterised and unidentifiable. In the error structure of rcov, this is taken care of such that rcov = ~ar1v(colf):ar1(rowf), rcov = ~ar1(colf):ar1v(rowf) and rcov = ~ar1(colf):ar1(rowf) will fit all the same model. It should be noted that this is not the case for separable covariance structures specified in random effects.

In comparison, the more restrictive API of lmer function does not allow the assumption on the random effects to be relaxed from $var(e) = \sigma^2 \mathbf{I}_n$. One may of course introduce a random effect, $b_e \sim N(\mathbf{0}, \sigma^2 \Sigma_c \otimes \Sigma_r)$, and assume $e \sim N(\mathbf{0}, \sigma^2 \mathbf{I}_n)$. However, this separable covariance structure also can not be specified within lmer function.

Known Covariance Structure. Often in plant breeding trials, the varieties of interest have some shared ancestry. This is captured in the form of pedigree data that contains 3 columns: individual ID, mother's ID and father's ID. The related structure is commonly captured by the use of a numerator relationship matrix, denoted here has $\mathbf{A}$ (Mrode 2014). For example, suppose that individuals i and j are full-siblings. Then the corresponding (i, j)-th entry in $\mathbf{A}$ is 0.5 (i.e., the average probability that a randomly drawn allele from individual i is identical by descent to the randomly drawn allele at the same autosomal locus from individual j).

With the additional information above, we may assume that $var(b) = \sigma_g^2 \mathbf{A}$ to exploit this *known* relatedness structure between varieties. The symbolic model formulae in lme4 alone is unable to specify this model and, an extension R package **pedigreemm** (Vazquez et al. 2010) is required. The pedigree data is parsed to make an object of **pedigree** class, which we refer to here as ped. This object `ped` is then included as part of the input in the main fitting function `pedigreemm`, as depicted below.

```
pedigreemm(yield ~ 0 + rep + (1 | gen),
pedigree = list(gen = ped))
```

In `asreml`, the fit is similar to the above, but the factor with the known covariance structure must be wrapped in `giv` with argument `ginverse` providing a named list with the inverse of the $\mathbf{A}$ in a sparse format, i.e. a data frame of 3 columns that consists the row and the column index of $\mathbf{A}$ and its corresponding value in $\mathbf{A}$ provided that the value is non-zero.

```
asreml(yield ~ 0 + rep, random = ~giv(gen),
ginverse = list(gen = Ainv))
```

4.3 Multi-environmental Trial: Separable Structure

In the final example, we consider CIMMYT Australia ICARDA Germplasm Evaluation (CAIGE) bread wheat yield 2016 data (CAIGE 2016), which consists of $t = 7$ sites across Australia, where the overall aim is to select the best genotype (`gen`). There were $m = 240$ genotype tested across seven trials and 252–391 plots, with a total of $n = 2127$ yield observations. Each trial employed a partially replicated (p-rep) design (Cullis et al. 2006), with p ranging from 0.23 to 0.39.

Fitting a model to a model should take into account the differential mean yield across sites, and allow for different genotypic variations by site. For simplicity, we ignore other variations for now. In turn, the LMM formulation in Eq. (2) may be used where $\boldsymbol{y}$ is the vector of yield (ordered rows within columns within sites); $\boldsymbol{\beta}$ is the t-vector of site effects; and $\boldsymbol{b}$ is the mt-vector of genotype-by-site effects (ordered by genotype within site). There are a number of distributions that may be considered for $\boldsymbol{b}$, as explained below.

We may consider a separable model such that $\boldsymbol{b} \sim N(\boldsymbol{0}, \boldsymbol{\Sigma}_s \otimes \boldsymbol{\Sigma}_g)$, where $\boldsymbol{\Sigma}_s$ and $\boldsymbol{\Sigma}_g$ are a $t \times t$ and $m \times m$ matrices, respectively. We may further assume that $\boldsymbol{\Sigma}_g$ has a known structure similar to Sect. 4.2, but for simple illustration here we will assume that the genotypes are independent, i.e. $\boldsymbol{\Sigma}_g = \mathbf{I}_m$. Also, we may assume that $\boldsymbol{\Sigma}_s = \mathrm{diag}\left(\sigma_{g1}^2, \sigma_{g2}^2, \cdots, \sigma_{gt}^2\right)$, i.e. a diagonal matrix with different variance paramterisation for each site, thus allowing for different genotypic variance at each site. This model can be fitted as below in `asreml`.

```
asreml(yield ~ 0 + site, random = ~ diag(site):id(gen))
```

The same model in `lmer` is somewhat more involved as shown in Table 1.

Table 1. The table lists the equivalent symbolic model formula in `lmer` and `asreml` for the site-by-genotype random effect, b and the corresponding mathematical form of the variance structure of b. Here, $\Sigma_{t \times t}$ is a $t \times t$ unstructured covariance matrix; $D_{t \times t} = \mathrm{diag}(\sigma_{g1}^2, ..., \sigma_{g7}^2)$, a $t \times t$ diagonal covariance matrix; m is the number of genotypes; t is the number of sites; and `S1` is a n-vector where the entry is one if the corresponding observation belongs to site 1 and zero otherwise (similar definitions hold for `S2`, ..., `S7`). The conversion of the factor `site` to numerical variables `S1`, ..., `S7` is required to have uncorrelated random effects in `lmer` via the || operator, as per the last row in the table. The || group separation in `lmer` is only effective when variables on LHS are numerical.

lmer	asreml	$var(b)$
(1 | site:geno)	idv(site):id(geno) id(site):idv(geno) site:geno	$\sigma_g^2 I_{tm}$
(0 + site | geno) (0 + site || geno) (0 + S1 + S2 + S3 + S4 + S5 + S6 + S7 | geno)	us(site):id(geno)	$\Sigma_{t \times t} \otimes I_t$
(0 + S1 + S2 + S3 + S4 + S5 + S6 + S7 || geno)	diag(site):id(geno)	$D_{t \times t} \otimes I_t$

The diagonal model assumes that genotype-by-site effects are uncorrelated across sites for the same genotype. However, a more realistic assumption is to assume that these effects are correlated, thus allowing for different correlation of genotype effects between pair of sites, i.e. we assume that Σ_s is an unstructured covariance matrix. The specification of such model for `lmer` and `asreml` is shown in Fig. 4.

A even more realistic model may consider including site-specific random row or column effects, and assuming an AR1×AR1 process for the error covariance at each site as in Sect. 4.2. These are easily included in `asreml` using the at function within the symbolic model formulae. For example, the inclusion of random row effect at site `S1` only and AR1×AR1 processes for the error covariance at each site is shown below.

```
asreml(yield ~ 0 + site,
random = ~us(site):id(gen) + at(site, "S1"):idv(rowf),
rcov = ~at(site):ar1(colf):ar1(rowf))
```

The above model cannot be specified using `lmer`.

5   Discussion

In fitting statistical models, the user may not necessary understand the full intricacies of model fitting process. However, it is essential that the user understands how to specify the model that they wish to fit and the interpretation from the fit. Symbolic model formulae is a way of bridging the gap between software and

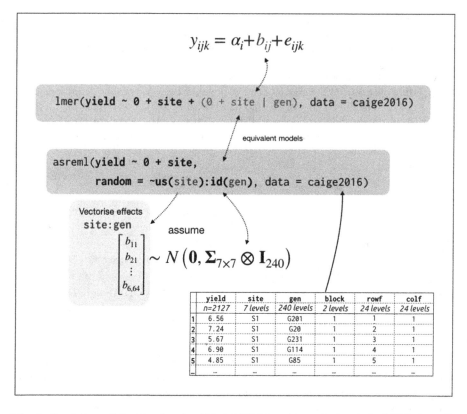

Fig. 4. Depiction of the fit of a simplified LMM for the analysis of the MET data. In modelling the `site-by-gen` random effect, the variance structure are specified differently using `lmer` and `asreml`, where latter shows resemblances of covariance structure written mathematically and when all random effects are vectorised and concatenated, while the former requires some additional computation.

mathematical representation of the model, and has been extensively employed in R for this purpose.

In this article, we have extensively compared two widely used LMM R-packages with contrasting model specification in functions: `lmer` and `asreml`. Both of these functions use symbolic model formulae to specify the model with `lmer` taking a more hierarchical approach to random effects specification, while `asreml` focuses on the covariance structure of the vectorised random effects (and the data for the matter). There are strength and weakness in both approaches as we discuss next.

It is clear from Sect. 4.1 that a random intercept and random slope model is verbose using the symbolic model formulae of `asreml`. Specifically, the random effect symbolic formula contains a function `str` that takes input of two other formula: the first input specifying the random effects, and second input specifying the covariance structure of the vectorised form of random effects specified in the first input. The second input also requires the dimension(s) of covariance structure as input. These number may need manual update when the data is subsequently updated, thus making this symbolic model formula clumsy to use.

On the other hand, the flexibility of `asreml` is evident in Sects. 4.2 and 4.3, where the LMMs fitted are less easy to pose hierarchically, but the vectorised version of the LMM remains straightforward provided one knows how to establish the set up the structure of the covariance matrices. Put another way, the vast set of in-built pre-defined covariance structures in `asreml` (e.g. scaled identity, diagonal structure, unstructured, autoregressive process), along with the capacity to modify the error covariance structure and incorporate separable structures makes the model specification embedded in `asreml` a superior choice here. There are many more pre-defined covariance structures not demonstrated in this paper, and interested readers may refer to Butler et al.(2009), (2018). By contrast, the lack of flexibility in `lme4` means that either a more obtuse workaround is required or the precise LMM can not be formulated at all.

That being said, the Bürkner (2017), (2018) (`brms`) make extensive discussion of symbolic model formula and extends on the framework built on `lme4`. The `brms` model formulae resembles `lme4` and many symbolic model formulae in our examples will be similar. The `brms` R-package uses a Bayesian approach to fit its models and model specification require further discussion on specifying priors. These discussions are left for future review, although we acknowledge that such extensions may well resolve some of the current limitations of `lme4` and bridge its gap in flexibility with `asreml`.

Symbolic model formulae in `R` is widely used and frameworks to specify mixed models by `lme4` and `asreml` (version 3) used for many years. This makes drastic changes difficult for these frameworks. Based on our review, we argue that ideally any new framework for symbolic model formulae should require intercepts to be specified explicitly. As discussed in Sect. 2.3, the default inclusion or explicit removal of intercepts removes the resemblance of symbolic model formulae to the model equation. Currently, the implicit inclusion of intercepts makes certain model formulation unclear and inconsistent across different LMM specifications, e.g. (`Time | Chick`) in `lmer` includes random intercept (and slope) for `Chick`, but the equivalent formulation `str(~Chick + Chick:Time, ~diag(2):id(50))` in `asreml` does not include the random intercept.

There is a trade-off between different types of symbolic model formulae: `lmer` syntax is no doubt less flexible and may be less intuitive to some, however, with a degree of familiarity pertains as a higher level language for symbolic model formula. For many hierarchical models, the formulation is more elegant and simpler than `asreml`. However, `asreml` is more flexible to specify variety of covariance matrices. This strength is predicated on having a deeper understanding of random effects and its covariance structure, and promotes the view of the LMM in a fully vectorised form.

Acknowledgement. This paper benefited from twitter conversation with Thomas Lumley. This paper is made using `R` Markdown (Xie et al. 2018). Huge thanks goes to the teams behind `lme4` and `asreml` R-packages that make fitting of general LMMs accessible to wider audiences. All materials used to produce this paper and its history of changes can be found on github https://github.com/emitanaka/paper-symlmm.

References

Aitkin, M., Dorothy, A., Francis, B., Hinde, J.: Statistical Modelling in GLIM. Oxford University Press, Oxford (1989)

Bates, D., Machler, M., Bolker, B., Walker, S.: Fitting linear mixed-effects models using lme4. J. Stat. Softw. **67**(1) (2015). https://doi.org/10.18637/jss.v067.i01

Buitinck, L., et al.: API design for machine learning software: experiences from the scikit-learn project. In: ECML PKDD Workshop: Languages for Data Mining and Machine Learning, pp. 108–122 (2013)

Butler, D.G., Cullis, B.R., Gilmour, A.R., Gogel, B.J.: Mixed models for s language environments ASReml-R reference manual (2009)

Butler, D.G., Gogel, B.J., Cullis, B.R., Thompson, R.: Navigating from ASReml-R version 3 to 4 (2018)

Bürkner, P.-C.: brms: an R package for Bayesian multilevel models using Stan. J. Stat. Softw. **80**(1), 1–28 (2017)

Bürkner, P.-C.: Advanced Bayesian multilevel modeling with the R package brms. R J. **10**(1), 395–411 (2018). https://doi.org/10.32614/RJ-2018-017

CAIGE: Caige project (2016). http://www.caigeproject.org.au

Chambers, J.M., Hastie, T.: Statistical models in S. Wadsworth & Brooks/Cole Computer Science Series. Wadsworth & Brooks/Cole Advanced Books & Software (1992). ISBN 9780534167646. http://books.google.fr/books?id=uyfvAAAAMAAJ

Crowder, M., Hand, D.: Analysis of Repeated Measures. Chapman and Hall, London (1990). http://www.python.org

Csárdi, G.: cranlogs: download logs from the 'RStudio' 'CRAN' mirror (2019). https://CRAN.R-project.org/package=cranlogs. R package version 2.1.1

Cullis, B.R., Smith, A.B., Coombes, N.E.: On the design of early generation variety trials with correlated data. J. Agric. Biol. Environ. Stat. **11**(4), 381–393 (2006). https://doi.org/10.1198/108571106X154443. ISSN 1085–7117

Gilmour, A.R., Cullis, B.R., Verbyla, A.P.: Accounting for natural and extraneous variation in the analysis of field experiments. J. Agric. Biol. Environ. Stat. **2**(3), 269–293 (1997). https://doi.org/10.2307/1400446

Gilmour, A.R., Gogel, B.J., Cullis, B.R., Thompson, R.: ASReml user guide release 3.0 (2009)

Kuhn, M.: parsnip: a common API to modeling and analysis functions (2018). https://topepo.github.io/parsnip. R package version 0.0.0.9003

Mrode, R.A.: Linear Models for the Prediction of Animal Breeding Values, 3rd edn. CABI, Wallingford (2014). https://doi.org/10.1017/CBO9781107415324.004. ISBN 1780643918, 9781780643915

Pedregosa, F., et al.: Scikit-learn: machine learning in Python. J. Mach. Learn. Res. **12**, 2825–2830 (2011)

Pinheiro, J., Bates, D., DebRoy, S., Sarkar, D., R Core Team: Nlme: linear and nonlinear mixed effects models (2019). https://CRAN.R-project.org/package=nlme. R package version 3.1-140

R Core Team: R: A language and environment for statistical computing. R Foundation for Statistical Computing, Vienna, Austria (2018). https://www.R-project.org/

Ryan, T.A., Joiner, B.L., Ryan, B.F.: The Minitab Student Handbook. Duxbury Press, London (1976)

Seabold, S., Perktold, J.: Statsmodels: econometric and statistical modeling with python. In: 9th Python in Science Conference (2010)

Smith, N.J., et al.: pydata/patsy: v0.5.1, October 2018. https://doi.org/10.5281/zenodo.1472929

Stan Development Team: RStan: the R interface to Stan (2019). http://mc-stan.org/. R package version 2.19.2

Van Rossum, G., Drake Jr, F.L.: Python tutorial. Centrum voor Wiskunde en Informatica Amsterdam, The Netherlands (1995). http://www.python.org

Vazquez, A.I., Bates, D.M., Rosa, G.J.M., Gianola, D., Weigel, K.A.: Technical note: an R package for fitting generalized linear mixed models in animal breeding. J. Anim. Sci. **88**, 497–504 (2010)

VSN International: Genstat for Windows 19th Edition. VSN International, Hemel Hempstead, UK (2017). Genstat.co.uk

Welham, S.J., Gezan, S.A., Clark, S.J., Mead, A.: Statistical Methods in Biology: Design and Analysis of Experiments and Regression. Chapman and Hall, London (2015)

Wickham, H., FranÃğois, R., Henry, L., MÃijller, K.: dplyr: a grammar of data manipulation (2019). https://CRAN.R-project.org/package=dplyr. R package version 0.8.3

Wilkinson, G.N., Rogers, C.E.: Symbolic description of factorial models for analysis of variance. J. Roy. Stat. Soc.: Ser. C (Appl. Stat.) **22**(3), 392–399 (1973)

Wright, K.: agridat: agricultural datasets (2018). https://CRAN.R-project.org/package=agridat. R package version 1.16

Xie, Y., Allaire, J.J., Grolemund, G.: R Markdown: The Definitive Guide. Chapman and Hall/CRC, Boca Raton (2018). ISBN 9781138359338. https://bookdown.org/yihui/rmarkdown

code::proof: Prepare for *Most* Weather Conditions

Charles T. Gray$^{(\boxtimes)}$ ⓘ

La Trobe University, Melbourne, Australia
charlestigray@gmail.com

Abstract. Computational tools for data analysis are being released daily on repositories such as the Comprehensive R Archive Network. How we integrate these tools to solve a problem in research is increasingly complex and requiring frequent updates. To mitigate these *Kafkaesque* computational challenges in research, this manuscript proposes *toolchain walkthrough*, an opinionated documentation of a scientific workflow. As a practical complement to our proof-based argument (Gray and Marwick, arXiv, 2019) for reproducible data analysis, here we focus on the practicality of setting up reproducible research compendia, with unit tests, as a measure of code::proof, confidence in computational algorithms.

Keywords: Metaresearch · Metaprogramming · Statistical computing

1 The Kafkaesque Dystopia of DevOps

In Franz Kafka's 1925 novel *The Trial* [14], the fictional character Josef K. is prosecuted for crimes that are not clear, in proceedings brought forth by an unidentified authority. For the diligent scientist attempting to answer a mathematical question computationally, such as measuring the efficacy of a statistical estimator via simulation, the process of implementing a scientific workflow to achieve this aim can be a *Kafkaesque* tour of computational tools and systems. The scientist may feel as if they are locked in a dystopia, tested repeatedly for practices in which they have not been trained, such as shell scripts and computational architecture. Whilst there are detailed guides for specific computational tools, it is hard to tell what is still relevant, as code frequently slides into obsolescence [22], and identify the optimal place to begin [34]. Significant cultural

---

[1] Stack Overflow (https://stackoverflow.com/) is forum for asking tightly scoped programming questions.

Thank you to Ben Marwick, Hien Nguyen, Emily Kothe, James Goldie, Mathew Ling, J.D. Long, Kate Smith-Miles, Greg Wilson, Kerrie Mengersen, Jacinta Holloway, Alex Hayes, Noam Ross, Rowland Mosbergen, Luke Prendergast, Dale Maschette, Elio Campitelli, Thomas Lumley, and Daniel S. Katz for advising on particular aspects of this manuscript.

© Springer Nature Singapore Pte Ltd. 2019
H. Nguyen (Ed.): RSSDS 2019, CCIS 1150, pp. 22–41, 2019.
https://doi.org/10.1007/978-981-15-1960-4_2

barriers continue to exist in programming fora; for example, only one in seventeen contributors to Stack Overflow[1] identify as women [6].

For many an unfortunate scientist, the dystopian experience is not confined to the *DevOps*, the developmental operations of preparation for the implementation of an algorithm [13]. Just as Josef K. was tried multiple times, the labours of the scientist attempting to answer a mathematical question computationally have only just begun. Analogous to how a string will knot with mathematical predictability when jostled [2], an algorithm will reliably require *debugging*, the process of identifying and correcting code, either to incorporate a new feature, or to correct an error. This scientist finds themselves part of the first generation of *research software engineers* (RSEs), who use computational tools in discipline-specific research practices [35]. By virtue of pioneering, RSEs are inadvertently cast as metaresearchers[2], developing new methodologies for scientific technologies that hitherto did not exist [15]. With the aim of mitigating the dystopia of DevOps and debugging for RSEs, this manuscript proposes a *toolchain walkthrough*, an *opinionated* [21] documentation of a scientific workflow, towards a measure of `code::proof`, a *good enough* [34] effort to provide computational confidence through reproducible research compendia with unit tests.

2 Toolchain Walkthrough

We define a *toolchain* as a collection of computational tools and commands that forms a scientific workflow to achieve a specific research objective, such as test the efficacy of a statistical estimator in a particular context. The term *walkthrough*, we borrow from video game terminology [5], and is defined as a guide for other players of the game. Various walkthrough formats exist to optimise the narrative enjoyment of the gamer. For example, the Universal Hint System [26] interface provides the gamer with ever more revealing hints without spoiling other parts of the game. Next generation walkthroughs see in-game modifiers, in games such as World of Warcraft, where these provide an option for on-screen boss-specific warnings [25].

We define *toolchain walkthrough* as an *opinionated* [21] documentation of a scientific workflow, where opinionated is a term appropriated from software engineering that acknowledges that software guides the user to certain choices. In this manuscript, we describe a workflow for building a research compendium that is opinionated in privileging reproducibility. As with the hint systems of gaming, a workflow can and must be tailored to the skill and background of the

---

[2] Visit the discussion on metaresearch and RSEs on the research compendium associated with this manuscript as an example of why this paper, and its companion [8], have so many acknowledgements. Canonical literature is not yet established in the field of RSE, and thus leaders of RSE projects, such as Alex Hayes' maintenance of the `broom::` [23]. This has propelled Hayes rapidly to the level of expert, by virtue of the pioneering collaborative structure of the package, where hundreds of statistical modellers contribute integrated code.

user. Thus toolchain walkthroughs can be extended and adapted for different disciplines.

Toolchain walkthroughs have not only intrinsic value in terms of solving the intended research problem, but also extrinsic value, pedagogically and from a developmental perspective. Frequently those who are undertaking research software engineering on statistical projects are not the most senior member of the team; in the case of university faculty, these are often also lecturers and service teachers. There is value in seeing the minutiae of what the footsoldiers of research development undertake and how they instruct others. This can inform as to what skillsets are required in graduate courses, or are required for those who wish to optimise scientific workflow for researchers. Much of what is being implemented right now, in workflows recommended in texts[3] such as *R Packages* [29] and *Advanced R* [30], is being adopted from existing software engineering principles. Toolchain walkthroughs can contribute to the literature on the adoption of these procedures in a research context, in addition to programming fora and blog posts.

Blog posts and programming fora, as well as printed texts, are inevitably bound for obsolescence [22]. Vignettes, tool-specific long-form documentation [29], focus on one tool in the chain. As a counterpoint to the inadvertently implied redundancy of the academic manuscript in the theoretical companion manuscript [8], here we consider if the ephemerality of most-recent publications, and the chronological nature of academic publishing, may serve the breakneck speed of research development. The toolchain walkthrough provides a documentation of a specific scientific workflow constructed by an expert, or expert in training, in the field. Indeed an expert in training is perhaps best placed, as by virtue of inexperience must research in order to solve the problem. The challenge above, say, the standard one might expect from a blog post, is to provide a *good enough*[4] [34] effort to avoid questionable research practices [7] that privilege, say, convention over optimal scientific methodology.

3   Two Research Compendia Case Studies

For concrete examples of the benefits of adopting software research engineering principals in mathematical science, we consider two in-development research compendia, `varameta::` and `simeta::`. The primary purpose of these packages is to provide a comparative analysis of estimators for the variance of the sample median when quartiles are provided, rather than a measure of standard deviation, within the meta-analytic context. However, by structuring the packages as such, rather than within a single script file, there is scope for solving similar problems.

---

[3] As in the companion manuscript [8], we focus on R packages, but the reader is invited to consider these as examples rather than definitive guidance. The same arguments hold for other languages, such as Python, and associated tools.

[4] As opposed to ofttimes unattainable or impractical *best practices* [33] in scientific computing.

3.1   The varameta:: Package; a Comparative Analysis

In contemporary meta-analytic computational tools, such as the R package metafor:: [27], a measure of both an effect and its variance are required to estimate the population parameters of interest.

However, not all studies report a variance of effect; particularly when scientists suspect an underlying asymmetry in the distribution of the observed data, prompting them to report quartiles, rather than sample standard deviations. One solution to this is to approximate estimators for mean and variance from quartiles [3,12,28]. We wish to explore the comparative efficacy of an estimator for the variance of the sample median derived from the estimator of [24]:

$$\operatorname{var}(m) \approx \frac{1}{4nf(\nu)^2}$$

where m denotes the sample median, n the sample size, ν the population median, and f the population probability density function.

However, in an experimental setting, we do not know the true distribution, nor the true population median. Thu, our method proposes that we assume a distribution, and estimate the parameters of that characterized the assumed distribution from the sample size and sample quartiles. We provide estimators derived for different distributions, to assess the efficacy of this analysis framework. One of which is the exponential distribution, which this manuscript will focus on.

If we assume that f is an exponential probability density function, with unknown rate parameter λ, then we can estimate this rate parameter via the sample median. Since the true median is given by $\log 2/\lambda$, we can estimate the rate parameter,

$$\lambda \approx \log 2/m, \tag{1}$$

via the sample median, m.

Each proposed estimator requires a different set of reported values as inputs and different calculations. It is notable that a most optimal estimation method for the problem above is generally unknown. For example, in the comparative analysis Wan et al. [28], it was shown that the performance of different estimators varied with the simulated sample sizes.

Thus, there is merit to providing not only the practical functionality of our proposed solution, but also the existing solutions. By structuring this comparative analysis as a reproducible research compendium we achieve practical improvements on a self-contained computational script file. Via roxygen::ised [32] documentation, estimators are provided in a modular fashion, with a devoted script file for each estimator that is easily sourced from the package environment. In addition to the advantage of debugging a single script file, the comparative analysis also serves a practical purpose, providing a characterisation of the functionality of each estimator.

To compare these estimators for the variance of the sample median, we undertook *coverage probability* simulations. Here, the coverage probability refers to the probability that the true parameter of interest falls within its constructed confidence interval. In order to do so, we require simulated meta-analytic data, which has the added complexity of a random effect that governs the variation *between* studies. To solve this with confidence in the implementation of computational algorithms and mathematical derivations, we structure this as a package. In addition to building `code::proof`, by separating the simulation component, we begin to develop a computational solution to not only solving this problem, but the testing of **any** estimator for the variance of the sample mean or median.

3.2 The `simeta::` Package

A *coverage probability* simulation repeats several trials with the same simulation meta-parameters where the differing factor is the random sampling of data. In order to separate simulation meta-parameters from trial-level parameters, and delineate this algorithm, we begin by considering a single trial from a standard coverage probability simulation.

3.3 Coverage Probability Simulation

Each trial draws a random sample, for example `rnorm(n = 100, mean = 3, sd = 0.2)` will produce 100 values drawn randomly from a normal distribution with mean 3 and standard deviation 0.2. From this sample, we calculate summary statistics. Using these summary statistics, we can compute an estimate of the parameter of interest $\hat{\nu}$, and its variance $\hat{\gamma}$. With these estimates, we can produce a $(1 - \alpha) \times 100\%$ confidence interval $\hat{\nu} \pm z_{1-\alpha/2}\sqrt{\hat{\gamma}}$, where $z_a = \Phi^{-1}(a)$ is the ath quantile of the standard normal distribution, and Φ is the standard normal distribution function. Given we set the parameters for the random sample drawn, we know the true parameter, ν. Thus we can ask, does ν fall within the confidence interval produced? We summarise the steps of a trial as an algorithm:

1. Draw a random sample from the distribution that is characterised by the parameter of interest, ν;
2. Calculate summary statistics from the random sample;
3. Calculate an estimate of ν from the summary statistics;
4. Construct a confidence interval using the parameter estimate;
5. Check if ν falls within the confidence interval.

A coverage probability simulation performs multiple trials and returns the proportion $p \in [0, 1]$ of confidence intervals for ν that contain the generative parameter value.

3.4 Simulating Meta-analysis Data

For a meta-analysis simulation, however, these steps are significantly more involved. And with this complexity, as we shall see, nesting, of the algorithm, the advantages of the package structure begin to become apparent. In a single script file, it is hard to find at which step of the algorithm that the code has failed. In addition to human error introduced into code, there are also practical considerations. For example, the random effects maximum likelihood model, method = REML, employed by metafor::rma [27] does not always converge on estimates for the effect and its variance, in which case a fixed effects model, method = FE, can be employed to produce parameter estimates.

The other point of complexity is in the sampling of meta-analytic data. As meta-analytic data is a collection of summary statistics for K studies of control and intervention samples, the first step of a coverage probability simulation trial,

1. Draw a random sample from the distribution that is characterised by the parameter of interest, ν,

requires several substeps. For the kth ($k \in \{1, \ldots, K\}$) study, we assume there is variation γ_k associated with that study, and, in particular, the control, with parameter ν_k^C, and intervention, with parameter ν_k^I, samples with ratio, $\rho = \nu_k^C/\nu_k^I$.

Let us consider a practical example from the estimators provided in the comparative analysis, varameta::. Our estimator of interest is the variance of the log-ratio of sample medians for control, ν^C, and intervention, ν^I groups. Since our focus is on building the research compendium to undertake this analysis, rather than the estimators in question, we will take the simplest case, where there is one parameter λ associated with the distribution of interest. Let us assume an underlying exponential distribution: Exponential(λ).

At the simulation level, which is to say, across all trials, we set λ, the parameter of the distribution of interest. Also at the simulation level, we define a ratio $\rho := \nu^C/\nu^I$ of interest for the population medians, where $\rho = 1$ would indicate no true difference between control and intervention groups. We assume that the log-ratio of sample medians $\log(m_k^I/m_k^C)$ for the kth study, can be characterised in terms of the log-ratio of populations medians $\log(\nu^C/\nu^I)$, with some error $\gamma \sim N(0, \tau^2)$ association with that study, as well as sampling error, $\varepsilon \sim N(0, \sigma^2)$,

$$\log(m_k^I/m_k^C) = \log(\nu^I/\nu^C) + \gamma_k + \varepsilon_k.$$

Since the underlying distribution is exponential, we need to find λ_k^J for $J \in \{C, I\}$ in order to sample n values $x_1, \ldots, x_n \sim$ Exponential(λ_k^J). We also know the median of the exponential distribution with rate parameter λ is given by $\log 2/\lambda$. Then, assuming the sampling error will be attained through the random computational process, we have

$$\log(m_k^I/m_k^C) = \log(\nu^I/\nu^C) + \gamma_k$$
$$\implies \log(\lambda_k^C) - \log(\lambda_k^I) = \log(\lambda^C) - \log(\lambda^I) + \gamma_k$$
$$\implies \log(\lambda_k^C) - \log(\lambda_k^I) = (\log(\lambda^C) + \gamma_k/2) - (\log(\lambda_k^I) - \gamma_k/2)$$

If we then split the random effect associated with the variation between studies γ_k equally, and divide the terms by experimental group $J \in \{C, I\}$, we obtain the following system for the control C and intervention I groups' kth parameter, λ_k^J.

$$\lambda_k^C = \lambda^C \exp(\gamma_k/2)$$
$$\lambda_k^I = \lambda^I \exp(-\gamma_k/2)$$

1. Draw a measure of variation for the kth study from $N(0, \tau^2)$ and calculate λ_I from fixed values, the ratio of medians, ρ, and the control group's rate parameter λ_C;
2. Calculate the rate parameters for the control, λ_k^C, and intervention, λ_k^I, groups for the kth study;
3. Draw a random samples of size n_k^J from Exponential(λ_k^J), for $J \in \{C, I\}$.

The sample size n_k^J for the Jth group of the kth study can also be sampled, by assuming $N_k := n_k^C + n_k^I$ and drawing N_k from a uniform distribution Uniform(a, b), where the minimum a, and maximum b, reflect knowledge about the domain of interest. The proportion of N_k given to n_k^I can be drawn from a beta distribution. But we shall omit the derivations of these sampling distributions, in the interests of brevity.

In the sampling steps that have been outlined, there are random values drawn, but there are also set simulation-level parameters. We may wish to see how our estimator performs for different numbers of studies, K, different expected variability between the studies, τ^2, and whether or not there is a difference between the control and intervention groups, ρ.

And finally, if we consider other distributions, with a mix of symmetric, say, normal or Cauchy distribution, and asymmetric, say, exponential or log-normal, we require different derivations for the sampling parameters.

3.5 Complexity and Formalised Analysis Structures

Via the modular nature of a research compendium R package, we can separate each layer of the algorithm into functions. We can produce automated *unit tests* for these functions that, at the very least, check that each component of the algorithm returns an output of expected type. We cannot automate the mathematical derivations, but we can produce an algorithm structure that provides far more computational confidence in implementation than a single script file in which the entire algorithm is nested.

However, structuring an analyses in research compendia is more challenging than simply coding directly into a .R script. Thus, there is benefit to outlining the computational workflow. We now turn to the practical *toolchain walkthrough* for establishing these analyses as research compendia. We may not be able to prepare for all errors, but we can aim to weather *most* problems that arise in the computational implementation of mathematical algorithms.

4   Research Compendia Toolchain Walkthrough

We now aim to provide a practical guide to computational research compendia for the comparative analysis, `varameta::`, and the simulation algorithm, `simeta::`, that supports it. As this is a first effort at a toolchain walkthrough, there will likely be aspects that are overlooked or underdeveloped.

4.1   DevOps

The DevOps section of this toolchain walkthrough aims to cover computational tools, why they were chosen, as well as some guidance as to how to source them.

Intended Audience. A toolchain walkthrough is a documentation of a specific scientific workflow created by a scientist who utilised this workflow for research. We begin by identifying the audience targeted who may benefit from detailing the minutiae of this process. We do not seek to generalise, but rather to provide a workflow that reflects the author's knowledge of good enough practices in scientific computing for this task, optimised for efficiency, scientific rigour, and, in the spirit of the gaming walkthrough: *fun*.

This toolchain walkthrough assumes an R user whose expertise is not primarily in computing, but rather a researcher who employs R for analysis in a discipline such as statistics, psychology, archaeology, or ecology. We make an effort to cover some of the less familiar aspects of computational workflow, such as shell commands, that might be considered trivial to a formally trained computer scientist.

Although many R users have gaps in their formal computational science education, researchers who utilise R are often implementing complex algorithms, such as the one outlined in Sect. 3.2, which describes the simulation of meta-analysis data for coverage probability simulation.

Burn It Down. This section only applies for work that has already begun. However, this is often the case for the development of a scientific project. We frequently have work that begin as small scripts, that develop in complexity and requirements.

In recognition of the ofttimes overwhelming density of resources, we list a few bash shell commands here that are particularly useful for moving files around when setting up an analysis as a research compendium. We enclose user input in <> and describe the utility of the command after #. A directory is colloquially referred to as a folder. These can be executed from a terminal.

```
.  # here
..  # up one
cd <directory path> # change location of .
ls -a # list files in .
cp <file> <toplace> # copy
mv <file> <toplace> # move or rename
rm -rf <directory> # remove directory and its contents
locate <partoffilename> # find a file
mkdir <directory> # create a directory
```

How to Code. The R software environment can be downloaded from R: The R Project for Statistical Computing. There are several excellent resources for getting started with programming with R. We list an opinionated selection here, chosen for clarity and enjoyment, all of which are freely available online:

- Learning Statistics with R by Danielle Navarro [20],
- R for Data Science by Grolemund Garrett and Hadley Wickham [9],
- R Cookbook by J.D. Long and Paul Teetor [17].

We now assume a working knowledge of the R programming language, as the intended audience of this toolchain workflow are researchers who have a working level of programming proficiency in R.

Where to Code. In this toolchain walkthrough, we emphasise cross-platform open-source software. There is, of course, the immediate benefit of accessibility. Furthermore, open-source invites an evolutionary development community where many can contribute small solutions that integrate to solve larger problems. RStudio is an integrated development environment for writing in the statistical language R. RStudio is *cross-platform* in that it can be installed on Windows, Macintosh, and Linux operating systems. There are many further advantages to this widely-used environment. For example, the `citr::` add-in [1] modifies RStudio to enable a connection to the open-source reference manager Zotero. Another example is the `datapasta::` [19] add-in that enables copy-paste of tables into R-formatted script.

4.2   Create Compendium Architecture

As `varameta::` is a research compendium containing comparative analyses and `simeta::` a package to provide simulation tools, the creation process for these two compendia are different.

We make use of two R packages, `rrtools::` [18] and `usethis::` [31], to assist in automating these tasks.

Compendiumise `varameta::`.

1. Open RStudio and close project via the toolbar File menu,
2. In the Console, set the working directory to desired location; e.g.,

```
> getwd()
[1] "/home/charles"
> setwd("Documents/repos/")
> getwd()
[1] "/home/charles/Documents/repos",
```

3. and `rrtools::use_compendium("varameta")`,
4. and update `DESCRIPTION` file with author, title, etc.,
5. Create analysis file structure with `rrtools::use_analysis()`.

For `varameta::`, we will have several reproducible documents that will form the basis of the analysis, as well as figures to contribute to the associated publication. The final step above automates the creation of a directory structure for a paper, figures, data, and templates.

Compendiumise `simeta::`. In this case, the file structure is less involved, however the testing structure will be need to be considerably more robust because of the complexity of the simulation algorithm described in Sect. 3.2:

1. Create a package with `usethis::create_package()`,
2. Switch to the package directory with `usethis::project_activate()`.

4.3   Common Steps Across both Packages

1. Set open source licence, with

    `usethis::use_mit_license(name = "Charles Gray");`

    this 'simple and permissive' choice of licence [31] serves the purpose of a comparative analysis of estimators,
2. Set up documentation for functions with `usethis::use_roxygen_md()`,
3. Set up data for internal datasets and examples with `usethis::use_data()`.

Connecting to GitHub. There are benefits to implementing a version control system, such as via the Git language and GitHub online repository archive, beyond the ability to trace work back to an earlier iteration [4]. The added benefit, arguably even greater benefit, is that of collaborative science. Storing work on GitHub allows for instantaneous sharing of code and analyses, and collaborative work with advanced project planning features, enabling other scientists to make very specific comments on work in progress.

Data Ethics and Further Considerations. In the case of `varameta::`'s estimators for meta-analysing medians, and `simeta::` for simulating meta-analysis estimators, there are no ethics in data considerations beyond ensuring contributors are recognised and credited for their work by time of publication. For some disciplines, sharing geographic locations might be an ethical consideration, say, in preventing fossil hunters from exploiting palaeontology sites [11]. Personal details, must, too be considered, that might inadvertently identify people and violate privacy considerations. Furthermore, various allowances might need to be made for institutional workflow. We note these here as a possible considerations, but as our case studies do not have such requirements, we now consider our research compendia instantiated.

However, as this algorithm has significant complexity, we need to include unit tests to provide confidence in our results, as we argue in the companion computational metamathematics manuscript [8], which motivates the practical steps laid out here.

5   Testing

We now expand in a practical sense on unit testing, which, in the theoretical companion manuscript, we describe 'the software engineering tool that provides a key piece of the correspondence between scientific claim and programming' [8]. It is in this manuscript that we sought to answer the question: why test? In this toolchain walkthrough, we will focus on the practical implementation of first unit tests.

5.1   What Is a Test?

Tests are collected in contexts. Each test comprises congruous expectation functions.

In the *head* of the 'bug hunt' context (under `context("bug hunt")`), we find the loading of packages. A seed is then set for reproducibility of errors. The first test, `"metasim runs for different n"`, tests the `simeta::metasim()` function for different orders of magnitude of `trials`. As each trial samples new data, this is the most direct way to test the scalability of the function for large datasets. We then follow up with a test that checks that the exponential distribution can be passed to all levels in the algorithm.

```
context("bug hunt")

set.seed(38)
library(tidyverse)
library(metasim)

test_that("metasim runs for different n", {
  expect_is(metasim(), 'data.frame')
```

```
  expect_is(metasim(trials = 100) , "data.frame")
  # expect_is(metasim(trials = 1000) , "data.frame")
})


test_that("exponential is parsed throughout", {

  # check sample
  expect_equal(
  sim_sample(10, rdist = "exp",
    par = list(rate = 3)) %>% length, 10)
  # check samples
  ...
```

5.2 Non-empty Thing of Expected Type

Simply asking '*does a function produce the expected output?*', induces a surprising number of considerations. To illustrate this, we return to our case studies.

Testing a Collection of Estimators in varameta::. In the interests of mathematical and computational brevity, we focus on one distributional example: the simple case of the exponential distribution, which is characterised by a single parameter. We return to the estimator of the rate $\hat{\lambda} := \log 2/m$ derived for the exponential distribution, as discussed in Sect. 3.4 and defined in Equation (1), explicitly coded in R.

```
function(n, median) {

  # Estimate parameters.
  lambda <- log(2) / median

  # Approximate the standard error of the sample median.
  1 / (2 * sqrt(n) * dexp(median, rate = lambda))

}
```

We create a context file, **tests/testthat/test-exponential.R** and provide a short context description in the first line of the script.

```
context("exponential estimator")
```

As a starting point, we can write unit tests to automate a check that this function returns non-empty thing of expected type. We arbitrarily choose values, a sample size of 10, and a proposed sample median of 4, for instance. The function should return a numeric **double** value, and should be positive.

```
test_that("non-empty thing of expected type, for fixed values", {

  # returns numeric
  expect_type(g_exp(10, 4), "double")

  # returns positive number
  expect_gt(g_exp(10, 4), 0)

})
```

In addition to choosing explicit values, we can also randomly sample the sample size n, and sample median m. To ensure reproducibility of these testing results on any machine, we set a random seed, passing `set.seed` an arbitrary numeric value.

```
set.seed(39) # ensures reproducibility of test results
```

```
# sample fuzz testing parameters
n <- sample(seq(2, 100), 1)
m <- runif(1, 1, 100)
```

We can then use these random *fuzz* values [16] to produce analogous unit tests for non-empty thing of expected type.

```
test_that("non-empty thing of expected type, for random values", {
  expect_type(g_exp(n, m), "double")
  expect_gt(g_exp(n, m), 0)
})
```

We can extend these tests to cover expected input errors. For example, we wish this function to fail when passed negative numbers. The sample size cannot be less than or equal to 0, and due to the logarithm, the function only works for positive sample medians. Here, we include the fixed and randomised values in the same test.

```
test_that("negative numbers throw an error", {
  expect_error(g_exp(-3, 4))
  expect_error(g_exp(3, -4))
  # with fuzz testing
  expect_error(g_exp(-n, m))
  expect_error(g_exp(n, -m))
})
```

Running all tests in a context tells us if the function is behaving as expected. The more tests we write, the more confidence we will have that our function behaves as we intended it to.

```
==> Testing R file using 'testthat'

Loading varameta
|  OK F W S | Context
|   8       | exponential estimator

 Results
OK:        8
Failed:    0
Warnings:  0
Skipped:   0
```

Test complete

There is a tradeoff with tests, in terms of time taken by updating the tests themselves. Here a test requires updating from an expected output of a numeric vector, to a dataframe. The function that is being tested.

```
==> Testing R file using 'testthat'

Loading simeta
|  OK F W S | Context
|   6 1     | bug hunt [7.1 s]

test-bug-hunt.R:21: failure: exponential is parsed throughout
sim_stats(rdist = "exp", par = list(rate = 3)) inherits from
'tbl_df/tbl/data.frame' not 'numeric'.


 Results
Duration: 7.1 s

OK:        6
Failed:    1
Warnings:  0
Skipped:   0
```

Test complete

Testing a Nested Algorithm in simeta::. Our other case study provides an example of a nested algorithm. In addition to ensuring each function returns a non-empty thing of expected type, we can automate checks that the functions form a toolchain. In the first place, it is helpful to know that our functions continue to form a toolchain under default settings.

We begin by setting our context. In this case, as we are running our functions on default settings, we do not require randomly sampled fuzz parmeters.

```
context("default pipeline")
```

We now check that the algorithm runs 'upwards', by running a test from most granular function in the algorithm to most nested. We could write a similiarly inverted test, from most nested function, downwards to most granular.

```
test_that("work upwards through algorithm", {
  expect_is(sim_n(), "data.frame")
  expect_gt(sim_n() %>% nrow(), 1)
  # sim_df calls sim_n
  expect_is(sim_df(), "data.frame")
  expect_is(sim_stats(), "data.frame")
  # metasim calls metatrial
  expect_is(metatrial(), "data.frame")
  expect_is(singletrial(), "data.frame") # alternate trial
  expect_is(metasim(trials = 3), "data.frame")
  # metasims calls sim_df & metasim
  expect_is(metasims(
    single_study = FALSE,
    trials = 3,
    progress = FALSE
  ),
  "sim_ma")
})
```

Now, if this test fails, we will know the combination of functions fails at some point in the nested algorithm. We follow this upwards test with a series of small tests for each function set to defaults to identify at which point in the pipeline where the algorithm fails, if the 'work upwards' test fails.

```
# test each component on defaults

test_that("sim_n", {
  expect_is(sim_n(), "data.frame")
})

test_that("sim_df", {
  expect_is(sim_df(), "data.frame")
})

test_that("metatrial", {
  # metasim calls metatrial
  expect_is(metatrial(), "data.frame")
})

test_that("singletrial", {
  expect_is(singletrial(), "data.frame") # alternate trial
```

```
})

test_that("metasim", {
  expect_is(metasim(trials =  3), "data.frame")
})

test_that("metasims", {
  expect_is(metasims(
    single_study = FALSE,
    trials = 3,
    progress = FALSE
  ),
  "list")
})
```

And we can now run all tests, for a starting point of automating checks that our algorithm runs on default settings.

```
==> Testing R file using 'testthat'

Loading simeta
  |   OK F W S | Context
  |   14       | default pipeline [28.7 s]

 Results
Duration: 28.7 s

OK:       14
Failed:   0
Warnings: 0
Skipped:  0
```

Test complete

To demonstrate how informative testing can be in identifying where an algorithm breaks, we now modify the `simeta::metasim` function to return a character string, `"error"`. Testing the default pipeline reveals where the algorithm is broken. Debugging is where the advantage of testing is exposed, and thus, arguably the requirement for testing increases with complexity of algorithm. Detailed output have been omitted for brevity.

```
==> Testing R file using 'testthat'

Loading simeta
 |  OK F W S | Context
 |  10 4     | default pipeline [32.3 s]

test-default-pipeline.R:12: failure: work upwards through algorithm
metasim(trials = 3) inherits from 'character' not 'data.frame'.

test-default-pipeline.R:14: error: work upwards through algorithm
Argument 1 must have names
 ...

test-default-pipeline.R:43: failure: metasim
metasim(trials = 3) inherits from 'character' not 'data.frame'.

test-default-pipeline.R:47: error: metasims
Argument 1 must have names
 ...


 Results
Duration: 32.3 s

OK:        10
Failed:    4
Warnings:  0
Skipped:   0

Test complete
```

From this output, we can see not only where the algorithm fails, but also what other functions fail because of a reliance on the elements that have failed.

5.3   Test-Driven Development

As we build new features into our package, such as checking that the single-trial setting works in the simulation function from `simeta::`, we can focus on a writing new tests that ensure our feature works within the ecosystem of our algorithm as expected. We can develop our algorithm from a testing setting, rather than focusing on rewriting functions and script files.

Another overview check that we can incorporate is from the `covr::` package [10]. Using `covr::package_coverage()`, we can check what proportion of lines of code have been tested in each function.

For the `varmeta::` package, at the time of writing, we have the following test coverage.

```
varameta Coverage: 90.00%
R/g_cauchy.R: 44.44%
R/g_norm.R: 71.43%
R/hozo_se.R: 92.31%
R/bland_mean.R: 100.00%
R/bland_se.R: 100.00%
R/effect_se.R: 100.00%
R/g_exp.R: 100.00%
R/g_lnorm.R: 100.00%
R/hozo_mean.R: 100.00%
R/wan_mean_C1.R: 100.00%
R/wan_mean_C2.R: 100.00%
R/wan_mean_C3.R: 100.00%
R/wan_se_C1.R: 100.00%
R/wan_se_C2.R: 100.00%
R/wan_se_C3.R: 100.00%
```

This is enables us to target specific functions that may require further testing. Testing lines of code is somewhat a blunt instrument, as we are not ensuring tests for every combination of inputs. However, test coverage is still an informative measure of software reliability. For example, here we see not all code in the g_* estimators have been checked.

These notes on testing are not intended to be comprehensive, but only aim to give the user an starting point for the initialisation of summarising an analysis in a reproducible research compendia, with an informative level of automated checks. Given only one quarter of packages on the largest R package repository CRAN have unit tests at all [8], it is arguable that there is much further scope for discussion and development with respect to the adoption of automated tests in reproducible research compendia.

6   Prepare for *most* weather conditions

Computational proof may be unachievable, however, a measure of code::proof can be attained by structuring research compendia in a standardised reproducible format, such as produced by rrtools:: [18]. Perhaps we cannot prove our software in the traditional mathematical sense [8]. However, we could consider building confidence in the mathematics that we implement computationally, like waterproofing our shoes. If we step in a big enough puddle, our feet are still going to get wet, but at least we have prepared to weather *most* of the problems associated with the implementation of statistical algorithms.

References

1. Aust, F.: Citr: 'RStudio' add-in to insert markdown citations (2018). https://github.com/crsh/citr. R package version 0.3.0
2. Belmonte, A.: The tangled web of self-tying knots. Proc. Natl. Acad. Sci. **104**(44), 17243–17244 (2007). https://doi.org/10.1073/pnas.0708150104. https://www.pnas.org/content/104/44/17243
3. Bland, M.: Estimating mean and standard deviation from the sample size, three quartiles, minimum, and maximum. Int. J. Stat. Med. Res. **4**(1), 57–64 (2014). http://lifescienceglobal.com/pms/index.php/ijsmr/article/view/2688
4. Bryan, J.: Excuse me, do you have a moment to talk about version control? PeerJ PrePrints **5**, e3159 (2017)
5. Consalvo, M.: Zelda 64 and video game fans: a walkthrough of games, intertextuality, and narrative. Televis. New Media **4**(3), 321–334 (2003). https://doi.org/10.1177/1527476403253993
6. Ford, D., Smith, J., Guo, P.J., Parnin, C.: Paradise unplugged: identifying barriers for female participation on stack overflow. In: Proceedings of the 2016 24th ACM SIGSOFT International Symposium on Foundations of Software Engineering, FSE 2016, Seattle, WA, USA, pp. 846–857. ACM, New York (2016). https://doi.org/10.1145/2950290.2950331
7. Fraser, H., Parker, T., Nakagawa, S., Barnett, A., Fidler, F.: Questionable research practices in ecology and evolution. PLOS One **13**(7), e0200303 (2018). https://doi.org/10.1371/journal.pone.0200303. https://journals.plos.org/plosone/article?id=10.1371/journal.pone.0200303. Bibtex*[shortjournal=PLOS ONE]
8. Gray, C.T., Marwick, B.: Truth, proof, and reproducibility: there's no counterattack for the codeless. arXiv:1907.05947 [math], July 2019
9. Grolemund, G., Wickham, H.: R for data science (2017). https://r4ds.had.co.nz/
10. Hester, J.: covr: Test Coverage for Packages (2018). https://CRAN.R-project.org/package=covr
11. Hopkin, M.: Palaeontology journal will 'fuel black market'. Nature **445**, 234–235 (2007). https://doi.org/10.1038/445234b. https://www.nature.com/articles/445234b
12. Hozo, S.P., Djulbegovic, B., Hozo, I.: Estimating the mean and variance from the median, range, and the size of a sample. BMC Med. Res. Methodol. **5**(1), 13 (2005). https://doi.org/10.1186/1471-2288-5-13
13. Httermann, M.: DevOps for Developers. Apress, New York (2012). google-Books-ID: JfUAkB8AA7EC
14. Kafka, F.: The Trial, April 2005. http://www.gutenberg.org/ebooks/7849
15. Katz, D.S., McHenry, K.: Super RSEs: combining research and service in three dimensions of Research Software Engineering, July 2019. https://danielskatzblog.wordpress.com/2019/07/12/
16. Klees, G., Ruef, A., Cooper, B., Wei, S., Hicks, M.: Evaluating fuzz testing. In: Proceedings of the 2018 ACM SIGSAC Conference on Computer and Communications Security, CCS 2018, Toronto, Canada, pp. 2123–2138. ACM, New York (2018). https://doi.org/10.1145/3243734.3243804. http://doi.acm.org/10.1145/3243734.3243804
17. Long, J.J., Teetor, P.: R Cookbook, 2nd edn. (2019). https://rc2e.com/
18. Marwick, B.: rrtools: creates a reproducible research compendium (2018). https://github.com/benmarwick/rrtools

19. McBain, M., Carroll, J.: Datapasta: R tools for data copy-pasta (2018). R package version 3.0.0. https://CRAN.R-project.org/package=datapasta
20. Navarro, D.: Learning statistics with R: A tutorial for psychology students and other beginners. (Version 0.6.1) (2019). https://learningstatisticswithr.com/book/
21. Parker, H.: Opinionated analysis development. preprint (2017). https://doi.org/10.7287/peerj.preprints.3210v1
22. Ragkhitwetsagul, C., Krinke, J., Paixao, M., Bianco, G., Oliveto, R.: Toxic code snippets on stack overflow. IEEE Trans. Softw. Eng. 1 (2019). https://doi.org/10.1109/TSE.2019.2900307
23. Robinson, D., Hayes, A.: broom: convert statistical analysis objects into tidy tibbles (2019). https://CRAN.R-project.org/package=broom
24. Serfling, R.J.: Approximation Theorems of Mathematical Statistics. Wiley, Hoboken (2009)
25. skyisup: Deadly Boss Mods Addon Guide (2019). https://www.wowhead.com/deadly-boss-mods-addon-guide
26. UHS: Universal Hint System: Not your ordinary walkthrough. Just the hints you need (2019). http://www.uhs-hints.com/
27. Viechtbauer, W.: Conducting meta-analyses in R with the metafor package. J. Stat. Softw. **36**(3), 1–48 (2010). http://www.jstatsoft.org/v36/i03/
28. Wan, X., Wang, W., Liu, J., Tong, T.: Estimating the sample mean and standard deviation from the sample size, median, range and/or interquartile range. BMC Med. Res. Methodol. **14**(1), 135 (2014). https://doi.org/10.1186/1471-2288-14-135
29. Wickham, H.: R Packages: Organize, Test, Document, and Share Your Code. O'Reilly Media, Newton (2015). https://books.google.com.au/books?id=DqSxBwAAQBAJ. Bibtex*[lccn=2015472811]
30. Wickham, H.: Advanced R. Routledge, Boca Raton (2014)
31. Wickham, H., Bryan, J.: usethis: automate package and project setup (2019). https://CRAN.R-project.org/package=usethis
32. Wickham, H., Danenberg, P., Eugster, M.: Roxygen2: in-line documentation for R (2019). R package version 6.1.1.9000. https://github.com/klutometis/roxygen
33. Wilson, G., et al.: Best Practices for scientific computing. PLoS Biol. **12**(1), e1001745 (2014). https://doi.org/10.1371/journal.pbio.1001745
34. Wilson, G., Bryan, J., Cranston, K., Kitzes, J., Nederbragt, L., Teal, T.K.: Good enough practices in scientific computing. PLOS Comput. Biol. **13**(6), e1005510 (2017). https://doi.org/10.1371/journal.pcbi.1005510
35. Wyatt, C.: Research Software Engineers Association (2019). https://rse.ac.uk/

Regularized Estimation and Feature Selection in Mixtures of Gaussian-Gated Experts Models

Faïcel Chamroukhi[1(⊠)], Florian Lecocq[2], and Hien D. Nguyen[3]

[1] Department of Mathematics, University of Queensland,
Brisbane, QLD 4072, Australia
`faicel.chamroukhi@unicaen.fr`
[2] Laboratory of Mathematics Nicolas Oresme LMNO - UMR CNRS,
University of Caen, Unicaen Campus 2, 14000 Caen, France
[3] Department of Mathematics and Statistics, La Trobe University,
Melbourne, VIC 3086, Australia

Abstract. Mixtures-of-Experts models and their maximum likelihood estimation (MLE) via the EM algorithm have been thoroughly studied in the statistics and machine learning literature. They are subject of a growing investigation in the context of modeling with high-dimensional predictors with regularized MLE. We examine MoE with Gaussian gating network, for clustering and regression, and propose an ℓ_1-regularized MLE to encourage sparse models and deal with the high-dimensional setting. We develop an EM-Lasso algorithm to perform parameter estimation and utilize a BIC-like criterion to select the model parameters, including the sparsity tuning hyperparameters. Experiments conducted on simulated data show the good performance of the proposed regularized MLE compared to the standard MLE with the EM algorithm.

Keywords: Mixtures-of-experts · Clustering · Feature selection · EM algorithm · Lasso · High-dimensional data

1 Introduction

Mixture-of-experts (MoE), originally introduced in [12,13], form a class of conditional mixture models [16] for modeling, clustering and prediction in the presence of heterogeneous data. Their construction rely on conditional mixture models [16] in which both the gating network, formed by the mixing proportions, and the experts network formed by the mixture components, depend on the predictors or the inputs. The most popular choices for the gating network are the softmax gating functions [12] or the Gaussian gating functions; the latter is a particular case of the exponential family gating functions introduced in [24].

Different choices are now common for the expert network model, depending on the type of the observed responses. For instance, a model for normal observations for regression and clustering was introduced in [5] or non-normally

© Springer Nature Singapore Pte Ltd. 2019
H. Nguyen (Ed.): RSSDS 2019, CCIS 1150, pp. 42–56, 2019.
https://doi.org/10.1007/978-981-15-1960-4_3

distributed expert models like in [1] to deal with skewed data distributions [3], to ensure robustness to outliers [2,19], or to accommodate both skewness and robustness as in [4]. A detailed review on MoE models can be found in [17].

Fitting MoE is generally performed by maximum-likelihood estimation (MLE) via the EM algorithm or its variants [8,15]. In a high-dimensional setting, the regularization of the MLE, to perform parameter estimation under a sparsity hypothesis and hence to simultaneously perform feature selection, has been studied in [14] and more recently in [7,11]. These approaches consider ℓ_1 and ℓ_2 penalties for the log-likelihood function, and are constructed upon softmax gating functions.

In this paper, we consider MoE with Gaussian gated functions, and propose an ℓ_1-regularized MLE via and EM-Lasso algorithm. We study the performance of the proposal on an experimental setup. The remainder of this paper is organized as follows. Section 2 describes the MoE modeling framework, and the Gaussian-gated MoE and its MLE with the EM algorithm. Then, Sect. 3 presents the proposed regularized MLE and the EM-Lasso algorithm. Finally, Sect. 4 is dedicated to numerical experiments.

2   Gaussian-Gated Mixture-of-Experts

2.1   MoE Modeling Framework

We consider mixtures-of-experts model to relate a high-dimensional predictor $X \in \mathbb{R}^p$ to a response $Y \in \mathbb{R}^d$, potentially multivariate $d \geq 1$. We assume that the pair (X, Y) is generated from a heterogeneous population governed by a hidden structure represented by a latent categorical variable $Z \in [K] = \{1, \ldots, K\}$. Assume that we observe a random sample $\{(X_i, Y_i)\}_{i=1,\ldots,n}$ of n independently and identically distributed (i.i.d) pairs (X_i, Y_i) from (X, Y), and let $\mathcal{D} = ((x_1, y_1), \ldots, (x_n, y_n))$ be an observed data sample. Assume that the pair (X, Y) follows a MoE distribution, then the MoE model can be defined as

$$f(y_i|x_i; \Psi) = \sum_{k=1}^{K} g_k(x_i; \mathbf{w}) f(y_i|x_i; \theta_k) \tag{1}$$

where $g_k(x; \mathbf{w}) = \mathbb{P}(Z = k|X = x; \mathbf{w})$ is the distribution of the hidden variable Z given the predictor x with parameters $\mathbf{w}$, which represents the gating network, and the conditional component densities $f(y|x; \theta_k) = f(y_i|X = x, Z = k; \theta)$ represent the experts network whose parameters are θ_k.

2.2   Gaussian-Gated Mixture-of-Experts

Let us define by $\phi_m(v; m, \mathbf{C}) = (2\pi)^{-m/2} |\mathbf{C}|^{-1/2} \exp\left(-\frac{1}{2}(v - m)^\top \mathbf{C}^{-1}(v - m)\right)$ the probability density function of a Gaussian random vector V of dimension m with mean m and covariance matrix $\mathbf{C}$. We consider mixture-of-experts for clustering and regression of heterogeneous data. In this case, the mixture of

Gaussian-gated experts models, we abbreviate as MoGGE, for multivariate real responses, is defined by (1) where the experts are (multivariate) Gaussian regressions, given by

$$f(\boldsymbol{y}_i|\boldsymbol{x}_i;\boldsymbol{\theta}_k) = \phi_d(\boldsymbol{y}_i;\mathbf{a}_k + \mathbf{B}_k^T\boldsymbol{x}_i,\boldsymbol{\Sigma}_k) \tag{2}$$

and the gating network $g(\boldsymbol{x}_i;\mathbf{w})$ is defined by Gaussian gating function of the form:

$$g_k(\boldsymbol{x}_i;\mathbf{w}) = \frac{\mathbb{P}(Z_i = k)f(\boldsymbol{x}_i|Z_i = k;\boldsymbol{w}_k)}{\sum_{\ell=1}^K \mathbb{P}(Z_i = \ell)f(\boldsymbol{x}_i|Z_i = \ell;\boldsymbol{w}_\ell)} = \frac{\alpha_k\phi_p(\boldsymbol{x}_i;\boldsymbol{\mu}_k,\mathbf{R}_k)}{\sum_{\ell=1}^K \alpha_\ell\phi_p(\boldsymbol{x}_i;\boldsymbol{\mu}_\ell,\mathbf{R}_\ell)} \tag{3}$$

with $\mathbb{P}(Z_i = k) = \alpha_k$, $f(\boldsymbol{x}_i|Z_i = k;\mathbf{w}) = \phi_p(\boldsymbol{x}_i;\boldsymbol{\mu}_k,\mathbf{R}_k)$ $(k = 1,\ldots,K)$. This Gaussian gating network was introduced in [24] to sidestep the need for a non-linear optimization routine in the inner loop of the EM algorithm in the case of a softmax function for the gating network. The MoGGE model is thus parameterized by the parameter vector $\boldsymbol{\Psi} = (\mathbf{w}^T,\boldsymbol{\theta}^T)^T$ where $\mathbf{w} = (\boldsymbol{w}_1^T,\ldots,\boldsymbol{w}_K^T)^T$ is the parameter vector of the gating network and $\boldsymbol{\theta} = (\boldsymbol{\theta}_1^T,\ldots,\boldsymbol{\theta}_K^T)^T$ is the parameter vector of experts network, with $\boldsymbol{w}_k = (\alpha_k,\boldsymbol{\mu}_k^T,\text{vech}(\mathbf{R}_k)^T)^T$ and $\boldsymbol{\theta}_k = (\mathbf{a}_k^T,\mathbf{B}_k^T,\text{vech}(\boldsymbol{\Sigma}_k)^T)^T$ for $k = 1,\ldots,K$. The approximation capabilities of this model have been studied very recently in [18].

2.3   Maximum Likelihood Estimation via the EM Algorithm

Mixtures-of-experts of the form (1) with softmax gating functions are in general estimated by maximizing the (conditional) log-likelihood $\sum_{i=1}^n \log f(\boldsymbol{y}_i|\boldsymbol{x}_i;\boldsymbol{\Psi})$ by using the EM algorithm , in which the M-step requires an internal iterative numerical optimization procedure (eg. a Newton-Raphson algorithm) to update the softmax parameters. We follow the approach of estimating MoGGE in [24], which relies on maximizing the joint loglikelihood, and in the MLE, the M-Step can then be solved in a closed form. Indeed, based on Eqs. (1), (2), and (3), we the MoGGE conditional density is given by:

$$f(\boldsymbol{y}_i|\boldsymbol{x}_i;\boldsymbol{\Psi}) = \sum_{k=1}^K \frac{\alpha_k\phi_p(\boldsymbol{x}_i;\boldsymbol{\mu}_k,\mathbf{R}_k)}{\sum_{\ell=1}^K \alpha_\ell\phi_p(\boldsymbol{x}_i;\boldsymbol{\mu}_\ell,\mathbf{R}_\ell)}\phi_d(\boldsymbol{y}_i;\mathbf{a}_k + \mathbf{B}_k^T\boldsymbol{x}_i,\boldsymbol{\Sigma}_k)\cdot \tag{4}$$

Then we can write the joint density as:

$$f(\boldsymbol{y}_i,\boldsymbol{x}_i;\boldsymbol{\Psi}) = f(\boldsymbol{x}_i;\mathbf{w})f(\boldsymbol{y}_i|\boldsymbol{x}_i;\boldsymbol{\theta}) = \sum_{k=1}^K \mathbb{P}(Z_i = k)f(\boldsymbol{x}_i;\boldsymbol{w}_k)f(\boldsymbol{y}_i|\boldsymbol{x}_i;\boldsymbol{\theta}_k)$$

$$= \sum_{k=1}^K \alpha_k\phi_p(\boldsymbol{x}_i;\boldsymbol{\mu}_k,\mathbf{R}_k)\phi_d(\boldsymbol{y}_i;\mathbf{a}_k + \mathbf{B}_k^T\boldsymbol{x}_i,\boldsymbol{\Sigma}_k)\cdot \tag{5}$$

The joint log-likelihood to be maximized by EM is therefore given by:

$$L(\boldsymbol{\Psi}) = \sum_{i=1}^n \log f(\boldsymbol{y}_i,\boldsymbol{x}_i;\boldsymbol{\Psi}) = \sum_{i=1}^n \log \sum_{k=1}^K \alpha_k\phi_p(\boldsymbol{x}_i;\boldsymbol{\mu}_k,\mathbf{R}_k)\phi_d(\boldsymbol{y}_i;\mathbf{a}_k + \mathbf{B}_k^T\boldsymbol{x}_i,\boldsymbol{\Sigma}_k)\cdot \tag{6}$$

2.4   The EM Algorithm for the MoGGE Model

The complete-data log-likelihood upon which the EM principle is constructed is then defined by

$$L_c(\boldsymbol{\Psi}) = \sum_{i=1}^{n} \sum_{k=1}^{K} Z_{ik} \log \left[\alpha_k \phi_p(\boldsymbol{x}_i; \boldsymbol{\mu}_k, \mathbf{R}_k) \phi_d(\boldsymbol{y}_i; \mathbf{a}_k + \mathbf{B}_k^T \boldsymbol{x}_i, \boldsymbol{\Sigma}_k) \right] \qquad (7)$$

where Z_{ik} being an indicator binary-valued variable such that $Z_{ik} = 1$ if $Z_i = k$ (i.e., if the ith pair $(\boldsymbol{x}_i, \boldsymbol{y}_i)$ is generated from the kth expert and $Z_{ik} = 0$ otherwise. The EM algorithm, after starting with an initial solution $\boldsymbol{\Psi}^{(0)}$, alternates between the E- and the M- Steps until convergence (when there is no longer a significant change in the log-likelihood (6)).

E-step: Compute the expectation of the complete-data log-likelihood (7), given the observed data $\mathcal{D}$ and the current parameter vector estimate $\boldsymbol{\Psi}^{(q)}$:

$$Q(\boldsymbol{\Psi}; \boldsymbol{\Psi}^{(q)}) = \mathbb{E}\left[L_c(\boldsymbol{\Psi}) | \mathcal{D}; \boldsymbol{\Psi}^{(q)} \right]$$
$$= \sum_{i=1}^{n} \sum_{k=1}^{K} \tau_{ik}^{(q)} \log \left[\alpha_k \phi_p(\boldsymbol{x}_i; \boldsymbol{\mu}_k, \mathbf{R}_k) \phi_d(\boldsymbol{y}_i; \mathbf{a}_k + \mathbf{B}_k^T \boldsymbol{x}_i, \boldsymbol{\Sigma}_k) \right], \quad (8)$$

where:

$$\tau_{ik}^{(q)} = \mathbb{P}(Z_i = k | \boldsymbol{y}_i, \boldsymbol{x}_i; \boldsymbol{\Psi}^{(q)}) = \frac{\alpha_k^{(q)} \phi_p(\boldsymbol{x}_i; \boldsymbol{\mu}_k^{(q)}, \mathbf{R}_k^{(q)}) \phi_d(\boldsymbol{y}_i; \mathbf{a}_k^{(q)} + \mathbf{B}_k^{(q)^T} \boldsymbol{x}_i, \boldsymbol{\Sigma}_k^{(q)})}{f(\boldsymbol{x}_i, \boldsymbol{y}_i; \boldsymbol{\Psi}^{(q)})},$$
$$(9)$$

is the posterior probability that the observed pair $(\boldsymbol{x}_i, \boldsymbol{y}_i)$ is generated by the kth expert. This step therefore only requires the computation of the posterior component membership probabilities $\tau_{ik}^{(q)}$ $(i = 1, \ldots, n)$, for $k = 1, \ldots, K$.

M-step: Calculate the parameter vector update $\boldsymbol{\Psi}^{(q+1)}$ by maximizing the Q-function (8), i.e, $\boldsymbol{\Psi}^{(q+1)} = \arg\max_{\boldsymbol{\Psi}} Q(\boldsymbol{\Psi}; \boldsymbol{\Psi}^{(q)})$. By decomposing the Q–function (8) as

$$Q(\boldsymbol{\Psi}; \boldsymbol{\Psi}^{(q)}) = \sum_{k=1}^{K} Q(\boldsymbol{w}_k; \boldsymbol{\Psi}^{(q)}) + Q(\boldsymbol{\theta}_k; \boldsymbol{\Psi}^{(q)}) \qquad (10)$$

where

$$Q(\boldsymbol{w}_k; \boldsymbol{\Psi}^{(q)}) = \sum_{i=1}^{n} \tau_{ik}^{(q)} \log \left[\alpha_k \phi_p(\boldsymbol{x}_i; \boldsymbol{\mu}_k, \mathbf{R}_k) \right] \qquad (11)$$

and

$$Q(\boldsymbol{\theta}_k; \boldsymbol{\Psi}^{(q)}) = \sum_{i=1}^{n} \tau_{ik}^{(q)} \log \phi_d(\boldsymbol{y}_i; \mathbf{a}_k + \mathbf{B}_k^T \boldsymbol{x}_i, \boldsymbol{\Sigma}_k), \qquad (12)$$

the maximization can then be done by performing K separate maximizations w.r.t the gating network parameters and the experts network parameters.

Updating the Gating Networks' Parameters: Maximizing (11) w.r.t $\boldsymbol{w}_k$'s corresponds to the M-Step of a Gaussian Mixture Model [16]. The closed-form expressions for updating the parameters are given by:

$$\alpha_k^{(q+1)} = \sum_{i=1}^{n} \tau_{ik}^{(q)} \Big/ n, \tag{13}$$

$$\boldsymbol{\mu}_k^{(q+1)} = \sum_{i=1}^{n} \tau_{ik}^{(q)} \boldsymbol{x}_i \Big/ \sum_{i=1}^{n} \tau_{ik}^{(q)}, \tag{14}$$

$$\mathbf{R}_k^{(q+1)} = \sum_{i=1}^{n} \tau_{ik}^{(q)} (\boldsymbol{x}_i - \boldsymbol{\mu}_k^{(q+1)})(\boldsymbol{x}_i - \boldsymbol{\mu}_k^{(q+1)})^T \Big/ \sum_{i=1}^{n} \tau_{ik}^{(q)}. \tag{15}$$

Updating the Experts' Network Parameters. Maximizing (12) w.r.t $\boldsymbol{\theta}_k$'s corresponds to the M-Step of standard MoE with multivariate Gaussian regression experts, see e.g [6]. The closed-form updating formulas are given by:

$$\mathbf{a}_k^{(q+1)} = \sum_{i=1}^{n} \tau_{ik}^{(q)} (\boldsymbol{y}_i - \mathbf{B}_k^{(q)T} \boldsymbol{x}_i) \Big/ \sum_{i=1}^{n} \tau_{ik}^{(q)}, \tag{16}$$

$$\mathbf{B}_k^{(q+1)} = \Big[\sum_{i=1}^{n} \tau_{ik}^{(q)} \boldsymbol{x}_i \boldsymbol{x}_i^T \Big]^{-1} \sum_{i=1}^{n} \tau_{ik}^{(q)} \boldsymbol{x}_i (\boldsymbol{y}_i - \mathbf{a}_k^{(q+1)})^T, \tag{17}$$

$$\boldsymbol{\Sigma}_k^{(q+1)} = \sum_{i=1}^{n} \tau_{ik}^{(q)} (\boldsymbol{y}_i - (\mathbf{a}_k^{(q+1)} + \mathbf{B}_k^{(q+1)T} \boldsymbol{x}_i))(\boldsymbol{y}_i - (\mathbf{a}_k^{(q+1)} + \mathbf{B}_k^{(q+1)T} \boldsymbol{x}_i))^T \Big/ \sum_{i=1}^{n} \tau_{ik}^{(q)}. \tag{18}$$

However, in a high dimensional setting, MLE may be unstable or even unfeasible. One possible way to proceed in such a context is the regularization of the objective function. In the context of MoE models, this has been studied namely in [7,11,14] where ℓ_1 and ℓ_2 regularization for the log-likelihood function of the standard MoE model with softmax gating network. This penalized MLE allow an efficient estimation for simultaneous parameter estimation and feature selection.

3   Penalized Maximum Likelihood Parameter Estimation

Here we study the regularized estimation of the MoGGE model. We first consider the case when $d = 1$ (univariate response $\boldsymbol{y}_i$). The expert densities are thus defined by $f(y_i|\boldsymbol{x}_i; \boldsymbol{\theta}_k) = \phi(y_i; \beta_{k,0} + \boldsymbol{\beta}_k^T \boldsymbol{x}_i, \sigma_k^2)$ with $\boldsymbol{\theta}_k = (\beta_{k,0}, \boldsymbol{\beta}_k^T, \sigma_k^2)^T$.

In our proposed approach, rather than maximizing the joint log-likelihood (6), we attempt to maximize its ℓ_1-regularized version, to encourage sparse models and to perform estimation and feature selection. The resulting penalized log-likelihood can then be defined by:

$$\mathcal{L}(\boldsymbol{\Psi}) = L(\boldsymbol{\Psi}) - \text{Pen}_{\lambda,\gamma}(\boldsymbol{\Psi}) \tag{19}$$

where $L(\boldsymbol{\Psi})$ is the observed-data log-likelihood of $\boldsymbol{\Psi}$ defined by (6) and $\text{Pen}_{\lambda,\gamma}(\boldsymbol{\Psi})$ is a Lasso [22] regularization term encouraging sparsity for the expert network parameters and the gating network parameters, with λ and γ positive real values representing tuning hyperparameters. For regularizing the expert parameters, the penalty is naturally applied to the regression coefficient vectors $\boldsymbol{\beta}_k$. For the gating network, since the estimates are those of a Gaussian mixture, we then follow the strategy of feature selection in model-based clustering in [20] in which we apply the penalty to the Gaussian mean vectors $\boldsymbol{\mu}_k$ and assume that the Gaussian covariance matrices of the gating network are diagonal, ie. $\mathbf{R}_k = \mathrm{diag}(\nu_1^2, \ldots, \nu_K^2)$. The penalty function is then given by:

$$\mathrm{Pen}_{\lambda,\gamma}(\boldsymbol{\Psi}) = \lambda \sum_{k=1}^{K} \|\boldsymbol{\beta}_k\|_1 + \gamma \sum_{k=1}^{K} \|\boldsymbol{\mu}_k\|_1. \qquad (20)$$

We now derive an EM-Lasso algorithm to maximize (19).

3.1   The EM-Lasso Algorithm for the MoGGE Model

Lets first define the penalized joint complete-data log-likelihood, which is given by

$$\mathcal{L}_c(\boldsymbol{\Psi}) = L_c(\boldsymbol{\Psi}) - \mathrm{Pen}_{\lambda,\gamma}(\boldsymbol{\Psi}) \qquad (21)$$

where $L_c(\boldsymbol{\Psi})$ is the non-regularized joint complete-data log-likelihood defined by (7). The EM-Lasso algorithm then alternates between the two following steps until convergence (when there is no significant change in (19).

E-step. This step computes the expectation of the complete-data log-likelihood (21), given the observed data $\mathcal{D}$, using the current parameter vector $\boldsymbol{\Psi}^{(q)}$:

$$\mathcal{Q}_{\lambda,\gamma}(\boldsymbol{\Psi}; \boldsymbol{\Psi}^{(q)}) = \mathbb{E}\left[\mathcal{L}_c(\boldsymbol{\Psi})|\mathcal{D}; \boldsymbol{\Psi}^{(q)}\right] = Q(\boldsymbol{\Psi}; \boldsymbol{\Psi}^{(q)}) - \mathrm{Pen}_{\lambda,\gamma}(\boldsymbol{\Psi}) \qquad (22)$$

which only requires the computation of the posterior probabilities of component membership $\tau_{ik}^{(q)}$ ($i = 1, \ldots, n$), for each of the K experts as defined by (9).

M-step. This step updates the value of the parameter vector $\boldsymbol{\Psi}$ by maximizing the Q-function (8) with respect to $\boldsymbol{\Psi}$, that is, by computing the parameter vector update $\boldsymbol{\Psi}^{(q+1)} = \arg\max_{\boldsymbol{\Psi}} \mathcal{Q}_{\lambda,\gamma}(\boldsymbol{\Psi}; \boldsymbol{\Psi}^{(q)})$. Now we have this decomposition

$$\mathcal{Q}_{\lambda,\gamma}(\boldsymbol{\Psi}; \boldsymbol{\Psi}^{(q)}) = \sum_{k=1}^{K} \mathcal{Q}_{\gamma}(\boldsymbol{w}_k; \boldsymbol{\Psi}^{(q)}) + \mathcal{Q}_{\lambda}(\boldsymbol{\Psi}_k; \boldsymbol{\Psi}^{(q)}) \qquad (23)$$

and the maximization is performed by K separate maximizations of the penalized Q-functions $\mathcal{Q}_{\gamma}(\boldsymbol{w}_k; \boldsymbol{\Psi}^{(q)})$ and $\mathcal{Q}_{\lambda}(\boldsymbol{\Psi}_k; \boldsymbol{\Psi}^{(q)})$.

Coordinate Ascent for Updating the Gating Network. Updating the gating network parameters consists of maximizing w.r.t $\boldsymbol{w}_k$ the following penalized Q-function

$$\mathcal{Q}_\gamma(\boldsymbol{w}_k;\boldsymbol{\Psi}) = \sum_{i=1}^{n}\tau_{ik}^{(q)}\log\left[\alpha_k\phi_p(\boldsymbol{x}_i;\boldsymbol{\mu}_k,\mathbf{R}_k)\right] - \gamma\sum_{j=1}^{p}|\mu_{k,j}|$$

$$= \sum_{i=1}^{n}\tau_{ik}^{(q)}\log\alpha_k + \sum_{i=1}^{n}\tau_{ik}^{(q)}\log\phi_p(\boldsymbol{x}_i;\boldsymbol{\mu}_k,\mathbf{R}_k) - \gamma\sum_{j=1}^{p}|\mu_{k,j}|.$$

It can be seen that the updates of the α_k's are unchanged compared to the standard algorithm and are given by (13). For the mean vectors, updating the coefficients $\mu_{k,j}$ corresponds to weighted version or and ℓ_1-regularized maximum likelihood estimation a Gaussian mean; The coefficients $\mu_{k,j}$ can then be updated in a cyclic way by using a Coordinate ascent algorithm until (24) is maximized. Coordinate ascent (CA) [10, sect. 5.4] [9,23] is indeed an efficient way to solve Lasso-regularization problems. For each coefficient index $j = 1,\ldots,p$, it can be easily shown that, after starting with the previous EM-Lasso estimate as initial value, i.e, $\mu_{kj}^{(0,q)} = \mu_{kj}^{(q)}$, each iteration t of the CA algorithm updates are given by the following updating formulas (see eg. [20]), written in a scalar and a vector form:

$$\mu_{kj}^{(t+1,q)} = sign(\tilde{\mu}_{kj}^{(q+1)})\left(|\tilde{\mu}_{kj}^{(q+1)}| - \frac{\gamma}{\sum_{i=1}^{n}\tau_{ik}^{(q)}}\nu_{kj}^{2(q)}\right)_+$$

$$= \mathcal{S}\left(\sum_{i=1}^{n}\tau_{ik}^{(q)}x_{ij};\gamma\nu_{kj}^{2(q)}\right)\bigg/\sum_{i=1}^{n}\tau_{ik}^{(q)}$$

$$= \mathcal{S}\left(\mathbf{X}_j^T\boldsymbol{\tau}_k^{(q)};\gamma\nu_{kj}^{2(q)}\right)\bigg/\mathbf{1}_n^T\boldsymbol{\tau}_k^{(q)} \tag{24}$$

with, $\tilde{\mu}_{kj}^{(q+1)} = \sum_{i=1}^{n}\tau_{ik}^{(q)}x_{ij}\big/\sum_{i=1}^{n}\tau_{ik}^{(q)}$ is the usual non-regularized MLE update for μ_k (Eq. (14)), $\mathbf{X}_j$ the jth column of $\mathbf{X}$, $\mathbf{1}_n$ is a vector of ones of size n, $\boldsymbol{\tau}_k^{(q)} = (\tau_{1k}^{(q)},\ldots,\tau_{nk}^{(q)})^T$, and $\mathcal{S}(u;\eta) := sign(u)(|u|-\eta)_+$ is the soft-thresholding operator with $(.)_+ = \max\{.,0\}$. The CA procedure is iterated until no significant change in (24) is observed. We then take the update at convergence of the CA algorithm, i.e $\mu_{kj}^{(q+1)} = \mu_{kj}^{(t+1,q)}$. Finally, the updates of the diagonal elements of the co-variance matrices are given by:

$$\nu_{kj}^{2(q+1)} = \sum_{i=1}^{n}\tau_{ik}^{(q)}(x_{ij} - \mu_{kj}^{(q+1)})^2\bigg/\sum_{i=1}^{n}\tau_{ik}^{(q)}. \tag{25}$$

Coordinate Ascent for Updating the Experts Network. The maximization step for updating the expert parameters $\boldsymbol{\theta}_k$ consists of maximizing the function $\mathcal{Q}_\lambda(\boldsymbol{\theta}_k;\boldsymbol{\Psi}^{(q)})$ given by:

$$Q_\lambda(\boldsymbol{\theta}_k; \boldsymbol{\Psi}^{(q)}) = Q(\boldsymbol{\Psi}_k; \boldsymbol{\Psi}^{(q)}) - \lambda \sum_{j=1}^{p} |\beta_{k,j}|$$

$$= -\frac{1}{2\sigma_k^2} \sum_{i=1}^{n} \tau_{ik}^{(q)} \left(y_i - (\beta_{k,0} + \boldsymbol{\beta}_k^T \boldsymbol{x}_i) \right)^2 - \frac{n_k^{(q)}}{2} \log(2\pi\sigma_k^2) - \lambda \sum_{j=1}^{p} |\beta_{k,j}|.$$

Updating $\boldsymbol{\beta}_k$, for each component k, consists of solving an independent weighted Lasso problem where the weights are the posterior component membership probabilities $\tau_{ik}^{(q)}$. Each of these weighted Lasso problems is then separately solved by Coordinate Ascent. The CA algorithm, after starting from the previous EM-Lasso estimate as initial values, i.e $\beta_{kj}^{(0,q)} = \beta_{kj}^{(q)}$, calculates, at each iteration t, the following coordinate updates, until no significant change in (26):

$$\beta_{kj}^{(t+1,q)} = \mathcal{S} \left(\sum_{i=1}^{n} \tau_{ik}^{(q)} r_{ikj}^{(t,q)} x_{ij}; \lambda \sigma_k^{(q)2} \right) / \sum_{i=1}^{n} \tau_{ik}^{(q)} x_{ij}^2 \qquad (26)$$

$$= \mathcal{S} \left(\mathbf{X}_j^T \mathbf{W}_k^{(q)} \boldsymbol{r}_{kj}^{(q)}; \lambda \sigma_k^{(q)^2} \right) / (\mathbf{X}_j^T \mathbf{W}_k^{(q)} \mathbf{X}_j), \qquad (27)$$

with $r_{ikj}^{(t,q)} = y_i - \beta_{k0}^{(q)} - \boldsymbol{x}_i^T \boldsymbol{\beta}_k^{(t,q)} + \beta_{kj}^{(t,q)} x_{ij}$, $\boldsymbol{r}_{kj}^{(t,q)} = \boldsymbol{y} - \beta_{k0}^{(q)} \mathbf{1}_n - \mathbf{X}\boldsymbol{\beta}_k^{(t,q)} + \beta_{kj}^{(t,q)} \mathbf{X}_j$ is the residual without considering the contribution of the j-th coefficient, and $\mathbf{W}_k^{(q)} = \text{diag}(\boldsymbol{\tau}_k^{(q)})$. The parameter vector update is then taken at convergence of the CA algorithm, i.e $\boldsymbol{\beta}_k^{(q+1)} = \boldsymbol{\beta}_k^{(t+1,q)}$. Then, the intercept and the variance, have the following standard updates:

$$\beta_{k,0}^{(q+1)} = \sum_{i=1}^{n} \tau_{ik}^{(q)} (y_i - \boldsymbol{x}_i^T \boldsymbol{\beta}_k^{(q+1)}) / \sum_{i=1}^{n} \tau_{ik}^{(q)} = \boldsymbol{\tau}_k^{(q)^T} (\boldsymbol{y} - \mathbf{X}\boldsymbol{\beta}_k^{(q+1)}) / \mathbf{1}_n^T \boldsymbol{\tau}_k^{(q)} \quad (28)$$

$$\sigma_k^{2(q+1)} = \sum_{i=1}^{n} \tau_{ik}^{(q)} \left(y_i - (\beta_{k,0}^{(q+1)} + \boldsymbol{x}_i^T \boldsymbol{\beta}_k^{(q+1)}) \right)^2 / \sum_{i=1}^{n} \tau_{ik}^{(q)} \qquad (29)$$

$$= \| \sqrt{\mathbf{W}_k^{(q)}} \left(\boldsymbol{y} - \beta_{k,0}^{(q+1)} \mathbf{1}_n - \mathbf{X}\boldsymbol{\beta}_k^{(q+1)} \right) \|_2^2 / \mathbf{1}_n^T \boldsymbol{\tau}_k^{(q)}. \qquad (30)$$

3.2 Algorithm Tuning and Model Selection

In practice, appropriate values of the tuning parameters (λ, γ) as well as the number of experts K should be chosen. In order to select them, we use a modified BIC based on a grid of candidate values for K, λ and γ. This modified BIC is an extension of the criterion used in [21] for regularized mixture of regressions and was used in [7, 11] and is defined as:

$$\text{BIC}(K, \lambda, \gamma) = L(\widehat{\boldsymbol{\Psi}}_{K,\lambda,\gamma}) - \text{df}(K, \lambda, \gamma) \frac{\log n}{2}, \qquad (31)$$

where $\widehat{\boldsymbol{\Psi}}_{K,\lambda,\gamma}$ is the penalized log-likelihood estimator obtained by the EM-Lasso algorithm, and $\text{df}(K, \lambda, \gamma)$ is the estimated number of non-zero coefficients in the model, interpreted as the degrees of freedom. Let's assume that $K_0 \in \{K_1, \ldots, K_M\}$, whith K_0 the true number of expert components. For each value of K, we define grids of tuning parameters $\{\lambda_1, \ldots, \lambda_{M_1}\}$ and $\{\gamma_1, \ldots, \gamma_{M_2}\}$. For each triplet (K, λ, γ), we calculated the penalized log-likelihood estimators $\widehat{\boldsymbol{\Psi}}_{K,\lambda,\gamma}$ and compute $\mathrm{BIC}(K, \lambda, \gamma)$. Finally, the model with parameters (K, λ, γ) having the highest BIC value, is then selected.

4 Experimental Study

In this section, we study the performance of our approach on simulated data. The codes are written in Matlab and in R and will be made publicly available on https://github.com/fchamroukhi. Different evaluation criteria are used to assess the model's performance, including sparsity, estimation of parameters and clustering accuracy.

Sparsity Performance. In order to evaluate the sparsity of the model, we calculate the specificity/sensitivity defined by:

- Sensitivity: proportion of correctly estimated zero coefficients;
- Specificity: proportion of correctly estimated nonzero coefficients.

Clustering Performance. For measuring the clustering performance, we calculate the correct classification rate and the Adjusted Rate index (ARI) between the true simulated partition and the partition estimated by the EM algorithms. The estimated cluster labels are obtained by plugin the Baye's allocation rule for the estimated model, which consists of maximizing the posterior probabilities defined in 9 and calculated with the estimated parameters. That is, the estimated class label $\hat{z}_i$ for the i-th pair $(\boldsymbol{X}_i, \boldsymbol{Y}_i)$ is given by

$$\hat{z}_i = \arg \max_{k=1}^{K} \tau_{ik}(\hat{\boldsymbol{\Psi}}) \quad (i = 1, \ldots, n). \tag{32}$$

For calculating the classification rate, we evaluate all the possible permutations of the obtained partition, and the one giving the best rate is then retained.

4.1 Simulation Study

The data are generated according to the following generative hierarchical process:

$$Z_i \sim \mathrm{Mult}(1; \alpha_1, \ldots, \alpha_K)$$
$$\boldsymbol{X}_i | Z_i = z_i \sim \mathcal{N}_p(.; \boldsymbol{\mu}_{z_i}, \mathbf{R}_{z_i})$$
$$\boldsymbol{Y}_i | \boldsymbol{X}_i = \boldsymbol{x}_i, Z_i = z_i \sim \mathcal{N}_d(.; \beta_{z_i,0} + \boldsymbol{\beta}_{z_i}^T \boldsymbol{x}_i, \sigma_{z_i}^2).$$

We consider a MoGGE model of $K = 2$ expert components. The parameters of the Gaussian gating function, whose prior probabilities are $\alpha_1 = \alpha_2 = 0.5$, are $\mu_1 = (0, 1, -1, -1.5, 0, 0.5, 0, 0)^T$, $\mu_2 = (2, 0, 1, -1.5, 0, -0.5, 0, 0)^T$ and $\mathbf{R}_1 = \mathbf{R}_2 = \text{diag}(\nu_1^2, \ldots, \nu_K^2)$ with $\nu_1^2 = \ldots = \nu_K^2 = 1$. The parameters of the Gaussian expert regressors are $\beta_1 = (0, 1.5, 0, 0, 0, 1, 0, -0.5)^T$, $\beta_2 = (1, -1.5, 0, 0, 2, 0, 0, 0.5)$, and $\sigma_1 = \sigma_2 = 1$. For each data set, we sample $n = 300$ data pairs, and for each experiment, 100 datasets were generated to average the results and provide error bars. In order to get the best model for each sample in the sense of the BIC criterion, we estimated the penalized model with the following grids of values for the parameters: $\lambda = (0, 1, 2, \ldots, 25)$, $\gamma = (0, 1, 2, \ldots, 25)$; The minimum and maximum values selected for λ and γ are respectively 4, 20 and 3, 18. Then we selected the penalized model which maximizes the modified BIC value (31). The results will be provided in the parts below.

Obtained Results

Parameter Estimation Accuracy. Figure 1 shows the estimated parameters for the gating network, with the error bars, for the proposed approach and for the standard MoGGE model. Similarly, Fig. 2 shows the estimated parameters of the gating network. It can be seen on the two figures that, as expected, the proposed lasso-regularization approach with the proposed EM-Lasso algorithm, clearly provides models that are sparser, compared to the standard approach with EM, where the zero-coefficients are not precisely recovered. This is observed for both the gating function parameters, and the expert function parameters. While the penalized version we can see that it may be subject of a bias in estimating the non-zero coefficients, the parameter estimated and the bias are still reasonable. Hence, if one would to encourage sparsity, and to still have a good performance in density estimation, then the penalized MoGGE is a better choice, compared to the standard MLE of the MoGGE model.

Table 1. Sensitivity (S_1) and specificity (S_2) results.

Method	Expert 1		Expert 2		Gate	
	S_1	S_2	S_1	S_2	S_1	S_2
MoGGE-EM	0.000	1.000	0.000	1.000	0.000	1.000
MoGGE-EMLasso-BIC	0.790	1.000	0.785	1.000	0.779	1.000

Sensitivity/Specificity Results. Table 1 gives the sensitivity (S_1) and specificity (S_2) results for the two compared approaches. Note that here since we have two components, then only the estimation of one Gaussian gating function is considered, as the parameters of the other one are zeros. It can be seen that,

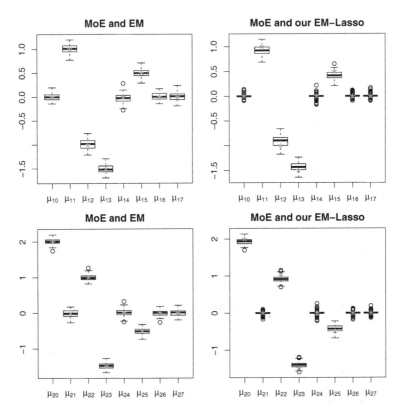

Fig. 1. Boxplots of the estimated gating network parameters $\mu_{k,j}$: component $k = 1$, top, and component $k = 2$, bottom. The red stars are the true values.

none of the parameters in the non penalized model has a null value. The penalized model provides naturally sparser models compared to the standard non-penalized one.

Clustering Results. We calculate the accuracy of clustering for each data set. The results in terms of correct classification rate and ARI values are provided in Table 2. We can see that the classification rate as well as as the Adjusted Rand Index are very close for the two methods, with a slight advantage to the proposed approach.

Table 2. Clustering results: correct classification rate and Adjusted Rand Index.

Model	C.rate	ARI
MoGGE - EM	$97.25\%_{(0.8770\%)}$	$89.28\%_{(3.325\%)}$
MoGGE-EMLasso-BIC	$97.43\%_{(0.8521\%)}$	$89.99\%_{(3.231\%)}$

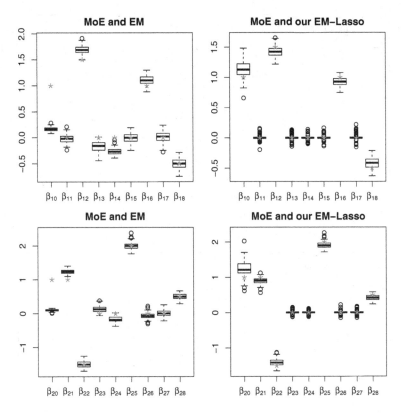

Fig. 2. Boxplots of the estimated expert network parameters $\beta_{k,j}$: component $k = 1$, top, and component $k = 2$, bottom. The red stars are the true values.

Selecting the Sparsity Tuning Parameters. We compute the Lasso path for a sample with same parameters as presented at the beginning of the section. On Fig. 3, we observe that even with very small values (null value as well, i.e. non penalized MoE) of γ, the true zero parameters have values very close to zero. We also note that for values of ratio close to 0.8 for both λ and γ, almost every true zero parameters have null values and the slight bias introduced in the true nonzero parameters is reasonable.

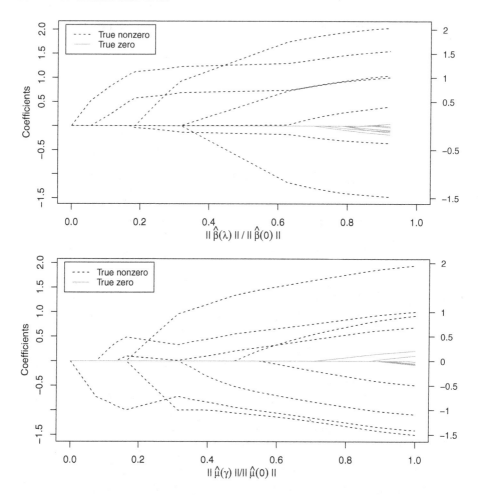

Fig. 3. Lasso paths of the estimated gating network parameters (top) and expert network parameters (bottom). The solid line represents the values of the true non-zero values, and the dashed line represents the true zero values.

5 Conclusion and Future Work

In this paper, the mixture of Gaussian-gated experts is studied towards modeling and clustering of heterogeneous regression data with high-dimensional predictors. A regularized MLE approach is proposed to simultaneously perform parameter estimation and feature selection. The developed EM-Lasso algorithm to fit the model relies on coordinate ascent updates of the regularized parameters, and its application in numerical experiments clearly shows it provides sparse models. Its performance is also compared to the state-of-the art fitting with the EM algorithm, shows its good performance, in particular in terms of sparsity. The diagonal hypothesis of the covariance matrix to derive the regularization

(19) is now being relaxed, so that the regularization is on the elements of the precision matrix, i.e a graphical Lasso regularization. A future extension will also consider multivariate response with dedicated sparsity on the matrices of regression coefficients.

Acknowledgments. This research is supported by Ethel Raybould Fellowship (Univ. of Queensland), ANR SMILES ANR-18-CE40-0014, and Région Normandie RIN AStERiCs.

References

1. Chamroukhi, F.: Non-normal mixtures of experts, July 2015. arXiv:1506.06707
2. Chamroukhi, F.: Robust mixture of experts modeling using the t-distribution. Neural Netw. **79**, 20–36 (2016)
3. Chamroukhi, F.: Skew-normal mixture of experts. In: The International Joint Conference on Neural Networks (IJCNN), Vancouver, Canada, July 2016
4. Chamroukhi, F.: Skew t mixture of experts. Neurocomputing **266**, 390–408 (2017)
5. Chamroukhi, F., Samé, A., Govaert, G., Aknin, P.: A regression model with a hidden logistic process for feature extraction from time series. In: International Joint Conference on Neural Networks (IJCNN), pp. 489–496 (2009)
6. Chamroukhi, F., Trabelsi, D., Mohammed, S., Oukhellou, L., Amirat, Y.: Joint segmentation of multivariate time series with hidden process regression for human activity recognition. Neurocomputing **120**, 633–644 (2013)
7. Chamroukhi, F., Huynh, B.T.: Regularized maximum likelihood estimation and feature selection in mixtures-of-experts models. J. Soc. Française Stat. **160**(1), 57–85 (2019)
8. Dempster, A.P., Laird, N.M., Rubin, D.B.: Maximum likelihood from incomplete data via the EM algorithm. JRSS B **39**(1), 1–38 (1977)
9. Friedman, J., Hastie, T., Höfling, H., Tibshirani, R.: Pathwise coordinate optimization. Technical report, Annals of Applied Statistics (2007)
10. Hastie, T., Tibshirani, R., Wainwright, M.: Statistical Learning with Sparsity: The Lasso and Generalizations. Chapman & Hall/CRC, London/Boca Raton (2015)
11. Huynh, T., Chamroukhi, F.: Estimation and feature selection in mixtures of generalized linear experts models. arXiv:1907.06994 (2019)
12. Jacobs, R.A., Jordan, M.I., Nowlan, S.J., Hinton, G.E.: Adaptive mixtures of local experts. Neural Comput. **3**(1), 79–87 (1991)
13. Jordan, M.I., Jacobs, R.A.: Hierarchical mixtures of experts and the EM algorithm. Neural Comput. **6**, 181–214 (1994)
14. Khalili, A.: New estimation and feature selection methods in mixture-of-experts models. Can. J. Stat. **38**(4), 519–539 (2010)
15. McLachlan, G.J., Krishnan, T.: The EM Algorithm and Extensions, 2nd edn. Wiley, New York (2008)
16. McLachlan, G.J., Peel, D.: Finite Mixture Models. Wiley, New York (2000)
17. Nguyen, H.D., Chamroukhi, F.: Practical and theoretical aspects of mixture-of-experts modeling: an overview. WIREs Data Min. Knowl. Discov. **8**, e1246-n/a (2018). https://doi.org/10.1002/widm.1246
18. Nguyen, H.D., Chamroukhi, F., Forbes, F.: Approximation results regarding the multiple-output mixture of linear experts model. Neurocomputing (2019). https://doi.org/10.1016/j.neucom.2019.08.014

19. Nguyen, H.D., McLachlan, G.J.: Laplace mixture of linear experts. Comput. Stat. Data Anal. **93**, 177–191 (2016)
20. Pan, W., Shen, X.: Penalized model-based clustering with application to variable selection. J. Mach. Learn. Res. **8**, 1145–1164 (2007)
21. Städler, N., Bühlmann, P., van de Geer, S.: Rejoinder: l1-penalization for mixture regression models. TEST **19**(2), 280–285 (2010)
22. Tibshirani, R.: Regression shrinkage and selection via the lasso. J. Roy. Stat. Soc. Ser. B **58**(1), 267–288 (1996)
23. Wu, T.T., Lange, K.: Coordinate descent algorithms for lasso penalized regression. Ann. Appl. Stat. **2**(1), 224–244 (2008). https://doi.org/10.1214/07-AOAS147
24. Xu, L., Jordan, M.I., Hinton, G.E.: An alternative model for mixtures of experts. In: Tesauro, G., Touretzky, D.S., Leen, T.K. (eds.) Advances in Neural Information Processing Systems, vol. 7, pp. 633–640. MIT Press, Cambridge (1995)

Flexible Modelling via Multivariate Skew Distributions

Geoffrey J. McLachlan[1]([⊠]) and Sharon X. Lee[2]

[1] University of Queensland, St. Lucia, Brisbane, QLD 4072, Australia
g.mclachlan@uq.edu.au
[2] University of Adelaide, Adelaide, SA, Australia
sharon.lee@adelaide.edu.au
https://people.smp.uq.edu.au/GeoffMcLachlan/,
https://www.adelaide.edu.au/directory/sharon.lee

Abstract. Mixtures of skew component distributions are being applied widely to model and partition data into clusters that exhibit non-normal features such as asymmetry and tails heavier than the normal. The number of contributions on skew distributions are now so many that it is beyond the scope of this paper to include them all here. However, many of these developments can be considered as special cases of a (location-scale variant) of the fundamental skew normal (CFUSN) distribution or of the fundamental skew t (CFUST) distribution. We therefore focus on mixtures of CFUSN and CFUST distributions, along with a recently proposed extension that can be viewed as a scale-mixture of the CFUSN distribution, namely the canonical fundamental skew (symmetric generalized) hyperbolic (CFUSH) distribution.

Keywords: Skew normal distribution · Skew t-distribution · Mixtures of skew components

1 Introduction

Finite mixtures of multivariate skew distributions have gained wide acceptance as a useful tool for analyzing a variety of heterogeneous datasets that exhibit non-normal features (McLachlan et al. 2019). These distributions provide flexible alternatives to traditional normal and t-mixture models, with additional features such as asymmetry and heavy tails, rendering them suitable for a wider range of applications. For a comprehensive survey of skew distributions, see, for example, the articles by Azzalini (2005), Arellano-Valle and Azzalini (2006), Arellano-Valle et al. (2006), the book edited by Genton (2004), and the recent monograph by Azzalini (2014). With various proposals appearing rapidly in recent years, which are similar but not identical, the connection between them and their relative performance becomes rather unclear. It led Lee and McLachlan (2013a) to provide a concise overview of these developments by presenting a systematic classification of the existing skew symmetric distributions into four

© Springer Nature Singapore Pte Ltd. 2019
H. Nguyen (Ed.): RSSDS 2019, CCIS 1150, pp. 57–67, 2019.
https://doi.org/10.1007/978-981-15-1960-4_4

types, restricted, unrestricted, extended, and generalized, thereby clarifying their close relationships.

In particular, the multivariate skew normal and skew t-distributions have been used extensively in model-based clustering. Their usefulness has also been exemplified by various applications in a range of scientific fields, including environmental science (Allard and Soubeyrand 2012; Tagle et al. 2019), flow cytometry (Pyne et al. 2009, 2014; Lee et al. 2016; Hejblum et al. 2019), financial risk analysis (McLachlan 2013b; Mousavi et al. 2019), fisheries science (Contreras-Reyes and Arellano-Valle 2013; Contreras-Reyes et al. 2018), astrophysics (Riggi and Ingrassia 2013; Voigt and Fried 2015), image segmentation (Lee and McLachlan 2013b), and the social sciences (Asparouhov and Muthén 2016; Hohmann et al. 2018). Some other notable contributions to multivariate nonnormal mixture models include the normal-inverse-Gaussian (NIG) mixture models (Karlis and Santourian 2009) and related densities arising from the family of generalized hyperbolic (GH) distributions, and mixtures of multiple-scaled distributions (Forbes and Wraith 2013; Wraith and Forbes 2015).

The number of contributions on skew distributions are now so many that it is beyond the scope of this paper to include them all here. However, many of these developments can be considered as special or limiting cases of a (location-scale variant) of the fundamental skew normal (FUSN) distribution or of the fundamental skew t (CFUST) distribution (Arellano-Valle and Genton 2005). Hence in this paper we shall focus on the CFUSN and CFUST distributions along with a recently proposed extension that can be viewed as a scale-mixture of the CFUSN distribution (Lee and McLachlan 2018). This distribution is referred to here as the canonical fundamental skew (symmetric generalized) hyperbolic (CFUSH) distribution. We shall present a comparison of mixtures of CFUST and CFUSH distributions in their application to the modelling and clustering of two data sets as considered in Murray et al. (2017). In this comparison, the performance of a mixture of CFUST distributions is found to be superior to the mixture model with CFUSH component distributions.

The CFUSN, CFUST, and CFUSH distributions belong to the family of skew symmetric distributions. Hence before defining these three distributions, we shall consider the family of skew symmetric distributions.

2   Skew Symmetric Distributions

To establish some notation, we let $\boldsymbol{Y}$ denote a p-dimensional random vector, $\boldsymbol{I}_p$ be the $p \times p$ identity matrix, and $\boldsymbol{0}$ be a vector/matrix of zeros of appropriate size.

The skew distributions to be considered here belong to the class of canonical fundamental skew symmetric (CFUSS) distributions proposed by Arellano-Valle and Genton (2005). The density of members of the class of CFUSS distributions can be expressed as

$$f(\boldsymbol{y};\, \boldsymbol{\theta}) = 2^r f_p(\boldsymbol{y};\, \boldsymbol{\theta})\, Q_r(\boldsymbol{y};\, \boldsymbol{\theta}), \tag{1}$$

where $f_p(\boldsymbol{y};\, \boldsymbol{\theta})$ is a symmetric density on R^p, $Q_r(\boldsymbol{y};\, \boldsymbol{\theta})$ is a skewing function that maps $\boldsymbol{y}$ into the unit interval, and $\boldsymbol{\theta}$ is the vector containing the parameters of

$\boldsymbol{Y}$. Let $\boldsymbol{U}$ be a $r \times 1$ random vector, where $\boldsymbol{Y}$ and $\boldsymbol{U}$ follow a joint distribution such that $\boldsymbol{Y}$ has marginal density $f_p(\boldsymbol{y}; \boldsymbol{\theta})$ and $Q_r(\boldsymbol{y}; \boldsymbol{\theta}) = P(\boldsymbol{U} > \boldsymbol{0} \mid \boldsymbol{Y} = \boldsymbol{y})$. If the latent random vector $\boldsymbol{U}$ (the vector of skewing variables) has its canonical distribution (that is, with mean $\boldsymbol{0}$ and scale matrix $\boldsymbol{I}_r$), we obtain the canonical form of (1), namely the CFUSS distribution. The class of CFUSS distributions encapsulates many existing distributions, including those to be considered here. We now proceed to define those members.

We are to focus on a finite mixture of canonical fundamental skew t (CFUST) distributions for a model based approach to clustering where the clusters are asymmetric and possibly long-tailed (Lee and McLachlan 2016). The CFUST distribution is an extension of the canonical fundamental skew normal (CFUSN) distribution through the addition of a scalar parameter ν representing the degrees of freedom as in the symmetric t-distribution. Hence we firstly define the CFUSN distribution.

3 CFUSN Distribution

The so-called canonical fundamental skew normal (CFUSN) distribution is a location-scale variant of the canonical fundamental skew normal distribution in Arellano-Valle and Genton (2005). If the $p \times 1$ random vector $\boldsymbol{Y}$ has a CFUSN distribution, its density is given by

$$f_{\text{CFUSN}}(\boldsymbol{y}; \boldsymbol{\mu}, \boldsymbol{\Sigma}, \boldsymbol{\Delta}) = 2^r \phi_p(\boldsymbol{y}; \boldsymbol{\mu}, \boldsymbol{\Omega})\, \Phi_r(c(\boldsymbol{y}); \boldsymbol{0}, \boldsymbol{\Lambda}), \qquad (2)$$

where

$$\boldsymbol{\Lambda} = \boldsymbol{I}_r - \boldsymbol{\Delta}^T \boldsymbol{\Omega}^{-1} \boldsymbol{\Delta},$$
$$c(\boldsymbol{y}) = \boldsymbol{\Delta}^T \boldsymbol{\Omega}^{-1}(\boldsymbol{y} - \boldsymbol{\mu}),$$
$$\boldsymbol{\Omega} = \boldsymbol{\Sigma} + \boldsymbol{\Delta}\boldsymbol{\Delta}^T,$$

and $\boldsymbol{\Delta}$ is a $p \times r$ matrix of skewness parameters with pr free parameters. They are identifiable up to a permutation of the columns of $\boldsymbol{\Delta}$. Also, $\phi_r(\boldsymbol{y}; \boldsymbol{\mu}, \boldsymbol{\Sigma})$ and $\Phi_r(\boldsymbol{y}; \boldsymbol{\mu}, \boldsymbol{\Sigma})$ denote the r-dimensional multivariate normal density and (cumulative) distribution, respectively, with mean $\boldsymbol{\mu}$ and covariance $\boldsymbol{\Sigma}$. The number of skewing variables r is not necessarily restricted to being less than or equal to p.

The convolution-type stochastic characterization of the CFUSN distribution is given by

$$\boldsymbol{Y} = \boldsymbol{\mu} + \boldsymbol{\Delta}|\boldsymbol{U}_0| + \boldsymbol{U}_1, \qquad (3)$$

where

$$\begin{bmatrix} \boldsymbol{U}_0 \\ \boldsymbol{U}_1 \end{bmatrix} \sim N_{r+p} \left(\begin{bmatrix} \boldsymbol{0} \\ \boldsymbol{0} \end{bmatrix}, \begin{bmatrix} \boldsymbol{I}_r & \boldsymbol{0} \\ \boldsymbol{0} & \boldsymbol{\Sigma} \end{bmatrix} \right). \qquad (4)$$

In the above, $|\boldsymbol{U}_0|$ denotes the vector whose ith element is the magnitude of the ith element of the vector $\boldsymbol{U}_0$.

3.1   Restricted Multivariate Skew Normal (rMSN) Distribution

In the univariate case $(p = 1)$ with a single skewing variable $(r = 1)$, the CFUSN density as defined by (2) reduces to the skew univariate normal distribution as pioneered by Azzalini (1985). Its extension to the multivariate case $(p > 1)$ with $(r = 1)$ was undertaken by Azzalini and Dalla Valle (1996).

Concerning the CFUSN Distribution in the multivariate case $(p > 1)$, Lee and McLachlan (2013a) referred to it as the restricted multivariate skew normal (rMSN) distribution if $r = 1$; that is, if only a single skewing variable is used. This is because with only a single skewing variable, the CFUSN density is restricted to modelling skewness in a single direction in the feature space.

Concerning the attempt in Azzalini et al. (2016) to show that the amount of skewness in the rMST distribution is unlimited, Mardia's measure of skewness for the rMST is indeed unbounded. But as subsequently explained by McLachlan and Lee (2016), since this measure is a squared Euclidean norm (Kollo and Srivastava 2007), the use of this property by Azzalini et al. (2016) to support the flexibility of the rMST distribution is similar to saying that rank-one covariance matrices will suffice for modelling multivariate scatter just because their Euclidean norm can be made as big as desired. Furthermore, Kollo (2008) showed some nice and simple examples of two multivariate distributions with very different shapes, but identical values of Mardia's measures of multivariate skewness and kurtosis.

3.2   Unrestricted MultivariateSkew Normal (uMSN) Distribution

Lee and McLachlan (2013a) referred to the CFUSN distribution in the particular case where $r = p$ and the matrix $\boldsymbol{\Delta}$ of skewness parameters is diagonal, as the unrestricted multivariate normal distribution (uMSN). They used the term "unrestricted" in the sense that the restricted version can be regarded as the unrestricted one with the restriction that all p skewing variables are the same. It should be noted, however, that the rMSN distribution is not nested within the unrestricted distribution, and so there can be situations where mixtures of rMSN distributions will be preferred over mixtures of uMSN distributions.

Although the uMSN distribution has p skewing variables, it is limited to modelling skewness in directions that are parallel to the axes of the feature space. Consequently, Lee and McLachlan (2014b; 2015; 2016) developed the methodology and algorithms for the fitting of mixtures of CFUSN and CFUST distributions with arbitrary skewness matrices so that they can handle skewness in multiple directions that are not necessarily parallel to the axes of the feature space.

We now define the CFUST distribution.

4   CFUST Distribution

The CFUST distribution provides a model for skew data with tails heavier than the CFUSN distribution. The density of the CFUST distribution is defined by

$$f\left(\boldsymbol{y}; \boldsymbol{\mu}, \boldsymbol{\Sigma}, \boldsymbol{\Delta}, \nu\right) = 2^r \, t_p\left(\boldsymbol{y}; \boldsymbol{\mu}, \boldsymbol{\Omega}, \nu\right) T_r\left(\mathbf{c}(\boldsymbol{y})\sqrt{\frac{\nu + p}{\nu + d(\boldsymbol{y})}}; \mathbf{0}, \boldsymbol{\Lambda}, \nu + p\right). \quad (5)$$

Here $t_p(\boldsymbol{y}; \boldsymbol{\mu}, \boldsymbol{\Omega}, \nu)$ denotes the p-dimensional t-distribution with location parameter $\boldsymbol{\mu}$, scale matrix $\boldsymbol{\Omega}$, and degrees of freedom ν, and $T_r(.)$ is the r-dimensional (cumulative) t-distribution function. As the degrees of freedom ν tend to infinity, the CFUST distribution tends to the CFUSN distribution.

The canonical fundamental skew t (CFUST) distribution can be characterized by

$$\boldsymbol{Y} = \boldsymbol{\mu} + \boldsymbol{\Delta}|\boldsymbol{U}_0| + \boldsymbol{U}_1 \tag{6}$$

where, conditional on $W = w$,

$$\begin{bmatrix} \boldsymbol{U}_0 \\ \boldsymbol{U}_1 \end{bmatrix} \sim N_{r+p}\left(\begin{bmatrix} \boldsymbol{0} \\ \boldsymbol{0} \end{bmatrix}, w\begin{bmatrix} \boldsymbol{I}_r & \boldsymbol{0} \\ \boldsymbol{0} & \boldsymbol{\Sigma} \end{bmatrix}\right) \tag{7}$$

and W is distributed according to the inverse gamma distribution $\mathrm{IG}(\frac{\nu}{2}, \frac{\nu}{2})$.

If we take the matrix $\boldsymbol{\Delta}$ of skewness parameters in the formulation (6) to be a p-dimensional vector (that is, $r = 1$), then we obtain the skew t-density as proposed by Azzalini and Capitanio (2003). Lee and McLachlan (2013a) referred to this distribution as the restricted multivariate skew t (rMST) distribution corresponding to its limiting rMSN distribution.

If we take $r = p$ with $\boldsymbol{\Delta}$ diagonal in (6), then we obtain the model considered by Sahu et al. (2003). It was termed the unrestricted multivariate skew t (uMST) distribution corresponding to its limiting uMSN distribution.

In an attempt to provide an automated approach to the clustering of flow cytometry data, Pyne et al. (2009) considered the fitting of mixtures of rMSN and rMST distributions. This paper and those of Lin (2009a; 2009b) would appear to be the first papers to consider the fitting of mixtures of multivariate skew distributions. Previously, Lin et al. (2007a; 2007b) had considered the fitting of mixtures of univariate skew normal and t-distributions.

5   Scale Mixture of CFUSN Distribution

A Scale Mixture of the CFUSN distribution (SMCFUSN) can be defined by the stochastic representation

$$\boldsymbol{Y} = \boldsymbol{\mu} + W^{\frac{1}{2}}\boldsymbol{Y}_0, \tag{8}$$

where $\boldsymbol{Y}_0$ follows a central CFUSN distribution and W is a positive (univariate) random variable independent of $\boldsymbol{Y}_0$. Thus, conditional on $W = w$, the density of $\boldsymbol{Y}$ has a CFUSN distribution with scale matrix $w\boldsymbol{\Sigma}$. The marginal density of $\boldsymbol{Y}$ is given by

$$\begin{aligned} &f_{\text{SMCFUSN}}(\boldsymbol{y}; \boldsymbol{\mu}, \boldsymbol{\Sigma}, \boldsymbol{\Delta}; F_\zeta) \\ &= 2^r \int_0^\infty \phi_p(\boldsymbol{y}; \boldsymbol{\mu}, w\boldsymbol{\Omega})\, \Phi_r\left(\frac{1}{\sqrt{w}}\boldsymbol{\Delta}^T\boldsymbol{\Omega}^{-1}(\boldsymbol{y} - \boldsymbol{\mu}); \boldsymbol{0}, \boldsymbol{\Lambda}\right) dF_\zeta(w), \end{aligned} \tag{9}$$

where F_ζ denotes the distribution function of W indexed by the parameter ζ. We shall use the notation $\boldsymbol{Y} \sim SMCFUSN_{p,r}(\boldsymbol{\mu}, \boldsymbol{\Sigma}, \boldsymbol{\Delta}; F_\zeta)$ if the density of $\boldsymbol{Y}$ can be expressed in the form of (9).

The CFUST distribution corresponds to taking W to have the inverse gamma distribution $\mathrm{IG}(\frac{\nu}{2}, \frac{\nu}{2})$.

6   CFUSH Distribution

The so-called hidden truncation hyperbolic distribution proposed by Murray et al. (2017) can be viewed as a member of the class of the canonical fundamental skew symmetric generalized hyperbolic (CFUSH) distributions (Lee, Lin, McLachlan 2018). To see this, we suppose now that the latent variable W in (9) follows a generalized inverse Gaussian (GIG) distribution (Seshadri 1997). The GIG density can be expressed as

$$f_{\text{GIG}}(w; \psi, \chi, \lambda) = \frac{\left(\frac{\psi}{\chi}\right)^{\frac{\lambda}{2}} w^{\lambda-1}}{2K_\lambda(\sqrt{\chi\psi})} e^{-\frac{\psi w + \frac{\chi}{w}}{2}}, \tag{10}$$

where $W > 0$, the parameters ψ and χ are positive, and λ is a real parameter. In the above, $K_\lambda(\cdot)$ denotes the modified Bessel function of the third kind of order λ. If we put $\chi = \nu$, $\lambda = -\frac{1}{2}\nu$, and let ψ tend to zero in (10), then it tends to the inverse gamma distribution $\text{IG}(\frac{1}{2}\nu, \frac{1}{2}\nu)$.

Taking the latent variable W in (9) to have a GIG distribution, we obtain the CFUSH distribution. It has the p-dimensional symmetric GH (generalized hyperbolic) density $h_p(\cdot)$ and the r-dimensional symmetric GH distribution function $H_r(\cdot)$, corresponding to the symmetric density $f_p(\cdot)$ and the distribution function $Q_r(\cdot)$, respectively, in (1).

The symmetric GH density is given by

$$h_p(\boldsymbol{y}; \boldsymbol{\mu}, \boldsymbol{\Sigma}, \psi, \chi, \lambda) = \left(\frac{\chi + \eta(\boldsymbol{y}; \boldsymbol{\mu}, \boldsymbol{\Sigma})}{\psi}\right)^{\frac{\lambda}{2}-\frac{p}{4}} \frac{\left(\frac{\psi}{\chi}\right)^{\frac{\lambda}{2}} K_{\lambda-\frac{p}{2}}(\sqrt{[\chi + \eta(\boldsymbol{y}; \boldsymbol{\mu}, \boldsymbol{\Sigma})]\psi})}{(2\pi)^{\frac{p}{2}} |\boldsymbol{\Sigma}|^{\frac{1}{2}} K_\lambda(\sqrt{\chi\psi})}, \tag{11}$$

where

$$\eta(\boldsymbol{y}; \boldsymbol{\mu}, \boldsymbol{\Sigma}) = (\boldsymbol{y} - \boldsymbol{\mu})^T \boldsymbol{\Sigma}^{-1}(\boldsymbol{y} - \boldsymbol{\mu}).$$

It is well known that the GH distribution has an identifiability issue in that the parameter vectors $\boldsymbol{\theta} = (\boldsymbol{\mu}, k\boldsymbol{\Sigma}, k\psi, \chi/k, \lambda)$ and $\boldsymbol{\theta}^* = (\boldsymbol{\mu}, \boldsymbol{\Sigma}, \psi, \chi, \lambda)$ both yield the same symmetric GH distribution (11) for any $k > 0$. It is therefore not surprising that the CFUSH distribution also suffers from such an issue. To handle this, restrictions are imposed on some of the parameters of the CFUSH distribution. An example is the HTH distribution considered in Murray et al. (2017) where the constraint $\psi = \chi = \omega$ is used, leading to the density

$$f_{\text{HTH}}(\boldsymbol{y}; \boldsymbol{\mu}, \boldsymbol{\Sigma}, \boldsymbol{\Delta}, \omega, \lambda)$$

$$= 2^r h_p(\boldsymbol{y}; \boldsymbol{\mu}, \boldsymbol{\Omega}, \omega, \omega, \lambda) H_r\left(\boldsymbol{\Delta}^T \boldsymbol{\Omega}^{-1}(\boldsymbol{y} - \boldsymbol{\mu})\left(\frac{\omega}{\omega + \eta}\right)^{\frac{1}{4}}; \boldsymbol{0}, \boldsymbol{\Lambda}, \gamma, \gamma, \lambda - \frac{p}{2}\right), \tag{12}$$

where $\gamma = \sqrt{\omega[\omega + \eta(\boldsymbol{y}; \boldsymbol{\mu}, \boldsymbol{\Sigma})]}$.

Note that in their terminology, Murray et al. (2017) are using 'hidden trunca-tion' to describe the latent skewing variable that follows a truncated distribution in the convolution-type characterization of the CFUSH distribution. Another alternative is to restrict the parameters of W so that, for example, $E(W) = 1$. A commonly used constraint on the GH distribution is to set $|\boldsymbol{\Sigma}| = 1$. This can be applied to the CFUSH distribution to achieve identifiability; see also the unrestricted skew normal generalized hyperbolic (SUNGH) distribution consid-ered by Maleki et al. (2019), which is equivalent to the CFUSH distribution on setting its scaling function equal to $W^{1/2}$.

Note that the CFUST distribution is not a special case of the CFUSH dis-tribution as stated in places in Murray et al. (2017). It can be obtained as a limiting case. One approach to obtain the limiting case is to put $\lambda = -\nu/2$ and to replace $\boldsymbol{\Sigma}$, $\boldsymbol{\Delta}$ and ω by $\frac{1}{k}\boldsymbol{\Sigma}, \frac{1}{k}\boldsymbol{\Delta}$, and $k\nu$ in the density (12) to give

$$f_{\text{HTH}}(\boldsymbol{y}; \boldsymbol{\mu}, \tfrac{1}{k}\boldsymbol{\Sigma}, \tfrac{1}{k}\boldsymbol{\Delta}, k\nu, -\tfrac{\nu}{2}),$$

and then to let k tend to zero; see also Murray et al. (2019).

7   Mixtures of CFUST Distributions Versus Mixtures of HTH Distributions

To demonstrate the performance of mixtures of HTH distributions in clustering data, they were compared by Murray et al. (2017) with mixtures of the two special cases of CFUST distributions defined in the previous section, namely the uMST (that is, CFUST($r = 1$)) and the rMST (CFUST(diag)) distributions, the latter two being referred to as the classical skew t and the SDM skew t distribution, respectively, by Murray et al. (2017). These three models were fitted to two real data sets referred to as the Seeds and HSCT (hematopoietic stem cell transplant) sets. However, a mixture of CFUST distributions was not fitted to these two data sets by Murray et al. (2017), who stated that "the SDB skew t mixture model is regarded by some as the state of the art approach (see [36])." The reference [36] is the paper by Lee and McLachlan (2013b) in which the CFUST distribution was not considered. In subsequent papers (Lee and McLachlan 2014b; 2015; 2016), which appeared before the submission by Murray et al. (2017) of their paper, it had been explained and demonstrated how the SDB skew t distribution, along with the classical skew t, are embedded in the CFUST distribution. The results in Murray et al. (2017) are reproduced here in Table 1, along with the results for mixtures of CFUST distributions as obtained by McLachlan and Lee (2019). The notation HTHu and HTHm in Table 1 as used in Murray et al. (2017) refer to the HTH component distributions having $r = 1$ and $r = p$ in forming the skewness matrix $\boldsymbol{\Delta}$ in the formulation of the HTH distribution as a member of the class of SMCFUSN distributions (9).

On fitting mixtures of CFUST distributions to the HSCT and Seeds data sets with $r = p$ skewing variables, McLachlan and Lee (2019) obtained higher values of the adjusted Rand Index (ARI), namely 0.991 and 0.916 relative to the values of 0.984 (0.976) and 0.877 (0.877) for mixtures of HTHm (HTHu) distributions fitted to the HSCT and Seeds data sets, respectively.

In commenting on their results in Table 1, Murray et al. (2017) stated that "The results (Table 1) show that the HTHu and HTHm mixture models outperform both the classical and SDB skew-t mixtures for both data sets. This is crucial when one considers that the HSCT and seeds data sets were used by [Lee and McLachlan (2014a)] to illustrate the excellent clustering performance of the SDB and classical skew-t mixture approaches". We wish to note here that Lee and McLachlan (2013b) did not consider the seeds data set to demonstrate the clustering performance of mixtures of rMST and uMST distributions. This is because they did not produce clusterings with higher ARI's than that obtained by mixtures of ordinary multivariate normal distributions.

In Murray et al. (2017), the very small value of 0.009 for the ARI for the SDB skew t mixture model in Table 1 is not interpreted as reflecting the failure of the algorithm to fit the model to the data. Rather it is taken at face value with the statement that "the SDB skew-t mixture performs better than the classical skew-t mixture approach for the Seeds data."

As stated in Murray et al. (2017), "the expected value of the ARI under random classification is zero". Thus the reported value of 0.009 for the SDB skew t-mixture model is implying that this model does no better than random classification! But this is unrealistic, particularly as an ARI value of 0.836 was obtained for the classical model; such a difference between the ARI's for these two models is highly unlikely as both embed the t-mixture model with just a few additional parameters to allow for any skewness in the data.

Table 1. ARI value of CFUST mixture model versus the ARI values for the mixture models in Murray et al. (2017)

	CFUST	HTHu	HTHm	Classical skew t	SDB skew t
HSCT	0.991	0.976	0.984	0.782	0.890
Seeds	0.916	0.877	0.877	0.836	0.009

To investigate this further, we followed the procedure adopted in Murray et al. (2017) for the fitting of this model. We first scaled the data so that each variable had mean zero and unit standard deviation before applying the EMMIXuskew program using the default options for the starting values for the skewness parameters. We found that the EM algorithm stopped after three iterations as essentially all the observations were being put into the one cluster. It was a consequence of the initial estimates of the skewness parameters not being scale invariant which for the Seeds data set after scaling caused problems. However, the EMMIXuskew algorithm also has optional starting strategies in addition to the default ones, such as those provided by k-means applied to the unscaled data and by the normal mixture model. When we used these latter options, we obtained a fit for the CFUST(diag) mixture model (that is, the SDB skew t model) with an ARI of 0.84. However, given the problems that our algorithm

EMMIXuskew encountered with the data as scaled by Murray et al. (2017), we have modified those default steps for the provision of starting values that were not scale invariant. But we recommend using our latest algorithm EMMIXcskew (Lee and McLachlan 2015; 2018), which also has the provision to fit mixtures of CFUST distributions.

8   Conclusions

Mixtures of CFUSN distrbutions provide a flexible approach to the clustering of heterogeneous data into clusters that exhibit skewness. For clusters with tails heavier than the normal, there is the mixture of CFUST distributions model. The CFUSN and CFUST distributions formally encompass other widely used models as special and limiting cases, including the restricted and unrestricted skew normal and t-distributions, and the normal and t-distributions.

Lee and McLachlan (2015; 2018) have developed an R package EMMIXcskew for the fitting of the CFUST distribution and finite mixtures of CFUST distributions via maximum likelihood (ML). An expectation–maximization (EM) algorithm is described for computing the ML estimates of the parameters of the FM-CFUST model, and different strategies for initializing the algorithm are discussed and illustrated. Concerning the implementation of the EM algorithm, the M-step can be carried out in closed form for CFUSN mixtures and also for CFUST mixtures apart for the degrees of freedom for each component which needs to be calculated iteratively on each M-step.

Recently, the so-called CFUSH distribution has been proposed from which the CFUST distribution can be obtained as a limiting case. For the two data sets considered in this paper, CFUSH mixtures did not perform as well as CFUST mixtures in providing a clustering.

References

Allard, A., Soubeyrand, S.: Skew-normality for climatic data and dispersal models for plant epidemiology: when application fields drive spatial statistics. Spat. Stat. **1**, 50–64 (2012)

Arellano-Valle, R.B., Azzalini, A.: On the unification of families of skew-normal distributions. Scand. J. Stat. **33**, 561–574 (2006)

Arellano-Valle, R.B., Branco, M.D., Genton, M.G.: A unified view on skewed distributions arising from selections. Can. J. Stat. **34**, 581–601 (2006)

Arellano-Valle, R.B., Genton, M.G.: On fundamental skew distributions. J. Multivar. Anal. **96**, 93–116 (2005)

Asparouhov, T., Muthén, B.: Structural equation models and mixture models with continuous non-normal skewed distributions. Struct. Equ. Model.:Multidisc. J. **23**, 1–19 (2016)

Azzalini, A.: A class of distributions which includes the normal ones. Scand. J. Stat. **12**, 171–178 (1985)

Azzalini, A.: The skew-normal distribution and related multivariate families. Scand. J. Stat. **32**, 159–188 (2005)

Azzalini, A.: The Skew-Normal and Related Families. Cambridge University Press, Cambridge (2014). Institute of Mathematical Statistics Monographs

Azzalini, A., Browne, R.P., Genton, M.G., McNicholas, P.D.: On nomenclature for, and the relative merits of, two formulations of skew distributions. Stat. Probab. Lett. **110**, 201–206 (2016)

Azzalini, A., Capitanio, A.: Distributions generated by perturbation of symmetry with emphasis on a multivariate skew t distribution. J. Roy. Stat. Soc. B **65**, 367–389 (2003)

Azzalini, A., Dalla Valle, A.: The multivariate skew-normal distribution. Biometrika **83**, 715–726 (1996)

Contreras-Reyes, J.E., Arellano-Valle, R.B.: Growth estimates of cardinalfish (Epigonus Crassicaudus) based on scale mixtures of skew-normal distributions. Fish. Res. **147**, 137–144 (2013)

Contreras-Reyes, J.E., López Quintero, F.O., Yáñez, A.A.: Towards age determination of Southern King crab (Lithodes Santolla) off Southern Chile using flexible mixture modeling. J. Marine Sci. Eng. **6**, 157 (2018)

Forbes, F., Wraith, D.: A new family of multivariate heavy-tailed distributions with variable marginal amounts of tailweight: application to robust clustering. Stat. Comput. **24**, 971–984 (2013)

Genton, M.G. (ed.): Skew-Elliptical Distributions and Their Applications: A Journey Beyond Normality. Chapman & Hall/CRC, Boca Raton/Florida (2004)

Hejblum, B.P., Alkhassim, C., Gottardo, R., Caron, F., Thiébaut, R.: Sequential Dirichlet process mixtures of multivariate skew t-distributions for model-based clustering of flow cytometry data. Ann. Appl. Stat. **13**, 638–660 (2019)

Hohmann, L., Holtmann, J., Eid, M.: Skew t mixture latent state-trait analysis: a Monte Carlo simulation study on statistical performance. Front. Psychol. **9**, 1323 (2018)

Karlis, D., Santourian, A.: Model-based clustering with non-elliptically contoured distributions. Stat. Comput. **19**, 73–83 (2009)

Kollo, T.: Multivariate skewness and kurtosis measures with an application in ICA. J. Multivar. Anal. **99**, 2328–2338 (2008)

Kollo, T., Srivastava, M.S.: Estimation and testing of parameters in multivariate Laplace distribution. Commun. Stat. - Theor. Methods **33**, 2363–2387 (2007)

Lee, S.X., Lin, T.-I., McLachlan, G.J.: Mixtures of factor analyzers with fundamental skew symmetric distributions. arXiv:1802.02467 (2018)

Lee, S.X., McLachlan, G.J.: On mixtures of skew normal and skew t-distributions. Adv. Data Anal. Classif. **7**, 241–266 (2013a)

Lee, S.X., McLachlan, G.J.: Model-based clustering and classification with non-normal mixture distributions (with discussion). Stat. Methods Appl. **22**, 427–479 (2013b)

Lee, S.X., McLachlan, G.J.: Finite mixtures of multivariate skew t-distributions: some recent and new results. Stat. Comput. **24**, 181–202 (2014a)

Lee, S.X., McLachlan, G.J.: Maximum likelihood estimation for finite mixtures of canonical fundamental skew t-distributions: the unification of the unrestricted and restricted skew t-mixture models. arXiv:1401.8182v1 [stat.ME] (2014b)

Lee, S.X., McLachlan, G.J.: EMMIXcskew: an R package for the fitting of a mixture of canonical fundamental skew t-distributions. arXiv:1509.02069.v1 [stat.CO] (2015)

Lee, S.X., McLachlan, G.J.: EMMIXcskew: an R Package for the fitting of a mixture of canonical fundamental skew t-distributions. J. Stat. Softw. **83**(3) (2018)

Lee, S.X., McLachlan, G.J.: Finite mixtures of canonical fundamental skew t-distributions: the unification of the restricted and unrestricted skew t-mixture models. Stat. Comput. **26**, 573–589 (2016)

Lee, S.X., McLachlan, G.J., Pyne, S.: Modelling of inter-sample variation in flow cytometric data with the joint clustering and matching (JCM) procedure. Cytometry Part A **89A**, 30–43 (2016)

Lin, T.-I.: Maximum likelihood estimation for multivariate skew normal mixture models. J. Multivar. Anal. **101**, 257–265 (2009a)

Lin, T.-I.: Robust mixture modeling using the multivariate skew t-distributions. Stat. Comput. **20**, 343–356 (2009b)

Lin, T.-I., Lee, J.C., Hsieh, W.: Robust mixture modeling using the skew t distribution. Stat. Comput. **17**, 81–92 (2007a)

Lin, T.-I., Lee, J.C., Yen, S.Y.: Finite mixture modelling using the skew normal distribution. Stat. Sinica **17**, 909–927 (2007b)

Maleki, M., Wraith, D., Arellano-Valle, R.B.: Robust finite mixture modeling of multivariate unrestricted skew-normal generalized hyperbolic distributions. Stat. Comput. **29**, 415–428 (2019)

McLachlan, G.J., Lee, S.X.: Comment "On the nomenclature for, and the relative merits of two formulations of skew distributions," by A. Azzalini, R. Browne, M. Genton, and P. McNicholas. Stat. Probab. Lett. **116**, 1–5 (2016)

McLachlan, G.J., Lee, S.X.: Comment on "Hidden truncation hyperbolic distributions, mixtures thereof, and their application for clustering" by Murray, Browne, and McNicholas. arXiv:1904.12057 (2019)

McLachlan, G.J., Lee, S.X., Rathnayake, S.I.: Finite mixture models. Ann. Rev. Stat. Appl. **6**, 355–378 (2019)

Mousavi, S.A., Amirzadeh, V., Rezapour, M., Sheikhy, A.: Multivariate tail conditional expectation for scale mixtures of skew-normal distribution. J. Stat. Comput. Simul. **89**, 3167–3181 (2019)

Murray, P.M., Browne, R.B., McNicholas, P.D.: Hidden truncation hyperbolic distributions, finite mixtures thereof, and their application for clustering. J. Multivar. Anal. **161**, 141–156 (2017)

Murray, P.M., Browne, R.B., McNicholas, P.D.: Note of Clarification on "Hidden truncation hyperbolic distributions, finite mixtures thereof, and their application for clustering", by Murray, Browne, and McNicholas, J. Multivariate Anal. 161 (2017) 141–156. J. Multivar. Anal. **171**, 475–476 (2019)

Pyne, S., et al.: Automated high-dimensional flow cytometric data analysis. Proc. Natl. Acad. Sci. USA **106**, 8519–8524 (2009)

Pyne, S., et al.: Joint modeling and registration of cell populations in cohorts of high-dimensional flow cytometric data. PLoS One **9**(7), e100334 (2014)

Riggi, S., Ingrassia, S.: Modeling high energy cosmic rays mass composition data via mixtures of multivariate skew-t distributions. arXiv:13011178 [astro-phHE] (2013)

Sahu, S.K., Dey, D.K., Branco, M.D.: A new class of multivariate skew distributions with applications to Bayesian regression models. Can. J. Stat. **31**, 129–150 (2003)

Seshadri, V.: Halphen's laws. In: Encyclopedia of Statistical Sciences, pp. 302–306. Wiley, New York (1997)

Tagle, F., Castruccio, S., Crippa, P., Genton, M.G.: A non-Gaussian spatio-temporal model for daily wind speeds based on a multi-variate skew-t distribution. J. Time Ser. Anal. **40**, 312–326 (2019)

Voigt, T., Fried, R.: Distance based feature construction in a setting of astronomy. In: Lausen, B., Krolak-Schwerdt, S., Böhmer, M. (eds.) Data Science, Learning by Latent Structures, and Knowledge Discovery. SCDAKO, pp. 475–485. Springer, Heidelberg (2015). https://doi.org/10.1007/978-3-662-44983-7_42

Wraith, D., Forbes, F.: Clustering using skewed multivariate heavy tailed distributions with flexible tail behaviour. Comput. Stat. Data Anal. **90**, 61–72 (2015)

Estimating Occupancy and Fitting Models with the Two-Stage Approach

Natalie Karavarsamis$^{(\boxtimes)}$

Department of Mathematics and Statistics, La Trobe University,
Melbourne, Australia
n.karavarsamis@latrobe.edu.au
https://www.natalie-karavarsamis.github.org

Abstract. The two-stage approach for occupancy modelling applies the
partial and conditional likelihood to occupancy data and is an alternative
to direct maximisation of the full likelihood that involves simultaneous
estimation of occupancy and detection probabilities. The two-stage approach resolves limitations with the full likelihood and allows full use of
GLM (generalised linear model) and GAM (generalised additive model)
computing functions in standard software such as R. It reduces computation time as it significantly reduces the number of models to be
assessed in model selection. The two-stage approach makes it easy to
include covariates for heterogeneous GLMs and GAMs and we present
these models for time dependent detection probabilities. For the basic
occupancy model we provide complete solutions for maximum likelihood
estimation at the boundaries of the sample space, where the score equations do not apply. We describe a region based on a convex hull within
which estimates are certain to exist and evaluate the bias of the occupancy estimator.

Keywords: Partial likelihood · Zero inflated binomial ZIB · GLM ·
GAM · Heterogeneous · Occupancy · Imperfect detection · IWLS

1 Introduction

Over recent years there has been a shift in focus from estimating population size
(or abundance) to mapping species dispersion and doing this through occupancy.
We give the two-stage approach for occupancy, specifically work in [12,14,15,
17]. We begin with the full likelihood for the homogeneous case and resolve
limitations. We present the maximum likelihood estimators for the boundaries
of the sample space, where the score equations do not apply and give a plausible
region that ensures estimation and give an overview of the evaluation of bias of
the occupancy estimator.

Methods for estimating occupancy of a species that account for imperfect
detection are necessary for sound species management. Systematic sampling
strategies are vital for species management when multiple visits to a site occur.

© Springer Nature Singapore Pte Ltd. 2019
H. Nguyen (Ed.): RSSDS 2019, CCIS 1150, pp. 68–80, 2019.
https://doi.org/10.1007/978-981-15-1960-4_5

Repeated visits to a site has the potential for introducing heterogeneity into the data. This heterogeneity may be accommodated either by covariates [19] or by assuming a distributional form for the detection probabilities [18]. Other methods for modelling heterogeneity include Bayesian approaches such as those of [1, 3, 5, 6, 10, 20–22, 25, 30]. However, an optimal outcome would be to design a study which minimises this potential heterogeneity at the outset.

This paper is outlined as follows. In Sect. 2, we describe the full likelihood function that describes the basic occupancy model. We describe limitations with estimation that exist at the boundary of the parameter space and identify another boundary problem and give the complete set of boundary solutions in Sect. 3 [7, 12]. Section 4 gives a plausible region defined by a convex hull on the parameter space as a function of the sufficient statistics, which ensures that estimates for occupancy and detection exist and are less biased. Limitations of the full likelihood provides motivation for the two-stage approach, presented in Sect. 6. We show our extensions to the basic occupancy model with a partial likelihood approach that leads to our two-stage approach for the homogeneous case. This includes a conditional likelihood to estimate detection. The homogeneous model is extended to allow covariates to be considered separately in each stage. Then, detection and occupancy are estimated with GLMs (generalised linear models) that also deals with restrictions of the full likelihood for site inhomogeneity. As noted elsewhere (for example, [7, 11, 16, 29]) the full likelihood may be numerically unstable and this is addressed with the two-stage approach. There is some loss of efficiency for detection that is outweighed by gains in computational efficiency. We extend the GLM for two-stage model to include covariates of time dependent detection. We show two methods for estimating occupancy for GLM that include direct maximisation and iterative weighted least squares (IWLS). The GLM model is extended to include nonlinear covariates with GAMS (generalised additive models) and occupancy is estimated with an iterative offset method. Details and further derivations are deferred to originating work as noted throughout. We provide a short summarising discussion in Sect. 7.

2   Full Likelihood

The most basic occupancy model has two parameters, occupancy ψ and detection p. It was proposed by [19] and is an adaptation of capture-recapture models [9]. Occupancy and detection probabilities are assumed to remain constant over all N sites and T survey occasions. Under this assumption, we may construct the likelihood by modelling the number of detections at each site $X_{i.} (= \sum_{i=1}^{T} x_{i.})$ as a zero-inflated Binomial random variable ZIB$(T, p; \psi)$ [8, 26, 27] and [20, p. 94]. The number of detections $X_{i.}$ at site i is distributed as $X_{i.} \stackrel{d}{=} 0$, with probability $(1 - \psi)$, and $X_{i.} \stackrel{d}{=} \text{Bi}(T, p)$, with probability ψ. Thus $P(X_{i.} = x_{i.}) = \psi(1 - p)^T + (1 - \psi)$ when $x_{i.} = 0$, and $P(X_{i.} = x_{i.}) = \psi \binom{T}{x_{i.}} p^{x_{i.}} (1 - p)^{T - x_{i.}}$ for $x_{i.} = 1, 2, \ldots, T$. There are two states and three possible outcomes; either the species *is* present and *is* detected ($x = 1$), or the species *is* present and is *not* detected ($x = 0$), or the species is *not* present and *not* detected ($x = 0$).

Occupancy is supposed to be permanent. Then, with an indicator for presence, $I(x_{i.} > 0)$ gives the likelihood of each site i, $L_i(\psi, p | \boldsymbol{x}_i) = P(X_{i.} = x_{i.} | \psi, p)$ for the observed history vector $\boldsymbol{x}_i = (x_{i1}, \ldots, x_{iT})^T$ for detections $x_{ij} = 1$ and nondetections $x_{ij} = 0$ at site i on occasion j. Then by taking the product over all site likelihoods $L_i(\psi, p | \boldsymbol{x}_i)$ results in the full likelihood

$$L(\psi, p | \boldsymbol{X}) = \prod_{i=1}^{N} L_i(\psi, p | x_{i.})$$
$$= \left(\psi(1-p)^T + (1-\psi) \right)^{N-k} \psi^k p^x (1-p)^{NT-x}, \qquad (1)$$

for the $N \times T$ history matrix $\boldsymbol{X} = [x_{ij}]$, where $k = \sum_{i=1}^{N} I(x_{i.} > 0)$ is number of sites at which any detection is made during the study and $x = \sum_{i=1}^{N} x_{i.}$ is the total number of detections. The first term of the product relates occupancy to nondetections (or 0s) and the remainder terms relates occupancy to detections (1s).

The conditional distribution of the data, given the statistic (X, K) from the observed (x, k) does not depend on the parameters (ψ, p), so (X, K) are sufficient statistics for (ψ, p). This was shown in [12]. All the information concerning (ψ, p) that is contained in the data is captured by (X, K) and hence $(\widehat{\psi}, \widehat{p}) = (\widehat{\psi}(x, k), \widehat{p}(x, k))$. This observation is important as it enables to specify the exact distribution of the maximum likelihood estimator, and hence to evaluate the exact expectation, exact variance and the bias of $\widehat{\psi}$.

3   Boundary Solutions

Solutions to the score equations are given by $\psi = k/N\theta$ and $p/\theta = x/kT$, where $\theta = g(p) = 1 - (1-p)^T$ is the probability of at least one detected site. The value of p is obtained numerically as there is no closed form solution. For given values of (x, k), and since N and T are known, solve $(1 - (1-\theta)^{1/T})/\theta = x/kT$ to obtain $\theta(x, k)$. Thus, the value for ψ may be calculated.

However, the score equations may or may not give the maximum likelihood estimates (mles). These give the mles if the maximum occurs at a turning point in the interior of the parameter space, the unit square $[0, 1]^2$. If the maximum of the likelihood occurs on the edge of the parameter space, then the maximum may not occur at a turning point and one (or both) of the score equations may not be satisfied. This will tend to happen when (x, k) is on the edge of the sample space ($k = N$ or $x = k$ or $x = kT$). This results in three sets of boundary solutions.

Values for $\widehat{\psi} < 1$ may be obtained when $N - k > N(1 - \widehat{\theta})$. As there are no closed form solutions for ψ, p and $\text{var}(\widehat{\psi})$, numerical maximisation is required. The full likelihood approach maximises ψ and p simultaneously via numerical maximisation, for example, with function optim in R to obtain estimates $\widehat{\psi}$ and $\widehat{p}$ [24]. However, direct maximisation does not always converge, as examined in [16]. These are known limitations with the full likelihood. Asymptotically, for N and T, the score equations will yield the mles, but otherwise this is not guaranteed. For small N and T, these will not apply.

Derivations for the boundary solutions are given in [12,17]. Any history matrices that give sufficient statistics that fall on an edge of the sample space do not fulfil the score equations. On the boundaries when $x = k$ or $k = N$, the maximum is obtained by assuming that occupancy is perfect (or complete, $\psi = 1$), and all sites are occupied ($\psi = 1$), and our failure to observe at all site-occasions is explained by imperfect (or incomplete) detectability ($p < 1$), where p is the proportion of all NT site-occasions where detections are made. The estimators on this boundary are given by the edge solution $\widehat{\psi} = 1$ and $\widehat{p} = x/NT$. On the boundary $x = kT$, detection is assumed to be perfect and leads to $\widehat{p} = 1$ and $\widehat{\psi} = k/N$. For example, see Figure 2.1 and Section 2.3.2 in [12].

4 Plausible Region

In [17] a region was proposed which the authors refer to as the 'plausible' region, within which it is guaranteed that the mles exist and which leads to estimates that are less biased. This region is a convex hull that covers the largest region of the sample space whilst mles exist.

An example of a sparse sample space for $N = 5$ and $T = 3$ is shown in Fig. 1. An example of a heavily populated sample space is given by $N = 27$ and $T = 4$ and the convex hull is shown in Fig. 2. The solid line connecting the outer bullets gives the convex hull $\mathbb{Q}_E$. The curved lines connecting the bullets mark the L_k-lines given by $\psi\theta = k/N$ for values of θ and $k = 1, \ldots, N$. Bullets along the L_k mark the mles corresponding to (x, k) where for each $k : x = k, k+1, \ldots, kT$. Boundary solutions are on the convex hull and internal points given by the score equations. The vertical and horizontal dashed lines indicate where mles do not exist, as given by $\psi = 1/N$ for $\theta = 1$ and $k = 1$. The lower bound for $\mathbb{Q}_E$ is shown with the curved dashed line defined by $\psi = 1/(N\theta)$, where $k = 1$, as seen in Figs. 2 and 1. But as line at $k = 1$, the L_1 line, has few points on it, at most $T - 2$ (i.e. $kT - 2$ with $k = 1$ and 2 boundary points for ψ and p) 'internal' points (i.e. points not on the boundary of the mles), so if T is small this lower bound curve may be well under the actual estimate points. The next curve L_2 at $k = 2$ is $\psi\theta = 2/N$ populated by (at most) $2T - 2$ internal points, which may be relatively few points for small T, for example when $N = 5$ and $T = 3$ in Fig. 1. Thus we want a curve close to $\mathbb{Q}_E$ and includes most of the points. So the approximation for the bottom edge of the convex hull of the set of mles $\mathbb{Q}_E(N, T)$ consists of the region that is not densely populated so that for points above this, unbiased estimation is plausible. In [12] the authors derive $\psi_{LB} = c_E/p + 1/N - c_E$ where $c_E = (1 - \epsilon^{1/N})/T$ for some small value of $\epsilon > 0$, for example $\epsilon = 0.01$. This approximate lower bound is intended to provide a reasonable practical lower bound for plausible estimation of ψ for a given p. They found that when N and T are small e.g. $N = 5$ and $T = 3$, $\epsilon = 0.01$ gives a good approximation.

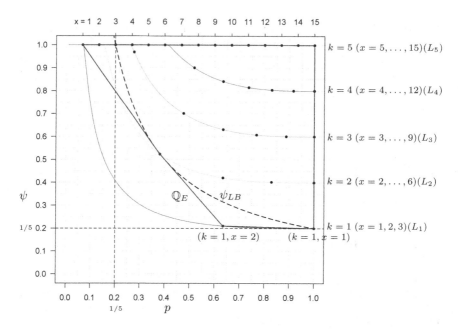

Fig. 1. Convex hull $\mathbb{Q}_E$ for $N = 5, T = 3$ marked by the solid line connecting the outer solid circles. The curved dashed line marks the lower bound for ψ, ψ_{LB}. The solid curved lines mark the L_k-lines for $k = 1, \ldots, N$ and the bullets along the L_k mark the mle corresponding to (x, k) where for each $k : x = k, k+1, \ldots, kT$, with the bottom curve starting at $k = 1$. The vertical and horizontal dashed lines are $1/N$.

5   Bias

To evaluate the bias of occupancy estimator $\widehat{\psi}$ requires the exact mean and variance of occupancy. In [17] the joint pmf of (X, K) was derived, and allows for definition of the exact expectation, exact variance and bias of occupancy estimator $\widehat{\psi}$.

In [17], the authors compared the exact variance to the asymptotic variance proposed by [20] and found that the asymptotic variance seriously underestimates the actual variance. For example, [19] recommended bootstrapping but they were not clear about which interval estimator to use and performance had not been assessed.

In [16] four bootstrap interval estimators for occupancy were examined including a normal approximation that uses the asymptotic variance of occupancy. They found that when N, T and p were not too small the studentized interval estimator was most consistent over the range of p and ψ. The asymptotic variance for occupancy $\widehat{\psi}$ was given in [20], and complete derivations given in [12],

$$\text{var}(\widehat{\psi}) = \frac{\psi}{N}\left((1 - \psi) + \frac{1 - \theta}{\theta - Tp(1-p)^{T-1}}\right).$$

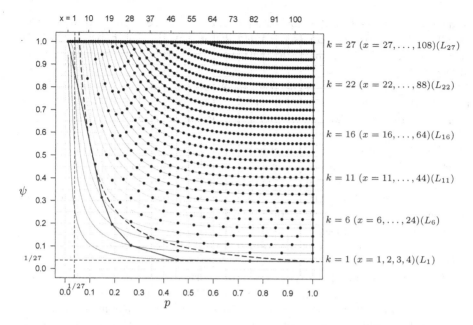

x = 1   10   19   28   37   46   55   64   73   82   91   100

$k = 27\ (x = 27, \ldots, 108)(L_{27})$

$k = 22\ (x = 22, \ldots, 88)(L_{22})$

$k = 16\ (x = 16, \ldots, 64)(L_{16})$

$k = 11\ (x = 11, \ldots, 44)(L_{11})$

$k = 6\ (x = 6, \ldots, 24)(L_6)$

$k = 1\ (x = 1, 2, 3, 4)(L_1)$

Fig. 2. Convex hull $\mathbb{Q}_E$ for $N = 27, T = 4$ marked by the solid line connecting the outer solid circles. The curved dashed line marks the lower bound for ψ, ψ_{LB}. The solid curved lines mark the L_k-lines for $k = 1, \ldots, N$ and the bullets along the L_k mark the mle corresponding to (x, k) where for each $k : x = k, k + 1, \ldots, kT$, with the bottom curve starting at $k = 1$. The vertical and horizontal dashed lines are $1/N$.

There is no closed form solution and a value for $\mathrm{var}(\widehat{\psi})$ is not always obtainable. Depending on the size of the study N and T, and depending on the values of the sufficient statistics (x, k) the Hessian may not be invertible [12,16,17]. The variance is obtained from the observed Fisher information $\mathrm{var}(\widehat{\psi}) = -\partial^2 l(\widehat{\psi}/x)/\partial\widehat{\psi}^2$.

A bias correction was examined, based on a conditional and an unconditional expectation for $\widehat{\psi}$ within the plausible region for estimation [17].

6   Two-Stage Approach and Modelling Occupancy

Modelling occupancy is possible by maximising the full likelihood. The function occu in the R package unmarked uses optim to maximise the full likelihood and simultaneously estimate ψ and p. Standard errors are possible with a bootstrap method or a Bayesian approach [3].

However there are limitations with the full likelihood, for example, that it may land outside the parameter space or simply will not converge, with multiple solutions for the mle as seen with the boundary solutions, or of identifiability with multiple local maxima, where numerical maximisation may converge to any of these or to none at all. The function may be too flat and estimates will be highly variable or variances may not be obtainable. And it may be too difficult to include covariates, for example as noted by [28]. Hence we would like an alternative approach. Bootstrap methods or Bayesian methods can include covariates but still have the issues with the basic occupancy model discussed above [17,28].

Presence-absence data results from recording capture information or detections. With the two-stage approach, we exploit the repeated visits to a site where there is more information on the detection probability than occupancy probabilities.

Thus [14,15] used a partial likelihood approach to propose a two-stage approach to estimating occupancy. This approach ignores information on the first capture and may be beneficial to gain computational efficiency at the cost of some small loss of efficiency in estimating the detection probabilities [14,15]. Often an investigator will wish to consider detection probabilities separately to occupancy probabilities, which is possible with the two-stage approach. Detection is estimated in the first stage then these values are used to estimate occupancy in the second stage.

We want to include this 'extra' covariate information into modelling relationships between ψ, of the species and its habitat. Occupancy and detection may be functions of, or influenced by covariates. With the partial likelihood approach, it is easy to include nonlinear functions of covariates for example with GAMs, whereas with the full likelihood approach there may be a large number of parameters to estimate. It resolves limitations and gives efficient closed form variance approximations [12,14,15]. And, it reduces the dimension of models in model selection. There may be time dependent covariates that affect detection that is a straight forward extension of the two-stage approach. [12,14,15] explored these scenarios in detail. The GAMs version is explored in [12,13].

6.1   Homogeneous Case

A two-stage approach to the homogeneous case is simple. In this case ψ and p are assumed constant across sites i and survey occasions j. It uses partial likelihoods to partition the full likelihood into one partial likelihood as a function of occupancy and detection $L_1(\psi, p)$ and another as a function of detection alone $L_2(p)$, as

$$L(\psi, p) \propto (1 - \psi + \psi(1-p)^T)^{N-k} \prod_{i=1}^{k} \psi p^{x_i \cdot} (1-p)^{T-x_i \cdot} \tag{2}$$

$$= (1 - \psi\theta)^{N-k} \psi^k \left(\prod_{i=1}^{k} (1-p)^{a_i-1} p \right) p^{x-k} (1-p)^{b-(x-k)}$$

$$\propto (1 - \psi\theta)^{N-k} \psi^k p^{x-k} (1-p)^{b-(x-k)}$$

$$= L_1(\psi, p) L_2(p),$$

where a_i are first detections for site i and b_i is the number of occasions remaining after the first detection at site i, then $b = \sum_{i=1}^{N} b_i$ are the total number of remaining occasions after a_i. This yields closed form estimators for p as a ratio of redetections over the number of remaining visits $\hat{p} = (x-k)/b$ and $\hat{\psi} = k/(N\hat{\theta})$ that are more stable than the full likelihood approach. Derivations are given in [14]. This model is equivalent to a conditional likelihood on detections, which yields orthogonal parameters as shown in [14] and can be computed using the VGAM package in R [24,33]. Variances are also given, and for detection, this is a simple binomial variance. Simulation studies show that the efficiency of $\hat{\psi}$ is above 90% for the two-stage approach [14]. With these positive results, the homogeneous case is extended and the heterogeneous case is developed to include covariates.

6.2 Heterogeneous Case

In the first instance, we explore site inhomogeneity, where occupancy and detection may vary between sites but remain constant within sites; in other words, remain constant over survey occasions. We give an overview here and derivations are given in [15].

The contribution of a single site i to the overall likelihood is

$$L_i(\psi_i, p_i) = \left\{ 1 - \psi_i + \psi_i(1 - p_i)^\tau \right\}^{1-w_i} \left\{ \binom{\tau}{x_{i.}} \psi_i p_i^{x_{i.}} \cdot (1 - p_i)^{\tau - x_{i.}} \right\}^{w_i}$$

$$\propto (1 - \psi_i\theta_i)^{1-w_i} \psi_i^{w_i} \left\{ p_i(1 - p_i)^{(a_i - 1)} \right\}^{w_i} \left\{ p_i^{(x_{i.} - 1)}(1 - p_i)^{b_i - x_{i.} + 1} \right\}^{w_i}$$

$$= (1 - \psi_i\theta_i)^{1-w_i} \psi_i^{w_i} \left\{ p_i^{(x_{i.} - 1)}(1 - p_i)^{b_i - x_{i.} + 1} \right\}^{w_i}, \tag{3}$$

$$= L_{1i}(\psi_i, p_i) L_{2i}(p_i)$$

where $w_i = 1$ is presence.

Let $\eta_s = \psi_s\theta_s$ for sites $s = 1, \ldots, S^1$ and then the likelihood can be re-written as a conditional likelihood for detection as in

$$L(\eta_s, p_s) = (1 - \eta_s)^{z_s} \eta_s^{1-z_s} \times \left\{ \frac{p_s^{y_s}(1 - p_s)^{1-y_s}}{\theta_s} \right\}^{1-z_s}$$

$$= L_1(\eta_s) L_2(p_s). \tag{4}$$

Modelling Detection. The conditional likelihood for detection $L_2(p_s)$ can be modelled with the vglm function in the VGAM package in R [32–34], and examples are given in [15]. Let p_{sj} be the detection probability at site s on occasion $j(= 1, \ldots, \tau)$ for an occupied site s then $\theta_s = 1 - \prod_{j=1}^{\tau}(1 - p_{sj})$ is the probability of at least one detection at site s for an occupied site. If there is no dependence on the survey occasion then $p_{sj} = p_s$ and $\theta_s = 1 - (1 - p_s)^\tau$. Then, detection can be modelled with a vector of covariates $\boldsymbol{u}_{sj}$ via $p_{sj} = h(\boldsymbol{u}_{sj}^T\boldsymbol{\beta})$ for a vector of

---

[1] For ease of derivations we make some slight modifications to notation.

coefficients $\boldsymbol{\beta} \in \mathbb{R}^q$. The function h usually will be the logistic function $h(x) = (1 + \exp(-x)^{-1})$. For time dependent covariates on detection the likelihood is

$$L(\eta_s, \boldsymbol{p}_s) = (1 - \eta_s)^{z_s} \eta_s^{1-z_s} \times \left\{ \frac{\prod_{j=1}^{\tau} p_{sj}^{y_{sj}} (1 - p_{sj})^{1-y_{sj}}}{\theta_s} \right\}^{1-z_s}$$

$$= L_1(\eta_s) L_2(\boldsymbol{p}_s),$$

where $\boldsymbol{p}_s = (p_{s1}, \ldots, p_{s\tau})^T$ for time dependent probabilities. For time independent case detection is constant over survey occasions, p_s and $p_s = p(\boldsymbol{u}_s, \boldsymbol{\beta}) = h(\boldsymbol{u}_s^T \boldsymbol{\beta})$ [15]. The contribution of site s to the log-likelihood is then

$$\ell(\eta_s, \boldsymbol{p}_s) = z_s \log(1 - \eta_s) + (1 - z_s) \log(\eta_s) \tag{5}$$

$$+ (1 - z_s) \left\{ \sum_{j=1}^{\tau} y_{sj} \log(p_{sj}) + \sum_{j=1}^{\tau} (1 - y_{sj}) \log(1 - p_{sj}) - \log(\theta_s) \right\}. \tag{6}$$

In the first stage, (6) is used to obtain estimated detection coefficients $\widehat{\boldsymbol{\beta}}$ and these yield the fitted values $\widehat{p}_s$. With these, calculate $\widehat{\theta}_s$ and define $\widetilde{\eta}_s = \psi_s \widehat{\theta}$. Then in the second stage, use $\widetilde{\eta}$ in (5) to estimate occupancy ψ. A large sample variance V_β is derived in [15]. Note that the conditional likelihood estimators $\widehat{p}$ will not be the mles, however [15] showed that estimation is not compromised by this fact.

Modelling Occupancy. In [15] explored three estimation methods for occupancy. Direct numeric maximisation of the first partial likelihood $L_{1s}(\psi_s, p_s)$ as a function of ψ_s from (3). Then $\psi_s = h(\boldsymbol{x}_s^T \boldsymbol{\alpha})$ where $\boldsymbol{x}_s$ is a vector of covariates associated with site s and $\boldsymbol{\alpha} \in \mathbb{R}^p$ is a vector of coefficients. To estimate $\boldsymbol{\alpha}$, maximise the partial likelihood $\prod_{s=1}^{S} L_{1s}(\widetilde{\eta}_s)$ where $\boldsymbol{p}_s$ and hence θ_s have been replaced by its estimator from the first stage $\widehat{\boldsymbol{p}}_s = \boldsymbol{p}_s(\widehat{\boldsymbol{\beta}})$. Then,

$$L_1(\boldsymbol{\alpha}) = \prod_{s=1}^{S} L_{1s}(\widetilde{\eta}_s) \propto \prod_{s=1}^{S} (1 - \psi_s \widehat{\theta}_s)^{z_s} \psi_s^{1-z_s},$$

where $w_s = 1 - z_s$ and the log-partial likelihood is

$$\ell(\boldsymbol{\alpha}) = \sum_{s=1}^{S} \left\{ (1 - w_s) \log(1 - \psi_s \widehat{\theta}_s) + w_s \log(\psi_s) \right\}.$$

This may be maximised numerically using the `optim` function in R.

The potential instability of the maximum likelihood estimates when computed using numerical optimization through the function `optim` in R motivated [15] to develop an iterative weighted least squares (IWLS) approach. This is quite straightforward for the logistic model in our two-stage approach.

For a logistic model, let matrix X have sth column x_s, $w = (w_1, \ldots, w_S)^T$, $E(w_s) = \eta_s = \theta_s \psi_s$, $\eta = (\eta_1, \ldots, \eta_S)^T$. Then, as θ_s is not a function of α, maximising the partial log-likelihood is equivalent to maximising $\ell(\eta) = \sum_{s=1}^{S} \{(1 - w_s) \log(1 - \eta_s) + w_s \log(\eta_s)\}$. Let $\eta(\alpha)$ be η evaluated at α and set $V = \text{diag}\{(1 - \eta)\eta\}$ and $U = \text{diag}\{\theta_s \psi_s (1 - \psi_s)\}$. Then $\alpha^{(k)}$ is estimate at the kth step and let $Z = UX\alpha^{(k)} + w - \eta(\alpha^{(k)})$. Then the estimate at the $(k+1)$th step is $\alpha^{(k+1)} = \left(XUV^{-1}UX^T\right)^{-1} XUV^{-1}UZ$ and this is repeated until convergence to yield the IWLS estimate for α. An estimate of the expected Fisher information corresponding to the partial likelihood, $E\{I(\alpha, \beta)\}$, is given by $\tilde{I}(\alpha, \beta) = XUV^{-1}UX^T$. Details and variances are derived in [15].

An iterative offset method was also explored in [15]. Their results show that the IWLS is preferred and recommend optim if IWLS does not converge. We defer further details and results with examples to their respective sources.

6.3   GAMs

Let u_s be the vector of covariates associated with detection whose effects will be modelled parametrically and $v_{s1}, \ldots, v_{sK}$ be those that will be modelled nonparametrically. Let the GAM for the linear predictor κ_s is $\kappa_s = u_s^T \beta + l_1(v_{s1}) + \cdots + l_K(v_{sK})$. Similarly, for occupancy let x_s be the parametric covariates and $r_{s1}, \ldots, r_{sJ}$ the nonparametric components. The GAM for the linear predictor is $\nu_s = x_s^T \alpha_1 + g_1(r_{s1}) + \cdots + g_J(r_{sJ})$.

Write $\alpha = (\alpha_1^T, \alpha_2^T)^T$, where α_1 are the parameters in the parametric component of the model and α_2 those in the nonparametric component.

For a given λ_S and penalty matrix $\mathcal{P}$, the penalised partial log-likelihood is $\ell_\lambda(\alpha, \beta) = \ell(\alpha_1, \beta) - \frac{1}{2}\lambda_S \alpha_2^T \mathcal{P} \alpha_2$,

$$\ell_\lambda(\alpha, \beta) = \sum_{s=1}^{S} \{(1 - w_s) \log(1 - \psi_s(\alpha_1)\theta_s) + w_s \log(\psi_s(\alpha_1))\} - \frac{1}{2}\lambda_S \alpha_2^T \mathcal{P} \alpha_2,$$

(7)

where $l(\alpha_1, \beta)$ is the likelihood from the GLM procedure. The penalised partial score function is $Q_\lambda(\alpha, \beta) = Q(\alpha_1, \beta) - \lambda_S \mathcal{P}^* \alpha_2$, where $Q(\alpha, \beta) = \partial \ell(\alpha, \beta)/\partial \alpha$ and $\mathcal{P}^*$ is a penalty matrix that penalises parameters in the nonparametric component of the model.

These models are fitted through penalised partial likelihoods in two stages, where, in the second stage, occupancy estimates are obtained through an iterative offset method. In the first stage, fit a GAM to the redetection data to yield estimates $\hat{p}_s$ for all sites as functions of u_s and $v_{s1}, \ldots, v_{sK}$.

In the second stage fit a GAM to the indicators of presence w_s with an offset using an iterative procedure. At the ith step the offset is $a_s^{(i)} = \log(\theta_s) - \log 1 + \exp(v^{(i-1)})(1 - \theta_s)$, for values of $\hat{\theta}_s$ for θ_s from the first step and $v^{(i-1)}$ is the predictor from the $(i - 1)$th step. Details of the iterative offset procedure are given in [15].

The GAM penalized likelihood is maximised, and coefficients estimated, by penalized iteratively re-weighted least squares (P-IRLS) [31, p. 169]. We use the gam function of the mgcv package to implement these methods from within R [24]. The penalty matrix may be extracted from fitting the gam function and computes an AIC (Aikaike information criterion) for the penalised likelihood using the effective degrees of freedom. The AIC is applied directly to model selection. To avoid overfitting with GAMs one way to control complexity is through the dimension of the basis of the smooth terms. Compute and minimize the $BIC_q = BIC - 2k \log(q/(1-q))$ using the bestglm package in R [4,23]. It is based on the BIC (Bayesian information criterion) for the fitted smooth model for a given k corrected for the number of parameters in the overall model, and $q \in (0,1)$. Then an appropriate k is chosen. This is done separately in each stage for detection and occupancy.

The variance approximation is an extension of the GLM variance approximation except for the penalty matrix, which penalises parameters in the nonparametric component of the model. Further details are given in [15].

7 Discussion

Occupancy models appear simple but their analysis is more difficult than expected. We have resolved problems with construction of estimators and interval estimators. Full likelihood approximations are possible but does not allow easy access to GLM machinery, [28] showed that problems with the full likelihood feed through to the covariate model. The two-stage approach with partial likelihood allows full access to GLM machinery at both stages and estimators from both stages are probabilities so are naturally constrained to be between 0 and 1. Mixed effects models and other maximisation method such as the EM (estimation-maximisation) algorithm can also be considered [2]. A natural next step is to resolve the boundary problems with covariates in the GLMs and GAMs cases, and is deferred to future work.

References

1. Aing, C., Halls, S., Oken, K., Dobrow, R., Fieberg, J.: A Bayesian hierarchical occupancy model for track surveys conducted in a series of linear, spatially correlated, sites. J. Appl. Ecol. **48**, 1508–1517 (2011)
2. Dempster, A., Laird, N., Rubin, D.: Maximum likelihood from incomplete data via the EM algorithm. J. R. Stat. Soc. B **39**(1), 1–38 (1977)
3. Fiske, I.J., Chandler, R.B.: unmarked: an R package for fitting hierarchical models of wildlife occurrence and abundance. J. Stat. Softw. **43**(10), 1–23 (2011). http://www.jstatsoft.org/v43/i10/
4. George, E., Foster, D.P.: Calibration and empirical Bayes variable selection. Biometrika **87**(4), 731–747 (2000)
5. Gimenez, O., et al.: WinBUGS for population ecologists: Bayesian modeling using Markov chain Monte Carlo methods. In: Thomson, D.L., Cooch, E.G., Conroy, M.J. (eds.) Modeling Demographic Processes in Marked Populations. Environmental and Ecological Statistics, vol. 3, pp. 883–915. Springer, Boston (2009). https://doi.org/10.1007/978-0-387-78151-8_41

6. Gimenez, O., et al.: State-space modelling of data on marked individuals. Ecol. Model. **206**, 431–438 (2007)

7. Guillera-Arroita, G., Ridout, M., Morgan, B.: Design of occupancy studies with imperfect detection. Methods Ecol. Evol. **1**(2), 131–139 (2010)

8. Hall, D.B.: Zero-inflated poisson and binomial regression with random effects: a case study. Biometrics **56**, 1030–1039 (2000)

9. Huggins, R.M.: On the statistical analysis of capture experiments. Biometrika **76**, 133–140 (1989)

10. Hui, C., Foxcroft, L.C., Richardson, D.M., MacFadyen, S.: Defining optimal sampling effort for large-scale monitoring of invasive alien plants: a Bayesian method for estimating abundance and distribution. J. Appl. Ecol. **48**(3), 768–776 (2011)

11. Hutchinson, R.A., Valente, J.J., Emerson, S.C., Betts, M.G., Dietterich, T.G.: Penalized likelihood methods improve parameter estimates in occupancy models. Methods Ecol. Evol. (2015). https://doi.org/10.1111/2041-210X.12368

12. Karavarsamis, N.: Methods for estimating occupancy. University of Melbourne, Victoria, Australia (2015)

13. Karavarsamis, N., Huggins, R.M.: The two-stage approach to the analysis of occupancy data iii: GAMs (2019, in preparation)

14. Karavarsamis, N., Huggins, R.M.: Two-stage approaches to the analysis of occupancy data I: the homogeneous case. Commun. Stat. - Theory Methods (2019). https://doi.org/10.1080/03610926.2019.1607385

15. Karavarsamis, N., Huggins, R.M.: Two-stage approaches to the analysis of occupancy data II. The heterogeneous model and conditional likelihood. Comput. Stat. Data Anal. **133**, 195–207 (2019)

16. Karavarsamis, N., Robinson, A.P., Hepworth, G., Hamilton, A., Heard, G.: Comparison of four bootstrap-based interval estimators of species occupancy and detection probabilities. Aust. N. Z. J. Stat. **55**(3), 235–252 (2013)

17. Karavarsamis, N., Watson, R.: Bias of the homogeneous occupancy estimator (2019, in preparation)

18. Mackenzie, D.I.: What are the issues with presence/absence data for wildlife managers? J. Wildl. Manag. **69**(3), 849–860 (2005)

19. MacKenzie, D.I., Nichols, J.D., Lachman, G.B., Droege, S., Royle, J., Langtimm, C.A.: Estimating site occupancy rates when detection probabilities are less than one. Ecology **83**(8), 2248–2255 (2002)

20. MacKenzie, D.I., Nichols, J., Royle, J., Pollock, K., Bailey, L., Hines, J.: Occupancy Estimation and Modeling Inferring Patterns and Dynamics of Species Occurrence. Elsevier, San Diego (2006)

21. MacKenzie, D.I., Nichols, J., Seamans, M.E., Gutiërrez, R.J.: Modeling species occurrence dynamics with multiple states and imperfect detection. Ecology **90**(3), 823–835 (2009)

22. Martin, J., Royle, J., MacKenzie, D., Edwards, H., Kéry, M., Gardner, B.: Accounting for non-independent detection when estimating abundance of organisms with a Bayesian approach. Methods Ecol. Evol. **2**(6), 595–601 (2011)

23. McLeod, A., Xu, C.: Best Subset GLM and Regression Utilities: Package 'bestglm' (2018)

24. R Development Core Team: R: A Language and Environment for Statistical Computing. R Foundation for Statistical Computing, Vienna, Austria (2018). http://www.R-project.org. ISBN 3-900051-07-0

25. Royle, J.A., Dorazio, R.: Hierarchical Modeling and Inference in Ecology: The Analysis of Data from Populations, Metapopulations, and Communities. Academic Press, San Diego (2008)

26. Royle, J.A.: Site occupancy models with heterogeneous detection probabiliies. Biometrics **62**, 97–102 (2006)
27. Royle, J.A., Nichols, J.D.: Estimating abundance from repeated presence-absence data or point counts. Ecology **84**(3), 777–790 (2003)
28. Welsh, A.H., Lindenmayer, D.B., Donnelly, C.F.: Fitting and interpreting occupancy models. PLoS One **8**(1), e52015 (2013). https://doi.org/10.1371/journal.pone.0052015.s001
29. Wintle, B., McCarthy, M., Parris, K., Burgman, M.: Precision and bias of methods for estimating point survey detection probabilities. Ecol. Appl. **14**, 703–712 (2004)
30. Wintle, B., McCarthy, M., Volinsky, C., Kavanagh, R.: The use of Bayesian model averaging to better represent uncertainty in ecological models. Conserv. Biol. **17**(6), 1579–1590 (2003)
31. Wood, S.N.: Generalized Additive Models: An Introduction with R. Chapman & Hall/CRC, Boca Raton (2006)
32. Yee, T.W.: The VGAM package for categorical data analysis. J. Stat. Softw. **32**, 1–34 (2010)
33. Yee, T.W., Stoklosa, J., Huggins, R.M.: The VGAM package for capture–recapture data using the conditional likelihood. J. Stat. Softw. **65**, 1–33 (2015)
34. Yee, T.W.: Vector Generalized Linear and Additive Models: With an Implementation in R. Springer, New York (2015). https://doi.org/10.1007/978-1-4939-2818-7

Component Elimination Strategies to Fit Mixtures of Multiple Scale Distributions

Florence Forbes[1](✉) (iD), Alexis Arnaud[1,2], Benjamin Lemasson[2], and Emmanuel Barbier[2]

[1] Univ. Grenoble Alpes, Inria, CNRS, Grenoble INP
(Institute of Engineering Univ. Grenoble Alpes), LJK, 38000 Grenoble, France
{florence.forbes,alexis.arnaud}@inria.fr
[2] Grenoble Institut des Neurosciences, Inserm U1216, Univ. Grenoble Alpes,
Grenoble, France
{alexis.arnaud,benjamin.lemasson,emmanuel.barbier}@univ-grenoble-alpes.fr

Abstract. We address the issue of selecting automatically the number of components in mixture models with non-Gaussian components. As a more efficient alternative to the traditional comparison of several model scores in a range, we consider procedures based on a single run of the inference scheme. Starting from an overfitting mixture in a Bayesian setting, we investigate two strategies to eliminate superfluous components. We implement these strategies for mixtures of multiple scale distributions which exhibit a variety of shapes not necessarily elliptical while remaining analytical and tractable in multiple dimensions. A Bayesian formulation and a tractable inference procedure based on variational approximation are proposed. Preliminary results on simulated and real data show promising performance in terms of model selection and computational time.

Keywords: Gaussian scale mixture · Bayesian analysis · Bayesian model selection · EM algorithm · Variational approximation

1 Introduction

A difficult problem when fitting mixture models is to determine the number K of components to include in the mixture. A recent review on the problem with theoretical and practical aspects can be found in [10]. Traditionally, this selection is performed by comparing a set of candidate models for a range of values of K, assuming that the true value is in this range. The number of components is selected by minimizing a model selection criterion, such as the Bayesian inference criterion (BIC), minimum message length (MML), Akaike's information criteria (AIC) to cite just a few [13,23]. Of a slightly different nature is the so-called slope heuristic [7], which involves a robust linear fit and is not simply based on criterion comparisons. However, the disadvantage of these approaches is that a whole set of candidate models has to be obtained and problems associated with

© Springer Nature Singapore Pte Ltd. 2019
H. Nguyen (Ed.): RSSDS 2019, CCIS 1150, pp. 81–95, 2019.
https://doi.org/10.1007/978-981-15-1960-4_6

running inference algorithms (such as EM) many times may emerge. When the components distributions complexity increases, it may then be desirable to avoid repetitive inference of models that will be discarded in the end. For standard Gaussian distributions however, this is not really a problem as efficient software such as Mclust [28] are available. Alternatives have been investigated that select the number of components from a single run of the inference scheme. Apart from the Reversible Jump Markov Chain Monte Carlo method of [26] which allows jumps between different numbers of components, two types of approaches can be distinguished depending on whether the strategy is to increase or to decrease the number of components. The first ones can be referred to as greedy algorithms (*e.g.* [30]) where the mixture is built component-wise, starting with the optimal one-component mixture and increasing the number of components until a stopping criterion is met. More recently, there seems to be an increase interest among mixture model practitioners for model selection strategies that start instead with a large number of components and merge them [18]. For instance, [13] proposes a practical algorithm that starts with a very large number of components, iteratively annihilates components, redistributes the observations to the other components, and terminates based on the MML criterion. The approach in [6] starts with an overestimated number of components using BIC, and then merges them hierarchically according to an entropy criterion, while [24] proposes a similar method that merges components based on measuring their pair-wise overlap. Another trend in handling the issue of finding the proper number of components is to consider Bayesian non-parametric mixture models. This allows the implementation of mixture models with an infinite number of components via the use of Dirichlet process mixture models. In [17,25] an infinite Gaussian mixture (IGMM) is presented with a computationally intensive Markov Chain Monte Carlo implementation. More recently, more flexibility in the cluster shapes has been allowed by considering infinite mixture of infinite Gaussian mixtures (I^2GMM) [32]. The flexibility is however limited to a cluster composed of sub-clusters of identical shapes and orientations, which may alter the performance of this approach. Beyond the Gaussian case, infinite Student mixture models have also been considered [31]. The Bayesian non-parametric approach is a promising technique. In this work, we consider a Bayesian formulation but in the simpler case of a finite number of components. We suspect all our Bayesian derivations could be easily tested in a non parametric setting with some minor adaptation left for future work. Following common practice that is to start from deliberately overfitting mixtures (*e.g.* [3,11,21,22]), we investigate 'component elimination strategies. Component elimination refers to a natural approach which is to exploit the vanishing component phenomenon that has been proved to occur in certain Bayesian settings [27]. This requires a Bayesian formulation of the mixture for the regularization effect due to the integration of parameters in the posterior distribution. This results in an implicit penalization for model complexity. Although this approach can be based on arbitrary mixture components, most previous investigation has been confined to Gaussian mixtures where the mixture components arise from multivariate Gaussian densities with component-specific parameters.

In this work, we address the issue of selecting automatically the number of components in a non-Gaussian case. We consider mixtures of so called multiple scale distributions for their ability to handle a variety of shapes not necessarily elliptical while remaining analytical and tractable. We propose a Bayesian formulation of these mixtures and a tractable inference procedure based on a variational approximation. We propose two different single-run strategies that make use of the component elimination property.

The rest of the paper is organized as follows. Mixture of multiple scale distributions, their Bayesian formulation and inference are specified in Sect. 2. The two proposed strategies are described in Sect. 3, illustrated with experiments on simulated data in Sect. 4.

2 Bayesian Mixtures of Multiple Scale Distributions

2.1 Multiple Scale Mixtures of Gaussians

A M-variate scale mixture of Gaussians is a distribution of the form:

$$p(\mathbf{y}; \boldsymbol{\mu}, \boldsymbol{\Sigma}, \boldsymbol{\theta}) = \int_0^\infty \mathcal{N}_M(\mathbf{y}; \boldsymbol{\mu}, \boldsymbol{\Sigma}/w) \, f_W(w; \boldsymbol{\theta}) \, \mathrm{d}w \tag{1}$$

where $\mathcal{N}_M(\,.\,; \boldsymbol{\mu}, \boldsymbol{\Sigma}/w)$ denotes the M-dimensional Gaussian distribution with mean $\boldsymbol{\mu}$, covariance $\boldsymbol{\Sigma}/w$ and f_W is the probability distribution of a univariate positive variable W referred to as hereafter as the weight variable. A common form is obtained when f_W is a Gamma distribution $\mathcal{G}(\nu/2, \nu/2)$ where ν denotes the degrees of freedom (we shall denote the Gamma distribution when the variable is X by $\mathcal{G}(x; \alpha, \gamma) = x^{\alpha-1} \Gamma(\alpha)^{-1} \exp(-\gamma x) \gamma^\alpha$ where Γ denotes the Gamma function). For this form, (1) is the density denoted by $t_M(\mathbf{y}; \boldsymbol{\mu}, \boldsymbol{\Sigma}, \nu)$ of the M-dimensional Student t-distribution with parameters $\boldsymbol{\mu}$ (real location vector), $\boldsymbol{\Sigma}$ ($M \times M$ real positive definite scale matrix) and ν (positive real degrees of freedom parameter).

The extension proposed by [14] consists of introducing a multidimensional weight. To do so, the scale matrix is decomposed into eigenvectors and eigenvalues. This spectral decomposition is classically used in Gaussian model-based clustering [5,9]. In a Bayesian setting, it is equivalent but more convenient to use matrix $\mathbf{T}$ the inverse of the scale matrix. We therefore consider the decomposition $\mathbf{T} = \mathbf{D}\mathbf{A}\mathbf{D}^T$ where $\mathbf{D}$ is the matrix of eigenvectors of $\mathbf{T}$ (equivalently of $\boldsymbol{\Sigma}$) and $\mathbf{A}$ is a diagonal matrix with the corresponding eigenvalues. The matrix $\mathbf{D}$ determines the orientation of the Gaussian and $\mathbf{A}$ its shape. Using this parameterization of $\mathbf{T}$, the scale Gaussian part in (1) is set to $\mathcal{N}_M(\mathbf{y}; \boldsymbol{\mu}, \mathbf{D}\boldsymbol{\Delta}_\mathbf{w}\mathbf{A}^{-1}\mathbf{D}^T)$, where $\boldsymbol{\Delta}_\mathbf{w} = \mathrm{diag}(w_1^{-1}, \ldots, w_M^{-1})$ is the $M \times M$ diagonal matrix whose diagonal components are the inverse weights $\{w_1^{-1}, \ldots, w_M^{-1}\}$. The multiple scale generalization consists therefore of:

$$p(\mathbf{y}; \boldsymbol{\mu}, \boldsymbol{\Sigma}, \boldsymbol{\theta}) = \int_0^\infty \ldots \int_0^\infty \mathcal{N}_M(\mathbf{y}; \boldsymbol{\mu}, \mathbf{D}\boldsymbol{\Delta}_\mathbf{w}\mathbf{A}^{-1}\mathbf{D}^T) \, f_\mathbf{w}(w_1 \ldots w_M; \boldsymbol{\theta}) \, \mathrm{d}w_1 \ldots \mathrm{d}w_M \tag{2}$$

where $f_{\mathbf{w}}$ is now a M-variate density depending on some parameter $\boldsymbol{\theta}$ to be further specified. In what follows, we will consider only independent weights, *i.e.* $\boldsymbol{\theta} = \{\boldsymbol{\theta}_1, \ldots, \boldsymbol{\theta}_M\}$ with $f_{\mathbf{w}}(w_1 \ldots w_M; \boldsymbol{\theta}) = f_{W_1}(w_1; \boldsymbol{\theta}_1) \ldots f_{W_M}(w_M; \boldsymbol{\theta}_M)$. For instance, setting $f_{W_m}(w_m; \theta_m)$ to a Gamma distribution $\mathcal{G}(w_m; \alpha_m, \gamma_m)$ results in a multivariate generalization of a Pearson type VII distribution (see *e.g.* [20] vol. 2 chap. 28 for a definition of the Pearson type VII distribution). For identifiability, this model needs to be further specified by fixing all γ_m parameters, for instance to 1. Despite this additional constraint, the decomposition of $\boldsymbol{\Sigma}$ still induces another identifiability issue due to invariance to a same permutation of the columns of $\boldsymbol{D}, \boldsymbol{A}$ and elements of $\boldsymbol{\alpha} = \{\alpha_1, \ldots, \alpha_M\}$. In a frequentist setting this can be solved by imposing a decreasing order for the eigenvalues in $\boldsymbol{A}$. In a Bayesian setting one way to solve the problem is to impose on $\boldsymbol{A}$ a non symmetric prior (see Sect. 2.2). An appropriate prior on $\boldsymbol{D}$ would be more difficult to set. The distributions we consider are therefore of the form,

$$\mathcal{MP}(\mathbf{y}; \boldsymbol{\mu}, \boldsymbol{D}, \boldsymbol{A}, \boldsymbol{\alpha}) = \prod_{m=1}^{M} \frac{\Gamma(\alpha_m + 1/2) A_m}{\Gamma(\alpha_m)(2\pi)^{1/2}} \left(1 + \frac{A_m[\boldsymbol{D}^T(\mathbf{y} - \boldsymbol{\mu})]_m^2}{2}\right)^{-(\alpha_m + 1/2)} \quad (3)$$

Let us consider an *i.i.d* sample $\mathbf{y} = \{\mathbf{y}_1, \ldots, \mathbf{y}_N\}$ from a K-component mixture of multiple scale distributions as defined in (3). With the usual notation for the mixing proportions $\boldsymbol{\pi} = \{\pi_1, \ldots, \pi_K\}$ and $\boldsymbol{\psi}_k = \{\boldsymbol{\mu}_k, \boldsymbol{A}_k, \boldsymbol{D}_k, \boldsymbol{\alpha}_k\}$ for $k = 1 \ldots K$, we consider,

$$p(\mathbf{y}; \boldsymbol{\Phi}) = \sum_{k=1}^{K} \pi_k \mathcal{MP}(\mathbf{y}; \boldsymbol{\mu}_k, \boldsymbol{A}_k, \boldsymbol{D}_k, \boldsymbol{\alpha}_k)$$

where $\boldsymbol{\Phi} = \{\boldsymbol{\pi}, \boldsymbol{\psi}\}$ with $\boldsymbol{\psi} = \{\boldsymbol{\psi}_1, \ldots \boldsymbol{\psi}_K\}$ denotes the mixture parameters. Additional variables can be introduced to identify the class labels: $\{Z_1, \ldots, Z_N\}$ define respectively the components of origin of $\{\mathbf{y}_1, \ldots, \mathbf{y}_N\}$. An equivalent modelling is therefore:

$$\forall i \in \{1 \ldots N\}, \quad \mathbf{Y}_i | \mathbf{W}_i = \mathbf{w}_i, Z_i = k, \boldsymbol{\psi} \sim \mathcal{N}(\boldsymbol{\mu}_k, \boldsymbol{D}_k \boldsymbol{\Delta}_{\mathbf{w}_i} \boldsymbol{A}_k^{-1} \boldsymbol{D}_k^T),$$

$$\mathbf{W}_i | Z_i = k, \boldsymbol{\psi} \sim \mathcal{G}(\alpha_{k1}, 1) \otimes \ldots \otimes \mathcal{G}(\alpha_{kM}, 1),$$

$$\text{and} \quad Z_i | \boldsymbol{\pi} \sim \mathcal{M}(1, \pi_1, \ldots, \pi_k),$$

where $\boldsymbol{\Delta}_{\mathbf{w}_i} = \text{diag}(w_{i1}^{-1}, \ldots, w_{iM}^{-1})$, symbol $\otimes$ means that the components of $\mathbf{W}_i$ are independent and $\mathcal{M}(1, \pi_1, \ldots, \pi_k)$ denotes the Multinomial distribution. In what follows, the weight variables will be denoted by $\mathbf{W} = \{\mathbf{W}_1, \ldots, \mathbf{W}_N\}$ and the labels by $\mathbf{Z} = \{Z_1, \ldots, Z_N\}$.

2.2   Priors on Parameters

In a Bayesian formulation, we assign priors on parameters in $\boldsymbol{\Phi}$. However, it is common (see *e.g.* [1]) not to impose priors on the parameters $\boldsymbol{\alpha}_k$ since no convenient conjugate prior exist for these parameters. Then the scale matrix decomposition imposes that we set priors on $\boldsymbol{\mu}_k$ and $\boldsymbol{D}_k, \boldsymbol{A}_k$. For the means $\boldsymbol{\mu}_k$, the standard Gaussian prior can be used:

$$\boldsymbol{\mu}_k \mid \boldsymbol{A}_k, \boldsymbol{D}_k \sim \mathcal{N}(\boldsymbol{m}_k, \boldsymbol{D}_k \boldsymbol{\Lambda}_k^{-1} \boldsymbol{A}_k^{-1} \boldsymbol{D}_k^T) \,, \tag{4}$$

where $\boldsymbol{m}_k$ (vector) and $\boldsymbol{\Lambda}_k$ (diagonal matrix) are hyperparameters and we shall use the notation $\boldsymbol{m} = \{\boldsymbol{m}_1, \ldots, \boldsymbol{m}_K\}$ and $\boldsymbol{\Lambda} = \{\boldsymbol{\Lambda}_1, \ldots \boldsymbol{\Lambda}_K\}$. For $\boldsymbol{A}_k$ and $\boldsymbol{D}_k$ a natural solution would be to use the distributions induced by the standard Wishart prior on $\boldsymbol{T}_k$ but this appears not to be tractable in inference scheme based on a variational framework. The difficulty lies in considering an appropriate and tractable prior for $\boldsymbol{D}_k$. There exists a number of priors on the Stiefel manifold among which a good candidate could be the Bingham prior and extensions investigated by [19]. However, it is not straightforward to derive from it a tractable E-$\boldsymbol{\Phi}^1$ step (see Sect. 2.3) that could provide a variational posterior distribution. Nevertheless, this kind of priors could be added in the M-$\boldsymbol{D}$-step. The simpler solution adopted in the present work consists of considering $\boldsymbol{D}_k$ as an unknown fixed parameter and imposing a prior only on $\boldsymbol{A}_k$, which is a diagonal matrix containing the positive eigenvalues of $\boldsymbol{T}_k$. It is natural to choose:

$$\boldsymbol{A}_k \sim \otimes_{m=1}^M \mathcal{G}(\lambda_{km}, \delta_{km}) \,, \tag{5}$$

where $\boldsymbol{\lambda}_k = \{\lambda_{km}, m = 1 \ldots M\}$ and $\boldsymbol{\delta}_k = \{\delta_{km}, m = 1 \ldots M\}$ are hyperparameters with $\boldsymbol{\lambda} = \{\boldsymbol{\lambda}_1, \ldots \boldsymbol{\lambda}_K\}$ and $\boldsymbol{\delta} = \{\boldsymbol{\delta}_1, \ldots \boldsymbol{\delta}_K\}$ as additional notation. It follows the joint prior on $\boldsymbol{\mu}_{1:K} = \{\boldsymbol{\mu}_1, \ldots, \boldsymbol{\mu}_K\}$, $\boldsymbol{A}_{1:K} = \{\boldsymbol{A}_1, \ldots, \boldsymbol{A}_K\}$ given $\boldsymbol{D}_{1:K} = \{\boldsymbol{D}_1, \ldots, \boldsymbol{D}_K\}$

$$p(\boldsymbol{\mu}_{1:K}, \boldsymbol{A}_{1:K}; \boldsymbol{D}_{1:K}) = \prod_{k=1}^K p(\boldsymbol{\mu}_k | \boldsymbol{A}_k; \boldsymbol{D}_k) \, p(\boldsymbol{A}_k) \tag{6}$$

where the first term in the product is given by (4) and the second term by (5).

Then a standard Dirichlet prior $\mathcal{D}(\tau_1, \ldots, \tau_K)$ is used for the mixing weights $\boldsymbol{\pi}$ with $\boldsymbol{\tau} = \{\tau_1, \ldots, \tau_K\}$ the Dirichlet hyperparameters.

For the complete model, the whole set of parameters is denoted by $\boldsymbol{\Phi}$. $\boldsymbol{\Phi} = \{\boldsymbol{\Phi}^1, \boldsymbol{\Phi}^2\}$ is decomposed into a set $\boldsymbol{\Phi}^1 = \{\boldsymbol{\Phi}_1^1, \ldots \boldsymbol{\Phi}_K^1\}$ with $\boldsymbol{\Phi}_k^1 = \{\boldsymbol{\mu}_k, \boldsymbol{A}_k, \pi_k\}$ of parameters for which we have priors and a set $\boldsymbol{\Phi}^2 = \{\boldsymbol{\Phi}_1^2, \ldots \boldsymbol{\Phi}_K^2\}$ with $\boldsymbol{\Phi}_k^2 = \{\boldsymbol{D}_k, \boldsymbol{\alpha}_k\}$ of unknown parameters considered as fixed. In addition, hyperparameters are denoted by $\boldsymbol{\Phi}^3 = \{\boldsymbol{\Phi}_1^3, \ldots \boldsymbol{\Phi}_K^3\}$ with $\boldsymbol{\Phi}_k^3 = \{\tau_k, \boldsymbol{m}_k, \boldsymbol{\Lambda}_k, \boldsymbol{\lambda}_k, \boldsymbol{\delta}_k\}$.

2.3   Inference Using Variational Expectation-Maximization

The main task in Bayesian inference is to compute the posterior probability of the latent variables $\boldsymbol{X} = \{\boldsymbol{W}, \boldsymbol{Z}\}$ and the parameter $\boldsymbol{\Phi}$ for which only the $\boldsymbol{\Phi}^1$ part is considered as random. We are therefore interested in computing the posterior $p(\boldsymbol{X}, \boldsymbol{\Phi}^1 \mid \boldsymbol{y}, \boldsymbol{\Phi}^2)$. This posterior is intractable and approximated here using a variational approximation $q(\boldsymbol{X}, \boldsymbol{\Phi}^1)$ with a factorized form

$q(\boldsymbol{X}, \boldsymbol{\Phi}^1) = q_X(\boldsymbol{X}) \, q_{\Phi^1}(\boldsymbol{\Phi}^1)$ in the set $\mathcal{D}$ of product probability distributions. The so-called variational EM procedure (VEM) proceeds as follows. At iteration (r), the current parameters values are denoted by $\boldsymbol{\Phi}^{2(r-1)}$ and VEM alternates between two steps,

$$\textbf{E-step:} \quad q^{(r)}(\boldsymbol{X}, \boldsymbol{\Phi}^1) = \arg\max_{q \in \mathcal{D}} \mathcal{F}(q, \boldsymbol{\Phi}^{2(r-1)})$$

$$\textbf{M-step:} \quad \boldsymbol{\Phi}^{2(r)} = \arg\max_{\boldsymbol{\Phi}^2} \mathcal{F}(q^{(r)}, \boldsymbol{\Phi}^2) \,,$$

where $\mathcal{F}$ is the usual free energy

$$\mathcal{F}(q, \boldsymbol{\Phi}^2) = E_q[\log p(\boldsymbol{y}, \boldsymbol{X}, \boldsymbol{\Phi}^1; \boldsymbol{\Phi}^2)] - E_q[\log q(\boldsymbol{X}, \boldsymbol{\Phi}^1)]. \tag{7}$$

The full expression of the free energy is not necessary to maximize it and to derive the variational EM algorithm. However, computing the free energy is useful. It provides a stopping criterion and a sanity check for implementation as the free energy should increase at each iteration. Then it can be used as specified in Sect. 3.1 as a replacement of the likelihood to provide a model selection procedure. Its detailed expression is given in a companion paper [2].

The E-step above divides into two steps. At iteration (r), denoting in addition by $q_X^{(r-1)}$ the current variational distribution for $\boldsymbol{X}$:

$$\textbf{E-}\boldsymbol{\Phi}^1\textbf{-step:} \quad q_{\Phi^1}^{(r)}(\boldsymbol{\Phi}^1) \propto \exp(E_{q_X^{(r-1)}}[\log p(\boldsymbol{\Phi}^1 | \boldsymbol{y}, \boldsymbol{X}; \boldsymbol{\Phi}^{2(r-1)})]) \tag{8}$$

$$\textbf{E-X-step:} \quad q_X^{(r)}(\boldsymbol{X}) \propto \exp(E_{q_{\phi^1}^{(r)}}[\log p(\boldsymbol{X} | \boldsymbol{y}, \boldsymbol{\Phi}^1; \boldsymbol{\Phi}^{2(r-1)})]) \,. \tag{9}$$

Then the M-step reduces to:

$$\textbf{M-step:} \quad \boldsymbol{\Phi}^{2(r)} = \arg\max_{\phi^2} E_{q_X^{(r)} q_{\phi^1}^{(r)}}[\log p(\boldsymbol{y}, \boldsymbol{X}, \boldsymbol{\Phi}^1; \boldsymbol{\Phi}^2)] \,.$$

The resulting variational EM algorithm is specified in [2] in two cases depending on the prior used for the mixing weights. For component elimination, the central quantity is $q_\pi^{(r)}(\boldsymbol{\pi})$ the approximate variational posterior of $\boldsymbol{\pi}$ that itself involves $q_Z^{(r)}(\boldsymbol{Z}) = \prod_i q_{Z_i}^{(r)}(Z_i)$ the variational posterior of the labels.

In what follows, we illustrate the use of this Bayesian formulation and its variational EM implementation on the issue of selecting the number of components in the mixture.

3   Single-Run Number of Component Selection

In this work, we consider approaches that start from an overfitting mixture with more components than expected in the data. In this case, as described by [16], identifiability will be violated in two possible ways. Identifiability issues can arise either because some of the components weights have to be zero (then component-specific parameters cannot be identified) or because some of the components have to be equal (then their weights cannot be identified). In practice, these two possibilities are not equivalent as checking for vanishing components is easier and is likely to lead to more stable behavior than testing for redundant components (see *e.g.* [27]).

Methods can be considered in a Bayesian and maximum likelihood setting. However, in a Bayesian framework, in contrast to maximum likelihood, considering a posterior distribution on the mixture parameters requires integrating out the parameters and this acts as a penalization for more complex models. The posterior is essentially putting mass on the sparsest way to approximate the true density, see *e.g.* [27]. Although the framework of [27] is fully Bayesian with priors on all mixture parameters, it seems that this penalization effect is also effective when only some of the parameters are integrated out. This is observed by [11] who use priors only for the component mean and covariance parameters. See [2] for details on the investigation of this alternative case with no prior on the mixing weights.

The idea of using overfitting finite mixtures with too many components K has been used in many papers. In a deliberately overfitting mixture model, a sparse prior on the mixture weights will empty superfluous components during estimation [21]. To obtain sparse solutions with regard to the number of mixture components, an appropriate prior on the weights π has to be selected. Guidelines have been given in previous work when the prior for the weights is a symmetric Dirichlet distribution $\mathcal{D}(\tau_1, \ldots, \tau_K)$ with all τ_k's equal to a value τ_0. To empty superfluous components automatically the value of τ_0 has to be chosen appropriately. In particular, [27] proposed conditions on τ_0 to control the asymptotic behavior of the posterior distribution of an overfitting mixture with respect to the two previously mentioned regimes. One regime in which a high likelihood is set to components with nearly identical parameters and one regime in which some of the mixture weights go to zero. More specifically, if $\tau_0 < d/2$ where d is the dimension of the component specific parameters, when N tends to infinity, the posterior expectation of the weight of superfluous components converges to zero. In practice, N is finite and as observed by [21], much smaller value of τ_0 are needed (*e.g.* 10^{-5}). It was even observed by [29] that negative values of τ_0 were useful to induce even more sparsity when the number of observations is too large with respect to the prior impact. Dirichlet priors with negative parameters, although not formally defined, are also mentioned by [13]. This latter work does not start from a Bayesian formulation but is based on a Minimum Message Length (MML) principle. [13] provide an M-step that performs component annihilation, thus an explicit rule for moving from the current number of components to a smaller one. A parallel is made with a Dirichlet prior with $\tau_0 = -d/2$ which according to [29] corresponds to a very strong prior sparsity.

In a Bayesian setting with symmetric sparse Dirichlet priors $\mathcal{D}(\tau_0, \ldots, \tau_0)$, the theoretical study of [27] therefore justifies to consider the posterior expectations of the weights $E[\pi_k|\boldsymbol{y}]$ and to prune out the too small ones. In practice this raises at least two additional questions: which expression to use for the estimated posterior means and how to set a threshold under which the estimated means are considered too small. The posterior means estimation is generally guided by the chosen inference scheme. For instance in our variational framework with a Dirichlet prior on the weights, the estimated posterior mean $E[\pi_k|\boldsymbol{y}]$ takes the following form (the (r) notation is removed to signify the convergence of the algorithm),

$$
\begin{aligned}
E[\pi_k|\boldsymbol{y}] \approx E_{q_\pi}[\pi_k] &= \frac{\tilde{\tau}_k}{\sum_{l=1}^{K} \tilde{\tau}_l} \\
&= \frac{\tau_k + n_k}{\sum_{k=1}^{K} \tau_k + N}
\end{aligned}
\tag{10}
$$

where $n_k = \sum_{i=1}^{N} q_{Z_i}(k)$ and the expression for $q_{Z_i}(k)$ is detailed in [2]. If we are in the no weight prior case, then the expectation simplifies to

$$
\pi_k \approx \frac{n_k}{N}
\tag{11}
$$

with the corresponding expression of $q_{Z_i}(k)$ also given in [2].

Nevertheless, whatever the inference scheme or prior setting, we are left with the issue of detecting when a component can be set as empty. There is usually a close relationship between the component weight π_k and the number of observations assigned to component k. This later number is itself often replaced by the sum $n_k = \sum_{i=1}^{N} q_{Z_i}(k)$. As an illustration, the choice of a negative τ_0 by [13] corresponds to a rule that sets a component weight to zero when $n_k = \sum_{i=1}^{N} q_{Z_i}(k)$ is smaller than $d/2$. This prevents the algorithm from approaching the boundary of the parameter space. When one of the components becomes too weak, meaning that it is not supported by the data, it is simply annihilated. One of the drawbacks of standard EM for mixtures is thus avoided. The rule of [13] is stronger than that used by [22] which annihilates a component when the sum n_k reduces to 1 or the one of [11] which corresponds to the sum n_k lower than a very small fraction of the sample size, i.e. $\sum_{i=1}^{N} q_{Z_i}(k)/N < 10^{-5}$ where N varies from 400 to 900 in their experiments. Note that [22] use a Bayesian framework with variational inference and their rule corresponds to thresholding the variational posterior weights (10) to $1/N$ because they set all τ_k to 0 in their experiments.

In addition to these thresholding approaches, alternatives have been developed that would worth testing to avoid the issue of setting a threshold for separating large and small weights. In their MCMC sampling, [21] propose to consider the number of non-empty components at each iteration and to estimate the number of components as the most frequent number of non-empty components. This is not directly applicable in our variational treatment as it would require to generate hard assignments to components at each iteration instead of dealing with their probabilities. In contrast, we could adopt techniques from the Bayesian non parametrics literature which seek for optimal partitions, such as the criterion of [12] using the so-called posterior similarity matrix ([15]). This matrix could be approximated easily in our case by computing the variational estimate of the probability that two observations are in the same component. However, even for moderate numbers of components, the optimization is already very costly.

In this work, we consider two strategies for component elimination. The first one is a thresholding approach while the second one is potentially more general as it is based on increasing the overall fit of the model assessed via the variational free energy at each iteration. Also it avoids the choice of a threshold for separating between large or small weights. The tested procedures are more specifically described in the next section.

3.1   Tested Procedures

We compare two single-run methods to estimate the number of components in a mixture of multiple scale distributions.

Thresholding Based Algorithm: A first method is directly derived from a Bayesian setting with a sparse symmetric Dirichlet prior likely to induce vanishing coefficients as supported by the theoretical results of [27]. This corresponds to the approach adopted in [21] and [22]. The difference between the later two being how they check for vanishing coefficients. Our variational inference leads more naturally to the solution of [22] which is to check the weight posterior means, that is whether at each iteration (r),

$$n_k^{(r)} < (K\tau_0 + N)\rho_t - \tau_0 \tag{12}$$

where ρ_t is the chosen threshold on the posterior means. When ρ_t is set such that (12) leads to $n_k^{(r)} < 1$, this method is referred to, in the next Section, as *SparseDirichlet+π*test. For comparison, the algorithm run with no intervention is called *SparseDirichlet*.

Free Energy Based Algorithm: We also consider a criterion based on the free energy (7) to detect components to eliminate. This choice is based on the observation that when we cannot control the hyperparameters ($e.g$ τ_k) to guide the algorithm in the vanishing components regime, the algorithm may as well go to the redundant component regime. The goal is then to test whether this alternative method is likely to handle this behavior. The proposal is to start from a clustering solution with too many components and to try to remove them using a criterion based on the gain in free energy. In this setting, the components that are removed are not necessarily vanishing components but can also be redundant ones. In the proposed variational EM inference framework, the free energy arises naturally as a selection criterion. It has been stated in [4] and [8] that the free energy penalizes model complexity and that it converges to the well known Bayesian Information Criterion (BIC) and Minimum Description Length (MDL) criteria, when the sample size increases, illustrating the interest of this measure for model selection.

The free energy expression used is given in [2]. The heuristic denoted by *SparseDirichlet+FEtest* can be described as follows (see the next section for implementation details).

1. Iteration $r = 0$: Initialization of the $K^{(0)}$ clusters and probabilities using for instance repetitions of k-means or trimmed k-means.
2. Iteration $r \geq 1$:
   (a) E and M steps updating from parameters at iteration $r - 1$
   (b) Updating of the resulting Free Energy value
   (c) In parallele, for each cluster $k \in \{1 \dots K^{(r-1)}\}$
      i. Re-normalization of the cluster probabilities when cluster k is removed from current estimates at iteration $r - 1$: the sum over the remaining $K^{(r-1)} - 1$ clusters must be equal to 1
      ii. Updating of the corresponding E and M steps and computation of the associate Free Energy value
   (d) Selection of the mixture with the highest Free Energy among the $K^{(r-1)}$-component mixture (step (b)) or one of the $(K^{(r-1)} - 1)$-component mixtures (step (c)).
   (e) Updating of $K^{(r)}$ accordingly, to $K^{(r-1)}$ or $K^{(r-1)} - 1$.
3. When no more cluster deletion occur (*eg.* during 5 steps), we switch to the EM algorithm (*SparseDirichlet*).

4   Experiments

In addition to the 3 methods *SparseDirichlet+π*test, *SparseDirichlet+FEtest* and *SparseDirichlet*, referred to below as $\mathcal{MP}$ single-run procedures, we consider standard Gaussian mixtures using the Mclust package [28] including a version with priors on the means and covariance matrices. The Bayesian Information Criterion (BIC) is used to select the number of components from $K = 1$ to 10. The respective methods are denoted below by *GM+BIC* and *Bayesian GM+BIC*. Regarding mixtures of $\mathcal{MP}$ distributions, we also consider their non Bayesian version, using BIC to select K, denoted below by *MMP+BIC*.

In practice, values need to be chosen for hyperparameters. These include the m_k that are set to 0, the Λ_k that are set to ϵI_M with ϵ small (set to 10^{-4}) so has to generate a large variance in (4). The δ_{km} are then set to 1 and λ_{km} to values $5 \times 10^{-4} = \lambda_1 < \lambda_2 < \dots < \lambda_M = 10^{-3}$. The τ_k's are set to 10^{-3} to favor sparse mixtures.

Initialization is also an important step in EM algorithms. For one data sample, each single-run method is initialized $I = 10$ times. These $I = 10$ initializations are the same for all single-run methods. Each initialization is obtained with $K = 10$ using trimmed k-means and excluding 10% of outliers. Each trimmed kmeans output is the one obtained after running the algorithm from $R = 10$ restarts and selecting the best assignment after 10 iterations. For each run of a procedure (data sample), the $I = 10$ initializations are followed by 5000 iterations maximum of VEM before choosing the best output. For Gaussian mixtures, the initialization procedure is that embedded in Mclust. For $\mathcal{MP}$ models, initial values of the α_{km}'s are set to 1.

Another important point for single-run procedures, is how to finally enumerate remaining components. For simplicity, we report components that are expressed by the maximum a posteriori (MAP) rule, which means components for which there is at least one data point assigned to them with the highest probability.

4.1   Simulated Data

We consider several models (more details can be found in [2]), 3 Gaussian mixtures and 10 $\mathcal{MP}$ mixtures, with 10 simulated samples each, for a total of 130 samples, K varying from 3 to 5, N from 900 to 9000, with close or more separated clusters. The results are summarized in Table 1 and the simulated samples illustrated in Fig. 1. Gaussian mixture models provide the right component number in 26% to 32% of the cases, which is higher than the number of Gaussian mixtures in the test (23%). All procedures hesitate mainly between the true number and this number plus 1. We observe a good behavior of the free energy heuristic with a time divided by 3 compared to the non Bayesian $\mathcal{MP}$ mixture procedure, although the later benefits from a more optimized implementation. For the first strategy, the dependence to the choice of a threshold value is certainly a limitation although some significant gain is observed over the cases with no component elimination (SparseDirichlet line in Table 1). Overall, eliminating components on the run is beneficial, both in terms of time and selection performance but using a penalized likelihood criterion (free energy) to do so avoid the commitment to a fix threshold and is more successful. A possible reason is that small components are more difficult to eliminate than redundant ones. Small components not only require the right threshold to be chosen but also they may appear at much latter iterations as illustrated in Fig. 2.

Table 1. 13 models simulated 10 times each: the true number of components is varying so the columns indicate the difference between the selection and the truth. The average time (for the total of the $I = 10$ repetitions, over the 130 samples) is indicated in the last column. The most frequent selection (in %) is indicated by a box while the true value is in green.

Procedures (10 restarts)	Difference between selected and true number of components								Average time (in seconds)
	0	1	2	3	4	5	6	7	
GM+BIC	26.1	33.0	8.4	3.8	19.2	1.5	2.3	5.3	177
Bayesian GM+BIC	31.5	34.6	3.0	3.0	20.7	3.8	1.5	1.5	92
MMP+BIC	94.6	5.3	.	.	.	.	.	.	9506
SparseDirichlet	54.6	39.2	5.3	.7	.	.	.	.	10355
SparseDirichlet+πtest	70.0	27.6	1.5	.7	.	.	.	.	4640
SparseDirichlet+FEtest	99.2	.	.	.7	.	.	.	.	3125

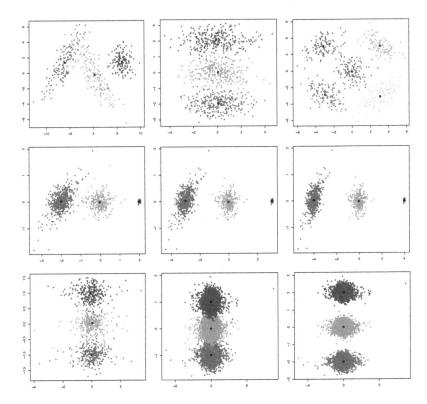

Fig. 1. Examples of simulated samples. First line: 3 Gaussian mixtures with 3 and 5 components. Second line: $\mathcal{MP}$ mixtures with different dof and increasing separation from left to right. Third line: $\mathcal{MP}$ mixtures with increasing separation, from left to right, and increasing number of points, $N = 900$ for the first plot, $N = 9000$ for the last two.

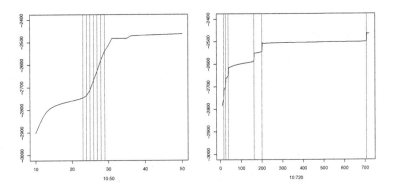

Fig. 2. Illustration of the two component elimination strategies: Free energy gain strategy, iterations 10 to 50 (left) and too small component proportion test, iterations 10 to 720 (right). Eliminations are marked with red lines. Most of them occur at earlier iterations when using the free energy test.

5   Discussion and Conclusion

We investigated, in the context of mixtures of non-Gaussian distributions, different single-run procedures to select automatically the number of components. The Bayesian formulation makes this possible when starting from an overfitting mixture, where K is larger than the expected number of components. The advantage of single run procedures is to avoid time consuming comparison of scores for each mixture model from 1 to K components. There are different ways to implement this idea: full Bayesian settings which have the advantage to be supported by some theoretical justification [27] and Type II maximum likelihood as proposed by [11] (not reported here but investigated in [2]). For further acceleration, we investigated component elimination which consists of eliminating components on the run. They are two main ways to do so: components are eliminated as soon as they are not supported by enough data points (their estimated weight is under some threshold) or when their removal does not penalize the overall fit. For the latest case, we proposed a heuristic based on the gain in free energy. The free energy acts as a penalized likelihood criterion and can potentially eliminate both too small components and redundant ones. Redundant components do not necessarily see their weight tend to zero and cannot be eliminated via a simple thresholding.

As non Gaussian components, we investigated in particular the case of multiple scale distributions [14], which have been shown to perform well in the modelling of non-elliptical clusters with potential outliers and tails of various heaviness. We proposed a Bayesian formulation of mixtures of such multiple scale distributions and derived an inference procedure based on a variational EM algorithm to implement the single-run procedures.

On preliminary experiments, we observed that eliminating components on the run is beneficial, both in terms of time and selection performance. Free energy based methods appeared to perform better than posterior weight thresholding methods: using a penalized likelihood criterion (free energy) avoids the commitment to a fix threshold and is not limited to the removal of small components. However, a fully Bayesian setting is probably not necessary as both in terms of selection and computation time, Type II maximum likelihood on the weights was competitive with the use of a Dirichlet prior with a slight advantage to the latter (results reported in [2]).

To confirm these observations, more tests on larger and real data sets would be required to better compare and understand the various characteristics of each procedure. Theoretical justification for thresholding approaches, as provided by [27], applies for Gaussian mixtures but may not hold in our case of non-elliptical distributions. A more specific study would be required and could provide additional guidelines as how to set the threshold in practice. Also time comparison in our study is only valid for the Bayesian procedures for which the implementation is similar while the other methods using BIC have been better optimized, but this does not change the overall conclusion as regards computational efficiency.

6   Supplementary Material

All details on the variational EM and free energy computations, plus additional illustrations can be found in a companion paper [2].

References

1. Archambeau, C., Verleysen, M.: Robust Bayesian clustering. Neural Netw. **20**(1), 129–138 (2007)
2. Arnaud, A., Forbes, F., Steele, R., Lemasson, B., Barbier, E.L.: Bayesian mixtures of multiple scale distributions, July 2019). https://hal.inria.fr/hal-01953393. Working paper or preprint
3. Attias, H.: Inferring parameters and structure of latent variable models by variational Bayes. In: UAI 1999: Proceedings of the Fifteenth Conference on Uncertainty in Artificial Intelligence, Stockholm, Sweden, 30 July–1 August 1999, pp. 21–30 (1999)
4. Attias, H.: A variational Bayesian framework for graphical models. In: Proceedings of Advances in Neural Information Processing Systems 12, pp. 209–215. MIT Press, Denver (2000)
5. Banfield, J., Raftery, A.: Model-based Gaussian and non-Gaussian clustering. Biometrics **49**(3), 803–821 (1993)
6. Baudry, J.P., Raftery, E.A., Celeux, G., Lo, K., Gottardo, R.: Combining mixture components for clustering. J. Comput. Graph. Stat. **19**(2), 332–353 (2010)
7. Baudry, J.P., Maugis, C., Michel, B.: Slope heuristics: overview and implementation. Stat. Comput. **22**(2), 455–470 (2012)
8. Beal, M.J.: Variational algorithms for approximate Bayesian inference. Ph.D. thesis, University of London (2003)
9. Celeux, G., Govaert, G.: Gaussian parsimonious clustering models. Pattern Recogn. **28**(5), 781–793 (1995)
10. Celeux, G., Fruhwirth-Schnatter, S., Robert, C.: Model selection for mixture models-perspectives and strategies. In: Handbook of Mixture Analysis. CRC Press (2018)
11. Corduneanu, A., Bishop, C.: Variational Bayesian model selection for mixture distributions. In: Proceedings Eighth International Conference on Artificial Intelligence and Statistics, p. 2734. Morgan Kaufmann (2001)
12. Dahl, D.B.: Model-based clustering for expression data via a Dirichlet process mixture model. In: Bayesian Inference for Gene Expression and Proteomics (2006)
13. Figueiredo, M.A.T., Jain, A.K.: Unsupervised learning of finite mixture models. IEEE Trans. Pattern Anal. Mach. Intell. **24**(3), 381–396 (2002)
14. Forbes, F., Wraith, D.: A new family of multivariate heavy-tailed distributions with variable marginal amounts of tailweights: application to robust clustering. Stat. Comput. **24**(6), 971–984 (2014)
15. Fritsch, A., Ickstadt, K.: Improved criteria for clustering based on the posterior similarity matrix. Bayesian Anal. **4**(2), 367–391 (2009)
16. Frühwirth-Schnatter, S.: Finite Mixture and Markov Switching Models. Springer, New York (2006). https://doi.org/10.1007/978-0-387-35768-3
17. Gorur, D., Rasmussen, C.: Dirichlet process Gaussian mixture models: choice of the base distribution. J. Comput. Sci. Technol. **25**(4), 653–664 (2010)

18. Hennig, C.: Methods for merging Gaussian mixture components. Adv. Data Anal. Classif. **4**(1), 3–34 (2010)
19. Hoff, P.D.: A hierarchical eigenmodel for pooled covariance estimation. J. R. Stat. Society. Ser. B (Stat. Methodol.) **71**(5), 971–992 (2009)
20. Johnson, N.L., Kotz, S., Balakrishnan, N.: Continuous Univariate Distributions, vol. 2, 2nd edn. Wiley, New York (1994)
21. Malsiner-Walli, G., Frühwirth-Schnatter, S., Grün, B.: Model-based clustering based on sparse finite Gaussian mixtures. Stat. Comput. **26**(1), 303–324 (2016)
22. McGrory, C.A., Titterington, D.M.: Variational approximations in Bayesian model selection for finite mixture distributions. Comput. Stat. Data Anal. **51**(11), 5352–5367 (2007)
23. McLachlan, G., Peel, D.: Finite Mixture Models. Wiley, Hoboken (2000)
24. Melnykov, V.: Merging mixture components for clustering through pairwise overlap. J. Comput. Graph. Stat. **25**(1), 66–90 (2016)
25. Rasmussen, C.E.: The infinite Gaussian mixture model. In: NIPS, vol. 12, pp. 554–560 (1999)
26. Richardson, S., Green, P.J.: On Bayesian analysis of mixtures with an unknown number of components (with discussion). J. R. Stat. Soc.: Ser. B (Stat. Methodol.) **59**(4), 731–792 (1997)
27. Rousseau, J., Mengersen, K.: Asymptotic behaviour of the posterior distribution in overfitted mixture models. J. R. Stat. Soc.: Ser. B (Stat. Methodol.) **73**(5), 689–710 (2011)
28. Scrucca, L., Fop, M., Murphy, T.B., Raftery, A.: mclust 5: clustering, classification and density estimation using Gaussian finite mixture models. R J. **8**(1), 205–233 (2016)
29. Tu, K.: Modified Dirichlet distribution: allowing negative parameters to induce stronger sparsity. In: Proceedings of the 2016 Conference on Empirical Methods in Natural Language Processing, EMNLP 2016, Austin, Texas, USA, 1–4 November 2016, pp. 1986–1991 (2016)
30. Verbeek, J., Vlassis, N., Kröse, B.: Efficient greedy learning of Gaussian mixture models. Neural Comput. **15**(2), 469–485 (2003)
31. Wei, X., Li, C.: The infinite student t-mixture for robust modeling. Signal Process. **92**(1), 224–234 (2012)
32. Yerebakan, H.Z., Rajwa, B., Dundar, M.: The infinite mixture of infinite Gaussian mixtures. In: Advances in Neural Information Processing Systems, pp. 28–36 (2014)

An Introduction to Approximate Bayesian Computation

Hien D. Nguyen[✉]

Department of Mathematics and Statistics, Latrobe University, Melbourne, Australia
h.nguyen5@latrobe.edu.au

Abstract. Many modern statistical settings feature the analysis of data that may arise from unknown generating processes, or processes for which the generative models are computationally infeasible to interact with. Conventional estimation and inference solution methods in such settings may be unwieldy or impossible to implement. The approximate Bayesian computation (ABC) approach is a potent method in such scenarios, since it does not require the knowledge of the underlying generative model in order to perform inference. Furthermore, when combined with sufficiently regular discrepancy measurements such as the energy statistic, ABC can be shown to have desirable asymptotic properties. We provide a concise introduction to the general ABC framework. To demonstrate the capabilities and usefulness of the ABC approach, we present the analyses of a number of artificial examples as well as one of a real-data example pertaining to circular statistics data.

Keywords: Approximate Bayesian computation · Bayesian statistics · Circular statistics · Energy statistic · Normal mixture model

1 Introduction

Suppose that we observe independent and identically distributed (IID) data $\mathbf{X}_n = \{\boldsymbol{X}_i\}_{i=1}^n$ ($n \in \mathbb{N}$), where $\boldsymbol{X}_1 \in \mathbb{X} \subseteq \mathbb{R}^d$ ($d \in \mathbb{N}$) arises from a data generating process (DGP) with probability density function or probability mass function (PDF or PMF) of form $f(\boldsymbol{x}|\boldsymbol{\theta})$, where $\boldsymbol{\theta} \in \mathbb{T}$ is some parameter vector. From hereon in, we shall refer to both PDF and PMF as PDF, when the nomenclature makes no difference.

As opposed to the classical *frequentist* framework of statistics, which considers that $\boldsymbol{\theta}$ is a constant, the *Bayesian* approach considers that $\boldsymbol{\theta}$ is a realization of some random variable $\boldsymbol{\Theta}$ with DGP characterized by the prior PDF $\pi(\boldsymbol{\theta})$ (cf. [19, Ch. 5]). Thus, whereas the frequentist goal is to produce some estimator $\hat{\boldsymbol{\Theta}}_n = \hat{\boldsymbol{\theta}}(\mathbf{X}_n)$ (function of the data) that is unbiased ($\mathbb{E}\left(\hat{\boldsymbol{\Theta}}_n\right) - \boldsymbol{\theta} = \mathbf{0}$) or that is consistent ($\hat{\boldsymbol{\Theta}}_n - \boldsymbol{\theta} \to \mathbf{0}$, in probability or almost surely, as $n \to \infty$), the goal Bayesian inference is to characterize the posterior distribution of $\boldsymbol{\Theta}|\mathbf{X}_n = \mathbf{x}_n$,

© Springer Nature Singapore Pte Ltd. 2019
H. Nguyen (Ed.): RSSDS 2019, CCIS 1150, pp. 96–108, 2019.
https://doi.org/10.1007/978-981-15-1960-4_7

where $\mathbf{x}_n$ is some realization of $\mathbf{X}_n$. That is, the Bayesian goal is to obtain the posterior PDF

$$\pi(\boldsymbol{\theta}|\mathbf{x}_n) = \frac{f(\mathbf{x}_n|\boldsymbol{\theta})\,\pi(\boldsymbol{\theta})}{\int_{\mathbb{T}} f(\mathbf{x}_n|\boldsymbol{\theta})\,\pi(\boldsymbol{\theta})\,\mathrm{d}\boldsymbol{\theta}}, \tag{1}$$

where $f(\mathbf{x}_n|\boldsymbol{\theta}) = \prod_{i=1}^{n} f(\boldsymbol{X}_i|\boldsymbol{\theta})$ is the likelihood of $\mathbf{X}_n$. Good introductions to the Bayesian statistics literature can be found in the works of [13,16], and [24].

In general, when conducting Bayesian inference, one has access to the functional forms of $f(\mathbf{x}_n|\boldsymbol{\theta})$ and $\pi(\boldsymbol{\theta})$, and can efficiently evaluate the functions for any inputs $\mathbf{x}_n$ and $\boldsymbol{\theta}$. If the PDFs $f(\mathbf{x}_n|\boldsymbol{\theta})$ and $\pi(\boldsymbol{\theta})$ are compatible, such as under conjugacy (cf. [24, Sec. 2.2]), then one may be able to evaluate the denominator $\int_{\mathbb{T}} f(\mathbf{x}_n|\boldsymbol{\theta})\,\pi(\boldsymbol{\theta})\,\mathrm{d}\boldsymbol{\theta}$ in closed form, and therefore obtain a closed form expression for the posterior PDF $\pi(\boldsymbol{\theta}|\mathbf{x}_n)$. If a closed form expression for $\int_{\mathbb{T}} f(\mathbf{x}_n|\boldsymbol{\theta})\,\pi(\boldsymbol{\theta})\,\mathrm{d}\boldsymbol{\theta}$ is not available, then it is still possible to use Markov chain Monte Carlo techniques to sample parameters from the posterior distribution, which can be used to estimate (1) and its related statistics (cf. [17]).

Unfortunately, one does not always have access to the likelihood function $f(\mathbf{x}_n|\boldsymbol{\theta})$, or also commonly, one does not have an efficient means of computing the likelihood function. In situations where access to the likelihood function is inhibited, an approximation of the posterior PDF is required. In this paper, we consider the so-called *approximate Bayesian computation* (ABC) method.

ABC is a simulation based method for approximation of the posterior PDF (1). The modern ABC method primarily arose from the need for intractable likelihood approximations in genetic settings, such as in the articles of [23] and [14]. Good modern introductions to the ABC method appear in [7,9], and [18].

There are now many variants on the theme of ABC algorithms. In order to keep the paper concise, we shall concentrate our attention on the so-called *rejection* ABC algorithm, which requires the notion of a *data discrepancy measurement*. Again, for conciseness, we shall concentrate our attention on the use of the so-called *energy statistic* (ES) of [20] and [22].

The remainder of the paper shall proceed as follows. In Sect. 2, we describe the basic rejection ABC algorithm. In Sect. 3, we present the ES and show how one can construct an ABC algorithm using the ES. In Sect. 4, a number of artificial examples are used to illustrate the ES-based ABC algorithm. In Sect. 5, a real-data example to circular statistics data is provided as a demonstration of the applicability of the presented methodology. A conclusion follows in Sect. 6.

2   Approximate Bayesian Computation

Let $\mathbf{X}_n$ be as in Sect. 1 and let $\mathbf{Y}_m = \{Y_i\}_{i=1}^{m}$ $(m \in \mathbb{N})$ be an IID random sample, where $Y_1 \in \mathbb{X}$. We call $\mathbf{Y}_m$ the *quasi-sample*. Define $D(\mathbf{X}_n, \mathbf{Y}_m) \geq 0$ to be some discrepancy measurement, that measures how different the sample $\mathbf{X}_n$ and $\mathbf{Y}_m$ are to one another. Here, we require the properties that when D is small, the samples $\mathbf{X}_n$ and $\mathbf{Y}_m$ are similar to one another in some sense, and similarly, when D is large, the two samples are different.

Using D, and for some $m \in \mathbb{N}$ and some *tolerance* $\epsilon > 0$, the goal of ABC is to approximate the posterior PDF (1), via the so-called *quasi-posterior* PDF

$$\pi_{\epsilon,m}(\boldsymbol{\theta}|\mathbf{x}_n) = \frac{\int_{\mathbb{X}} [\![D(\mathbf{x}_n, \mathbf{y}_m) \leq \epsilon]\!] f(\mathbf{y}_m|\boldsymbol{\theta}) \pi(\boldsymbol{\theta}) \, \mathrm{d}\mathbf{y}_m}{\int_{\mathbb{T}} \int_{\mathbb{X}} [\![D(\mathbf{x}_n, \mathbf{y}_m) \leq \epsilon]\!] f(\mathbf{y}_m|\boldsymbol{\theta}) \pi(\boldsymbol{\theta}) \, \mathrm{d}\mathbf{y}_m \mathrm{d}\boldsymbol{\theta}}, \tag{2}$$

where $[\![A]\!]$ is the Iverson bracket, which takes a value of 1 if the logical proposition A is true, and takes value of 0, otherwise (cf. [5]). Here, we can configure m and ϵ to tweak the quality of the approximation that (2) provides. For example, if we take ϵ to be small, then the approximation will be more accurate, whereas a larger value of ϵ will yield a more liberal approximation. As noted in [10], a degree of liberty may be suitable in situations where there may be potential for model misspecification. The value of m is usually taken to be equal to n. However, computational considerations may influence the choices of m and ϵ, as we shall discuss in the sequel.

Unfortunately, we cannot evaluate (2) explicitly. However it is possible to drawn an IID random sample $\mathbf{T}_N = \{\boldsymbol{\Theta}_j\}_{j=1}^N$ ($N \in \mathbb{N}$) from the quasi-posterior distribution that is characterized by (2), using the so-called rejection algorithm. Upon defining $[N] = \{1, \ldots, N\}$, we may present the rejection algorithm as Algorithm 1.

Input: Data $\mathbf{X}_n$. Data discrepancy function D. Accuracy parameter $\epsilon > 0$.
quasi-sample size $m \in \mathbb{N}$. Posterior sample size $N \in \mathbb{N}$.
For: $k \in [N]$;
    1. Generate $\boldsymbol{\Theta}$ from DGP with PDF $\pi(\boldsymbol{\theta})$;
    2. Generate $\mathbf{Y}_m$ from GDP with PDF $f(\mathbf{y}_m|\boldsymbol{\Theta})$;
    If: $D(\mathbf{X}_n, \mathbf{Y}_m) < \epsilon$;
        Set $\boldsymbol{\Theta}_k = \boldsymbol{\Theta}$;
    Else: Repeat Steps 1 and 2.
Output: Sample $\mathbf{T}_N = \{\boldsymbol{\Theta}_k\}_{k=1}^N$ from the quasi-posterior DGP with PDF (2).

Algorithm 1. Rejection algorithm for generating an IID sample $\mathbf{T}_N$ of size N from the DGP that is characterized by the PDF (2).

Provided that D is sufficiently regular, it is provable that (2) converges to a fixed asymptotic limiting distribution, as $n \to \infty$ and $m(n) \to \infty$, for any fixed $\epsilon > 0$, where m is a function of n. The following result was proved by [6].

Theorem 1. *Let $\mathbf{X}_n$ and $\mathbf{Y}_m$ be IID samples from DGPs that can be characterized likelihood functions $f(\mathbf{x}_n|\boldsymbol{\vartheta})$ and $f(\mathbf{y}_m|\boldsymbol{\theta})$, respectively, which are determined by parameters $\boldsymbol{\vartheta}$ and $\boldsymbol{\theta}$. Suppose that*

$$D(\mathbf{X}_n, \mathbf{Y}_m) \to D_\infty(\boldsymbol{\vartheta}, \boldsymbol{\theta}),$$

almost surely, as $n \to \infty$ and $m(n) \to \infty$. If $D \geq 0$, $\epsilon > 0$, and $D_\infty(\boldsymbol{\vartheta}, \boldsymbol{\theta}) \neq \epsilon$, then

$$\pi_{\epsilon,m}(\boldsymbol{\theta}|\mathbf{x}_n) \to \frac{[\![D_\infty(\boldsymbol{\vartheta}, \boldsymbol{\theta}) \leq \epsilon]\!] \pi(\boldsymbol{\theta})}{\int_{\mathbb{T}} [\![D_\infty(\boldsymbol{\vartheta}, \boldsymbol{\theta}) \leq \epsilon]\!] \pi(\boldsymbol{\theta}) \, d\boldsymbol{\theta}},$$

almost surely, as $n \to \infty$.

That is, the quasi-posterior PDF converges to a function that is proportional to the prior PDF $\pi(\boldsymbol{\theta})$, truncated to the set that is characterized by the relationship $D_\infty(\boldsymbol{\vartheta}, \boldsymbol{\theta}) \leq \epsilon$, for fixed $\boldsymbol{\vartheta}$ that is determined by $\mathbf{X}_n$. This result provides a theoretical guarantee that the quasi-posterior PDF has a limit as n gets large, and that the limit concentrates the density of the quasi-posterior PDF around a region that is determined by the sample $\mathbf{X}_n$.

3   The Energy Statistic

Let $f(\boldsymbol{x})$, $g(\boldsymbol{x})$, and $h(\boldsymbol{x})$ be a set of three PDFs over the domain $\mathbb{X} \subseteq \mathbb{R}^d$. We may define a *distance* $\Delta(f, g)$ between any two PDFs as a function that satisfies the following conditions:

1. $\Delta(f, g) \geq 0$ (non-negativity);
2. $\Delta(f, g) = 0$ if and only if $f = g$ (equivalence);
3. $\Delta(f, g) = \Delta(g, f)$ (symmetry);
4. $\Delta(f, g) + \Delta(g, h) \geq \Delta(f, h)$ (triangle inequality).

One such function Δ that satisfies the definition of a distance is the *energy distance* (ED) function. Suppose that $\boldsymbol{X}_1, \boldsymbol{X}_2$ are IID and arise from a DGP that is characterized by $f(\boldsymbol{x})$, and that $\boldsymbol{Y}_1, \boldsymbol{Y}_2$ are IID and arise from a DGP that is characterized by $g(\boldsymbol{y})$ $(\boldsymbol{x}, \boldsymbol{y} \in \mathbb{X})$. Then, the ED between f and g is defined as

$$\Delta(f, g) = 2\mathbb{E}\|\boldsymbol{X}_1 - \boldsymbol{Y}_1\|_2 - \mathbb{E}\|\boldsymbol{X}_1 - \boldsymbol{X}_2\|_2 - \mathbb{E}\|\boldsymbol{Y}_1 - \boldsymbol{Y}_2\|_2,$$

where $\|\cdot\|_2$ is the Euclidean distance. From Proposition 1 of [21], we have the following result.

Proposition 1. *Let $\boldsymbol{X}$ and $\boldsymbol{Y}$ arise from DGPs with PDFs f and g, respectively. If $f(\boldsymbol{x}) = f(\boldsymbol{x}|\boldsymbol{\vartheta})$ and $g(\boldsymbol{y}) = f(\boldsymbol{y}|\boldsymbol{\theta})$ are parametric with respective characteristic functions $\varphi(t; \boldsymbol{\vartheta})$ and $\varphi(t; \boldsymbol{\theta})$, and if the second moments of $\boldsymbol{X}$ and $\boldsymbol{Y}$ exist, in the sense that $\mathbb{E}\|\boldsymbol{X}\|_2 + \mathbb{E}\|\boldsymbol{Y}\|_2 < \infty$, then*

$$\Delta(f, g) = \frac{\Gamma\left(\frac{d+1}{2}\right)}{\pi^{(d+1)/2}} \int_{\mathbb{R}^d} \frac{|\varphi(t; \boldsymbol{\vartheta}) - \varphi(t; \boldsymbol{\theta})|^2}{\|t\|_2^{d+1}} \, dt. \tag{3}$$

Thus, Proposition 1 ensures that when the second moments of f and g exist, so does the distance $\Delta(f, g)$. Furthermore, if $f(\boldsymbol{x}) = f(\boldsymbol{x}|\boldsymbol{\vartheta})$ and $g(\boldsymbol{y}) = f(\boldsymbol{y}|\boldsymbol{\theta})$ are parametric PDFs from the same family, then the ED is a function of the two parameters $\boldsymbol{\vartheta}$ and $\boldsymbol{\theta}$. In the parametric case, we thus write $\Delta(f, g) = \Delta(\boldsymbol{\vartheta}, \boldsymbol{\theta})$, where $\Delta(\boldsymbol{\vartheta}, \boldsymbol{\theta})$ equals the right-hand side of (3).

Let $\mathbf{X}_n$ be an IID sample from a DGP with PDF f and let $\mathbf{Y}_m$ be an IID sample from a DGP with PDF g. We may estimate $\Delta(f, g)$ using $\mathbf{X}_n$ and $\mathbf{Y}_m$ using the ES defined via the *V-statistic*

$$\mathcal{E}\left(\mathbf{X}_n, \mathbf{Y}_m\right) = \frac{2}{mn} \sum_{i=1}^{n} \sum_{j=1}^{m} \|\mathbf{X}_i - \mathbf{Y}_j\|_2$$

$$-\frac{1}{n^2} \sum_{i=1}^{n} \sum_{j=1}^{n} \|\mathbf{X}_i - \mathbf{X}_j\|_2 - \frac{1}{m^2} \sum_{i=1}^{m} \sum_{j=1}^{m} \|\mathbf{Y}_i - \mathbf{Y}_j\|_2,$$

Under the assumption that $\mathbb{E}\|\mathbf{X}_1\|_2^2 + \mathbb{E}\|\mathbf{Y}_1\|_2^2 < \infty$, and provided that $\min\{m, n\} \to \infty$, we have the result that $\mathcal{E}\left(\mathbf{X}_n, \mathbf{Y}_m\right)$ converges almost surely to $\Delta(f, g)$ (cf. [11]). Upon setting $f(\boldsymbol{x}) = f(\boldsymbol{x}|\boldsymbol{\vartheta})$ and $g(\boldsymbol{y}) = f(\boldsymbol{y}|\boldsymbol{\theta})$, and letting $D_\infty(\boldsymbol{\vartheta}, \boldsymbol{\theta}) = \Delta(\boldsymbol{\vartheta}, \boldsymbol{\theta})$, we have the following corollary to Theorem 1.

Theorem 2. *Let $\mathbf{X}_n$ and $\mathbf{Y}_m$ be IID samples from DGPs that can be characterized likelihood functions $f(\mathbf{x}_n|\boldsymbol{\vartheta})$ and $f(\mathbf{y}_m|\boldsymbol{\theta})$, respectively, such that $\mathbb{E}\|\mathbf{X}_1\|_2^2 + \mathbb{E}\|\mathbf{Y}_1\|_2^2 < \infty$. If $D = \mathcal{E}$, $\epsilon > 0$, and $\Delta(\boldsymbol{\vartheta}, \boldsymbol{\theta}) \neq \epsilon$, then*

$$\pi_{\epsilon,m}\left(\boldsymbol{\theta}|\mathbf{x}_n\right) \to \frac{[\![\Delta(\boldsymbol{\vartheta}, \boldsymbol{\theta}) \leq \epsilon]\!]\pi(\boldsymbol{\theta})}{\int_{\mathbb{T}} [\![\Delta(\boldsymbol{\vartheta}, \boldsymbol{\theta}) \leq \epsilon]\!]\pi(\boldsymbol{\theta}) \, d\boldsymbol{\theta}},$$

almost surely, as $\min\{m, n\} \to \infty$.

Thus, Theorem 2 establishes that the ES can be used within the ABC framework in order to generate samples from DGPs quasi-posterior PDFs that are meaningful in some sense.

4   Artificial Examples

From hereon in, we shall assume that we use the ES as our choice of discrepancy measurement (i.e., we set $D = \mathcal{E}$). We note that in Algorithm 1, the specification of the accuracy parameter ϵ determines the runtime of the algorithm. That is, with all else held constant, we may anticipate that a larger value of ϵ will lead to a faster runtime of the algorithm, since Steps 1 and 2 will require less repetition, whereas smaller values of ϵ will lead to slower runtime. Furthermore, without knowledge of the probability of the event $D\left(\mathbf{X}_n, \mathbf{Y}_m\right) \leq \epsilon$, it is impossible to predict the runtime of the algorithm, as a function of ϵ.

In order to make the runtime more predictable, we shall consider an alternative version of Algorithm 1, which over samples a fixed number $\nu \in \mathbb{N}$ of parameters from the DGP characterized by $\pi(\boldsymbol{\theta})$, such that $\nu > N$. Then, a fraction $\gamma \leq N/\nu$ of the sample is used to construct $\mathbf{T}_N = \{\boldsymbol{\Theta}_k\}_{k=1}^{N}$, where the elements of $\mathbf{T}_N$ are chosen using the discrepancy measurement $\delta_k = D\left(\mathbf{X}_n, \mathbf{Y}_{m,k}\right)$, where $\mathbf{Y}_{m,k}$ is a sample from the DGP characterized by the likelihood $f\left(\mathbf{y}_k|\boldsymbol{\Theta}_k\right)$. Upon defining $\lceil \cdot \rceil$ to be the ceiling operator, we call this algorithm the quantile ABC algorithm, and elaborate upon it in Algorithm 2.

Input: Data $\mathbf{X}_n$. Data discrepancy function D. Quantile parameter $\gamma \in (0,1)$.
quasi-sample size $m \in \mathbb{N}$. Posterior sample size $N \in \mathbb{N}$.
Define: Over-sample size $\nu = \lceil N/\gamma \rceil$;
For: $k \in [\nu]$;
    1. Generate $\boldsymbol{\Theta}$ from DGP with PDF $\pi(\boldsymbol{\theta})$;
    2. Generate $\mathbf{Y}_m$ from GDP with PDF $f(\mathbf{y}_m|\boldsymbol{\Theta})$;
    3. Compute $\delta_k = D(\mathbf{X}_n, \mathbf{Y}_m)$;
Define:
    Order statistics $\delta_{(1)}, \ldots, \delta_{(\nu)}$, equal to the values of $\delta_1, \ldots, \delta_\nu$, ordered from
smallest to largest;
    Accuracy parameter $\epsilon = \delta_{(N)}$;
    Posterior sample $\mathbf{T}_N = \{\boldsymbol{\Theta}_k | \delta_k \leq \delta_{(N)}, k \in [\nu]\}$;
Output: Sample $\mathbf{T}_N$ from the quasi-posterior DGP with PDF (2).

Algorithm 2. γ-quantile algorithm for generating an IID sample $\mathbf{T}_N$ of size
N from the DGP that is characterized by the PDF (2).

We note that the quantile ABC algorithm has been suggested in the past in
articles such as [4]. In Algorithm 2, we may note that γ plays the role of the
accuracy parameter, and the runtime of the algorithm is inversely proportional
to its value. However, unlike ϵ, for any value of γ, one requires a predictable
number $\nu = \lceil N/\gamma \rceil$ of samples from the DGP that is characterized by the prior
$\pi(\boldsymbol{\theta})$. Thus, one may always predict the runtime of the algorithm for any choice
of γ. We shall use Algorithm 2 in all instances, for the remainder of the paper.

4.1   Normal Model

Let

$$\phi\left(x; \mu, \sigma^2\right) = \left(2\pi\sigma^2\right)^{1/2} \exp\left[-\frac{(x-\mu)^2}{2\sigma^2}\right],$$

denote the normal PDF with mean $\mu \in \mathbb{R}$ and variance $\sigma^2 > 0$. Suppose that Θ
is a mean parameter that is generated from a DGP with PDF $\pi(\theta) = \phi(\theta; \mu, 1)$.
Furthermore, suppose that we observe IID data $\mathbf{X}_n = \{X_i\}_{i=1}^n$ that arise from a
DGP with PDF $f(x|\theta) = \phi(x; \theta, 1)$. Via a conjugacy relationship (cf. [16, Sec.
3.3]), we may write the posterior PDF $f(\theta|\mathbf{X}_n)$ in closed form as

$$f(\theta|\mathbf{X}_n) = \phi\left(\theta; \frac{1}{n+1}\left[\mu + n\bar{X}_n\right], \frac{1}{n+1}\right), \tag{4}$$

where $\bar{X}_n = n^{-1}\sum_{i=1}^n X_i$ is the sample mean of $\mathbf{X}_n$.

Suppose that $\mu = 1$ and $n = 100$. Upon drawing a fixed sample $\mathbf{X}_n$ from
the prior DGP with PDF $\pi(\theta) = \phi(\theta; 1, 1)$, we implement Algorithm 2 with
$m = 100$, $\gamma = 1/100$, and $N = 100$, in order to obtain a quasi-posterior sample
$\mathbf{T}_N = \{\boldsymbol{\Theta}_k\}_{k=1}^N$. Using the `density()` function from the R programming language
[15], we plot the kernel density estimate (KDE) of the sample $\mathbf{T}_N$ along with
the true posterior PDF in Fig. 1. We notice that the two PDFs share centrality
and dispersion properties, and thus we can see that the ABC approach yields a
reasonable approximation to the posterior PDF in this situation.

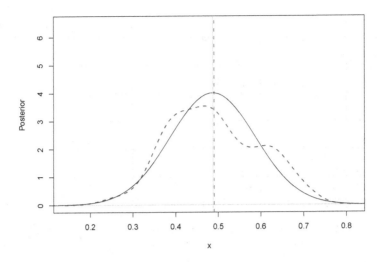

Fig. 1. Posterior and quasi-posterior PDFs for normal distribution example. The true posterior PDF is represented by the solid curve, with the solid line indicating its mean. The ABC quasi-posterior PDF is plotted as a dashed curve, with the dashed vertical line indicating its mean.

4.2   Normal Mixture Model

Let

$$\phi\left(\boldsymbol{x};\boldsymbol{\mu},\boldsymbol{\Sigma}\right)=\left|2\pi\boldsymbol{\Sigma}\right|^{-1/2}\exp\left[-\frac{1}{2}\left(\boldsymbol{x}-\boldsymbol{\mu}\right)^{\top}\boldsymbol{\Sigma}^{-1}\left(\boldsymbol{x}-\boldsymbol{\mu}\right)\right],$$

denote the multivariate normal PDF, where $\boldsymbol{\mu}\in\mathbb{R}^{d}$ and $\boldsymbol{\Sigma}\in\mathbb{R}^{d\times d}$ is a positive definite and symmetric matrix. Here, $(\cdot)^{\top}$ is the matrix transposition operator.

Suppose that $d=2$ and that data $\mathbf{X}_{n}=\left\{\boldsymbol{X}_{i}\right\}_{i=1}^{n}$ are generated IID, such that $\boldsymbol{X}_{1}$ arises from a DGP with two-component mixture density

$$f\left(\boldsymbol{x}|\boldsymbol{\theta}\right)=p\phi\left(\boldsymbol{x};\mu\mathbf{1},\mathbf{I}\right)+\left(1-p\right)\phi\left(\boldsymbol{x};-\mu\mathbf{1},\mathbf{I}\right),\tag{5}$$

where $\boldsymbol{\theta}^{\top}=\left(\mu,p\right)$ ($\mu\in\mathbb{R}$, and $p\in[0,1]$). Here $\mathbf{1}$ and $\mathbf{I}$ are the ones vector and identity matrix, respectively. We suppose for example that $\mathbf{X}_{n}$ is generated from a DGP with $\boldsymbol{\theta}^{\top}=\boldsymbol{\vartheta}^{\top}=(2,1/2)$, in particular. For $n=100$, a realization of $\mathbf{X}_{n}$ is visualized in Fig. 2.

We wish to recover the parameter vector $\boldsymbol{\theta}$ that defines the DGP of $\mathbf{X}_{n}$ via ABC. We utilize Algorithm 2 with $m=100$, $\gamma=1/100$, and $N=100$, in order to generate a quasi-posterior sample $\mathrm{T}_{N}=\left\{\Theta_{k}\right\}_{k=1}^{N}$ using (5) to simulate data, and using the prior PDF

$$\pi\left(\boldsymbol{\theta}\right)=\frac{\llbracket\mu\in[0,4]\rrbracket}{4}\times\llbracket p\in[0,1]\rrbracket,\tag{6}$$

which characterizes a distribution where the first coordinate of Θ is sampled uniformly between 0 and 4, and where the second coordinate is sampled uniformly, independent of the first coordinate, between 0 and 1. That is

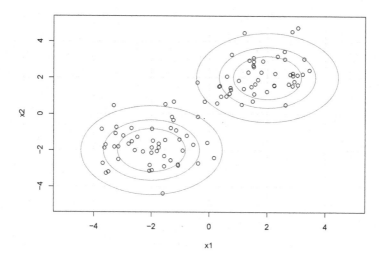

Fig. 2. A sample of $n = 100$ IID realizations from the DGP characterized by the PDF (5), with $\boldsymbol{\vartheta}^\top = (2, 1/2)$. Concentric circles indicate density contours of the two normal distributions that make up the mixture model (5).

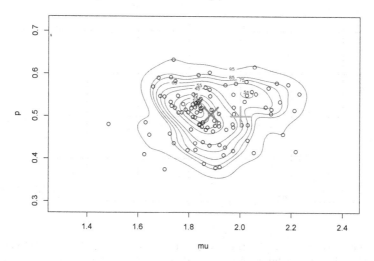

Fig. 3. A sample of $N = 100$ IID realizations $\mathbf{T}_N$ from the ABC quasi-posterior distribution, obtained using the mixture model (5) and prior distribution characterized by (6). The contours indicate the KDE of the sample $\mathbf{T}_N$. The plus sign indicates the location of $\boldsymbol{\vartheta}^\top = (2, 1/2)$ and the cross indicates the mean of $\mathbf{T}_N$.

$\boldsymbol{\Theta} \sim \text{Uniform}([0, 4] \times [0, 1])$. Figure 3 visualizes the quasi-posterior sample $\mathbf{T}_N$, along with its KDE, as estimated via the kde() function from the R package ks of [3]. We observe that the quasi-posterior sample is concentrated in a region that is very close to $\boldsymbol{\vartheta}$. In fact, the mean of the p coordinate is almost exactly equal to $1/2$.

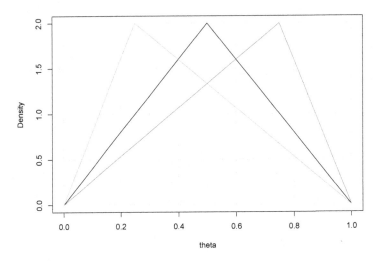

Fig. 4. PDFs of form (7), with $\theta \in \{1/4, 1/2, 3/4\}$.

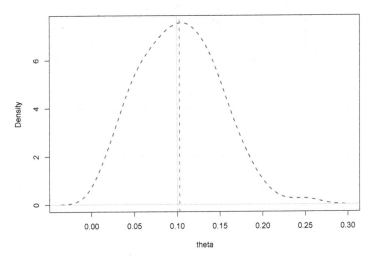

Fig. 5. Quasi-posterior PDF for the triangle distribution example, obtained via a KDE of the sample T_N. The solid vertical line indicates the value $\vartheta = 1/10$, and the dashed vertical like indicates the mean of the quasi-posterior sample T_N.

4.3   Triangle Distribution

We now suppose that the IID sample $X_n = \{X_i\}_{i=1}^{n}$ is such that X_1 is generated from a triangle distribution that is characterized by the PDF

$$f(x|\theta) = \frac{2x}{\theta} [\![x \in [0, \theta]]\!] + \frac{2(1-x)}{1-\theta} [\![x \in (\theta, 1]]\!], \tag{7}$$

where $\theta \in (0, 1)$. Visualizations of the PDF (7) with different values of θ appears in Fig. 4. A characterization of the difficulty in estimating the parameter θ of the triangle distribution can be found in [12].

We generate a $n = 100$ observations sample X_n from a triangle distribution with parameter $\theta = \vartheta = 1/10$. Upon setting $m = 100$, $\gamma = 1/100$, and $N = 100$, and using (7) to characterize the DGP of X_n and sampling Θ from the uniform prior distribution with PDF

$$\pi(\theta) = [\![\theta \in [0, 1]]\!],$$

we obtain a quasi-posterior sample $T_N = \{\Theta_k\}_{k=1}^{N}$. We plot the KDE for the quasi-posterior sample in Fig. 5. We observe that the mean of the quasi-posterior sample is a good estimator of the generative parameter value ϑ.

5   Application

The analysis of wind direction data plays an important role in meteorology. The data set wind from the R package circular [1]. The data set $X_n = \{X_i\}_{i=1}^{n}$ contains $n = 310$ measurements of wind direction, in radians anti-clockwise from due East, such that $X_i \in \mathbb{X} = [0, 2\pi)$ ($i \in [n]$). Each measurement is taken at the same location at the weather station of a place named *Col de la Roa* in the Italian Alps. The measurements are all acquired between 29 January and 31 March, in 2001. A visualization of the data is presented in Fig. 6.

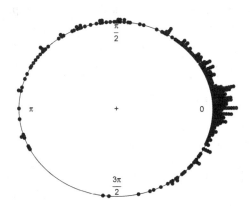

Fig. 6. Visualization of the wind data set from the R package circular.

Due to the circular and thus non-Euclidean nature of the domain $\mathbb{X}$ (e.g., the fact that 0 and 2π represent the same point), we cannot characterize the data via DGPs that are only applicable to Euclidean spaces, such as via the normal distribution. A particularly useful distribution for characterization of circular data is the *von Mises* distribution (cf. [8, Sec. 2.2]).

The von Mises distribution can be characterized via the PDF

$$f\left(x|\boldsymbol{\theta}\right) = \frac{\exp\left[\kappa\cos\left(x-\mu\right)\right]}{2\pi I_0\left(\kappa\right)}, \tag{8}$$

where $\mu \in \mathbb{X}$ and $\kappa > 0$ represent the centrality and scale of the data, respectively, and $\boldsymbol{\theta}^\top = (\mu, \kappa)$. Unfortunately, $I_0\left(\kappa\right)$ is the modified Bessel function of order zero, which takes the infinite sum form of

$$I_0\left(\kappa\right) = \sum_{j=0}^{\infty}\left(j!\right)^{-2}\left(\frac{\kappa}{2}\right)^{2j},$$

and thus makes optimization involving (8) difficult. This therefore inhibits the usual processes for obtaining parameter estimates. Fortunately, simulation from a DGP that is characterized by (8) is still possible via the use of algorithms such as that of [2].

We thus apply Algorithm 2 to the problem of estimating $\boldsymbol{\theta}$ for the `wind` data. We set $m = n$, $\gamma = 1/100$, and $N = 100$. The prior PDF was chosen to take the form

$$\pi\left(\boldsymbol{\theta}\right) = \frac{[\![\mu \in \mathbb{X}]\!]}{2\pi}\exp\left(-\kappa\right).$$

That is, $\boldsymbol{\Theta} \sim \text{Uniform}\left(\mathbb{X}\right) \times \text{Exponential}\left(1\right)$. The `rvonmises`() function from `circular` is then used to simulate data from a von Mises distribution with any particular parameter combination that is drawn from the prior DGP. In Fig. 7, we present the marginal histograms for the quasi-posterior sample $\mathbf{T}_N = \{\boldsymbol{\Theta}_k\}_{k=1}^{N}$.

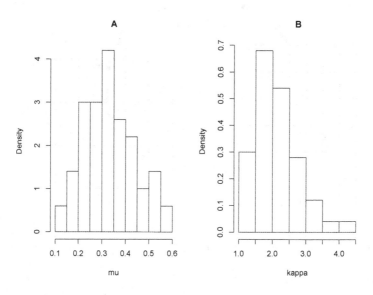

Fig. 7. Histograms A and B visualize the distribution of the μ and κ coordinates from the quasi-posterior sample $\mathbf{T}_N$, respectively, for the ABC algorithm applied to analyze the `wind` data.

We may also obtain the mean of $\mathbf{T}_N$, which we can write as $\bar{\boldsymbol{\Theta}}_N^\top =$ $(0.3297, 2.1404)$. We can then plot the estimated PDF $f\left(x|\bar{\boldsymbol{\Theta}}_N\right)$ along with the data in order to visualize the centrality and dispersion of the wind directions in Fig. 8. We can simply observe that the density is highest in East North-East direction with the scale parameter accounting for the variability around this centroid. Overall the model provides a reasonable fit to the data.

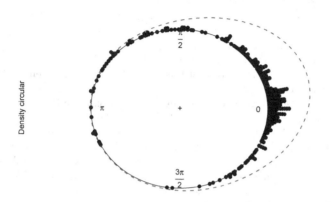

Fig. 8. Visualization of the ABC estimated PDF $f\left(x|\bar{\boldsymbol{\Theta}}_N\right)$ along with the `wind` data.

6 Conclusion

Modern data analysis can require the estimation of computationally intractable models. If data can be simulated from such models, then the ABC method provides a possible avenue for resolving the complex model estimation problem. The ABC framework along with the ES statistic approach yields a theoretically and practically sound solution for parameter estimation problems. We have demonstrated that such a framework can be applied to study both simple artificial estimation problems as well as interesting applied ones. Using our examples, we invite the reader to explore the many potential uses of the ABC method to estimation problems that they may face.

Acknowledgements. The author is supported by Australian Research Council grants DE170101134 and DP180101192.

References

1. Agostinelli, C., Lund, U.: R package 'circular': Circular Statistics (2017)
2. Best, D.J., Fisher, N.I.: Efficient simulation of the von Mises distribution. J. R. Stat. Soc. C **28**, 152–157 (1979)
3. Duong, T.: ks: kernel density estimation and kernel discriminant analysis for multivariate data in R. J. Stat. Softw. **21**(7), 1–16 (2007)

4. Frazier, D.T., Martin, G.M., Robert, C.P., Rousseau, J.: Asymptotic properties of approximate Bayesian computation. Biometrika **3**, 593–607 (2018)
5. Iverson, K.E.: A Programming Language. Wiley, Hoboken (1962)
6. Jiang, B., Wu, T.Y., Wong, W.H.: Approximate Bayesian computation with Kullback-Leibler divergence as data discrepancy. In: Proceedings of the 21st International Conference on Artificial Intelligence and Statistics (AISTATS) (2018)
7. Karabatsos, G., Leisen, F.: An approximate likelihood perspective on ABC methods. Stat. Surv. **12**, 66–104 (2018)
8. Ley, C., Verdebout, T.: Modern Directional Statistics. CRC Press, Boca Raton (2017)
9. Marin, J.M., Pudlo, P., Robert, C.P., Ryder, R.J.: Approximate Bayesian computation methods. Stat. Comput. **22**, 1167–1180 (2012)
10. Miller, J.W., Dunson, D.B.: Robust Bayesian inference via coarsening. J. Am. Stat. Assoc. **113**, 340–356 (2018)
11. Nguyen, H.D., Arbel, J., Lu, H., Forbes, F.: Approximate Bayesian computation via the energy statistic. ArXiv arXiv:1905.05884 (2019)
12. Nguyen, H.D., McLachlan, G.J.: Progress on a conjecture regarding the triangular distribution. Commun. Stat. - Theory Methods **46**, 11261011271 (2017)
13. Press, S.J.: Subjective and Objective Bayesian Statistics. Wiley, Hoboken (2003)
14. Pritchard, J.K., Seielstad, M.T., Perez-Lezaun, A., Feldman, M.W.: Population growth of human Y chromosomes: a study of Y chromosome microsatellites. Mol. Biol. Evol. **16**, 1791–1798 (1999)
15. R Core Team: R: A Language and Environment for Statistical Computing. R Foundation for Statistical Computing, Vienna (2019)
16. Robert, C.P.: The Bayesian Choice: From Decision-Theoretic Foundations to Computatoinal Implementation. Springer, New York (2007). https://doi.org/10.1007/0-387-71599-1
17. Robert, C.P., Casella, G.: Monte Carlo Statistical Methods. Springer, New York (1999). https://doi.org/10.1007/978-1-4757-3071-5
18. Sisson, S.A., Fan, Y., Beaumont, M.A. (eds.): Handbook of Approximate Bayesian Computation. CRC Press, Boca Raton (2019)
19. Spokoiny, V., Dickhaus, T.: Basics of Modern Mathematical Statistics. Springer, Berlin (2015). https://doi.org/10.1007/978-3-642-39909-1
20. Szekely, G.J., Rizzo, M.L.: Testing for equal distributions in high dimension. InterStat **5**, 1–16 (2004)
21. Szekely, G.J., Rizzo, M.L.: Energy statistics: a class of statistics based on distances. J. Stat. Plan. Inference **143**, 1249–1272 (2013)
22. Szekely, G.J., Rizzo, M.L.: The energy of data. Annu. Rev. Stat. Appl. **4**, 447–479 (2017)
23. Tavare, S., Balding, D.J., Griffiths, R.C., Donnelly, P.: Inferring coalescence times from DNA sequence data. Genetics **145**, 505–518 (1997)
24. Turkman, M.A.A., Paulino, C.D., Muller, P.: Computational Bayesian Statistics: An Introduction. Cambridge University Press, Cambridge (2019)

Contributing Papers

Truth, Proof, and Reproducibility: There's No Counter-Attack for the Codeless

Charles T. Gray[1]([⊠])(iD) and Ben Marwick[2](iD)

[1] La Trobe University, Melbourne, Australia
charlestigray@gmail.com
[2] University of Washington, Seattle, USA
bmarwick@uw.edu

Abstract. Current concerns about reproducibility in many research communities can be traced back to a high value placed on empirical reproducibility of the physical details of scientific experiments and observations. For example, the detailed descriptions by 17th century scientist Robert Boyle of his vacuum pump experiments are often held to be the ideal of reproducibility as a cornerstone of scientific practice. Victoria Stodden has claimed that the computer is an analog for Boyle's pump – another kind of scientific instrument that needs detailed descriptions of how it generates results. In the place of Boyle's hand-written notes, we now expect code in open source programming languages to be available to enable others to reproduce and extend computational experiments. In this paper we show that there is another genealogy for reproducibility, starting at least from Euclid, in the production of proofs in mathematics. Proofs have a distinctive quality of being necessarily reproducible, and are the cornerstone of mathematical science. However, the task of the modern mathematical scientist has drifted from that of blackboard rhetorician, where the craft of proof reigned, to a scientific workflow that now more closely resembles that of an experimental scientist. So, what is proof in modern mathematics? And, if proof is unattainable in other fields, what is due scientific diligence in a computational experimental environment? How do we measure truth in the context of uncertainty? Adopting a manner of Lakatosian conversant conjecture between two mathematicians, we examine how proof informs our practice of computational statistical inquiry. We propose that a reorientation of mathematical science is necessary so that its reproducibility can be readily assessed.

Keywords: Metaresearch · Reproducibility · Mathematics

Thank you to Kerrie Mengersen, Kate Smith-Miles, Mark Padgham, Hien Nguyen, Emily Kothe, Fiona Fidler, Mathew Ling, Luke Prendergast, Adam Sparks, Hannah Fraser, Felix SingletonThorn, James Goldie, Michel Penguin (Michael Sumner), in no particular order, with whom initial bits and pieces of this paper were discussed. Special thanks to Brian A. Davey for proofing the proofs and Alex Hayes for his edifying post [16].

H. Nguyen (Ed.): RSSDS 2019, CCIS 1150, pp. 111–129, 2019.
https://doi.org/10.1007/978-981-15-1960-4_8

In David Auburn's Pulitzer prize-winning 2000 play *Proof*, a young mathematician, Catherine, struggles to prove to another mathematician, Hal, that her argument is not a reproduction of the intellectual work of her deceased father, a professor [2]. Her handwriting similar to her father's, there is no way to discern her proof from his. But if Catherine were a computational scientist, we would have a very different story. We reimagine Hal challenging Catherine for different mathematical questions and the reproducibility of her solutions. We consider simple to complex mathematical questions that can be answered at the blackboard, and then consider the scenario where Catherine must use a combination of mathematical and computational tools to answer a question in mathematical science. Via these scenarios, we question to what extent proof methodology continues to inform our choices as mathematical scientists become as much research software engineers[1] as they are mathematicians.

Mathematical science is the compendium of research that binds the Catherine's methodology of work indistinguishably from her father's. However, in computational science, we not only do not have a common language in the traditional sense, with programming languages such as Python, R, and C++ performing overlapping tasks, but our research workflows comprise tools and platforms and operating systems, such as Linux or Windows, as well. Many inadvertent reasons conspire so that scientists are arriving at similar problems with different approaches to data management and version control. Code scripts, arguably the most immediately analogous to mathematical proof, are but one of the many components that make up the outputs of computational science.

If Catherine were a contemporary computational mathematician, she would not only struggle to reproduce another person's work, but she would likely struggle to reproduce her own. She may be overwhelmed by the diversity of research outputs [5], and find that she needs to rewrite her work to unpick what she did with specific computational functions under specific software package releases. The language of mathematical science has changed from something we write, to something we collect. In order to diligently answer scientific questions computationally, the mathematician must now consider her work within that of a research compendium. In this paper we ask: how can we extend the certainty afforded by a mathematical proof further down the research workflow into the 'mangle of practice' [31]? We show that communities of researchers in many scientific disciplines have converged on a toolkit that borrows heavily from software engineering to robustly provides many points to verify certainty, from transparency via version control, to stress testing of algorithms. We focus on unit testing as a strong measure of certainty.

---

[1] We might argue here we employ the term *research software engineer* (RSE) as Katz and McHenry would define *Super RSEs*, developers who 'work with and support researchers, and also work in teams of RSEs who research and develop their own software, support it, grow it, sustain it, etc.' [20]. Or choose the more ambiguous Research Software Engineers Association definition of RSEs as people in academia who 'combine expertise in programming with an intricate understanding of research' [45].

1   The Technological Shift in Mathematical Inquiry

The task of a mathematical scientist in the pre-computer age was largely that of a blackboard rhetorician, where the craft of proof reigned. For a proof such as that featured in Auburn's play, the argument can often be included in the article, or as a supplementary file. This allows the reader to fully reproduce the author's reasoning, by tracing the flow of argument through the notation. As computers have become ubiquitous in research, mathematical scientists have seen their workflow shift to one that now more closely resembles that of a generic scientist, concerned with diligent analysis of observational and experimental data, mediated by computers [30]. But the answer to the question of what constitutes a diligent attempt to answer a scientific question examined in a computationally intensive analysis, is unclear, and remains defined by the era of the blackboard mathematician.

So, what is proof in mathematics, when experimental and computer-assisted methods are common? And, beyond mathematics, in fields where literal proofs are unattainable, what counts as an equivalent form of scientific certainty in a computational experimental environment? How do we measure truth in the context of uncertainty? Among the histories of science we can trace three efforts to tackle these questions. First is the empirical effort, most prominently represented by Robert Boyle (1627–1691), known for his vacuum pump experiments [34]. Boyle documented his experiments in such detail and to an extent that was uncommon at the time. He was motivated by a rejection of the secrecy common in science at his time, and by a belief in the importance of written communication of experimental expertise (as a supplement to direct witnessing of experimental procedures). Boyle's distinctive approach of extensive documentation is often cited by modern advocates of computational reproducibility [35]. Making computer code openly available to the research community is argued to be the modern equivalent of Boyle's exhaustive reporting of his equipment, materials, and procedures [22].

A second effort to firming up certainty in scientific work, concerned with statistical integrity, can be traced at least as far back as Charles Babbage (1791–1871), mathematician and inventor of some of the first mechanical computers. In his 1830 book 'Reflections on the Decline of Science in England, and on Some of Its Causes' he criticised some of his contemporaries, characterising them as 'trimmers' and 'cooks' [14]. Trimmers, he wrote, were guilty of smoothing of irregularities to make the data look extremely accurate and precise. Cooks retained only those results that fit their theory and discarded the rest [26]. These practices are now called data-dredging, or p-hacking, where data are manipulated or removed from an analysis until a desirable effect or p-value is obtained [17].

A third effort follows the history of formal logic through to the time when an equivalence between philosophical logic and computation was noted. This observation is called the Curry-Howard isomorphism or the proofs-as-programs interpretation. First stated in 1959, this correspondence proposed that proofs in some areas of mathematics, such as type theory, are exactly programs from a particular programming language [37]. The bridging concepts come from intuitionistic logic and typed lambda calculi, which have lead to the design of computational formal proof management systems such as the Coq language. This language is designed to write mathematical definitions, execute algorithms and theorems, and check proofs [3]. This correspondence has not been extensively discussed in the context of reproducibility, but we believe it has relevance and is motivating beyond mathematics. Our view is that this logic-programming correspondence can be extended in a relaxed way beyond mathematics in proofs to scientific claims in general, such that computational languages can express those claims in ways that can establish a high degree of certainty.

Questions of confidence in scientific results are far from restricted to the domains of mathematics or computers; indeed, science is undergoing a broad reexamination under what is categorised as a crisis of inference [12]. How we reproduce scientific results is being examined across a range of disciplines [6,38]. An early answer to some of these questions is that authors should make available the code that generated the results in their paper [11,36]. These recommendations mark the emergence of a concern for computational reproducibility in mathematics. This paper extends this argument for computational reproducibility further into the workflow of modern statistical inquiry, expanding and drawing on solutions proposed by methods that privilege computational reproducibility.

Systemic problems are now being recognised in the practice of conventional applied statistics, with a tendency towards *dichotomania* [1] that reduces complex and nuanced questions to Boolean statements of TRUE or FALSE. This has diluted the trust the can be placed in scientific results, and led to a crisis of replication, where results can not easily be reproduced [12] and questionable research practices [13] proliferate.

As the conventions of statistics are called into question, it stands to reason that the research practices of the discipline of statistics itself require examination. For those practicing statistical computing, a conversation is emerging about what constitutes best practice [43]. But best practice may be unrealistic, especially for those applying statistics from fields where their background has afforded limited computational training. And thus the question is becoming reframed in terms of *good enough* standards [44] we can reasonably request of statistical practitioners. By extension, we must reconsider how we prepare students in data-analytic degree programs.

Proofs, derivations, verification, all form the work of mathematics. How do we make mathematical arguments in a computational[2] environment? In constructing mathematical arguments, we posit that we require an additional core element: unit testing for data analysis. We propose an expansion of the spectrum of reproducibility, Fig. 1, to include unit testing for data analytic algorithms facilitated by a tool such as testthat:: [41], for answering mathematical research

---

[2] We focus in this manuscript on R packages, but the reader is invited to consider these as examples rather than definitive guidance. The same arguments hold for other languages, such as Python, and associated tools.

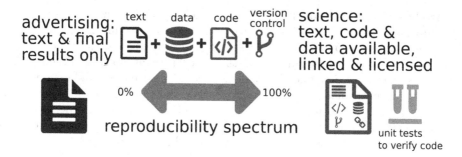

Fig. 1. We propose updating this spectrum of reproducibility [25] with unit tests for data analysis. In addition to the advertising, the **formal** scientific argument put forward, many *informal* and traditionally hidden scientific outputs comprise the compendium of research that produces the results. Given the underutilised nature of unit tests, we suggest there is further work to be done to facilitate the adoption of *good enough* [44] research software engineering practices for answering mathematical questions computationally. The informal components of mathematical research compendium are shaded grey. This figure has been adapted with permission [33] and is licensed under CC-BY 2.0.

questions computationally. In order to motivate this practice, we turn to the purest of sciences, mathematical proof.

2   Truth in Mathematics

The titular proof [2] of Auburn's play is a mathematical argument, a formalised essay in mathematical science. The creator of the proof, Catherine, is questioned by Hal, who is capable of following the argument; that is, Hal can *replicate* an approximation of the type of thought process that leads to a *reproduction* of the argument presented in the proof.

In Fig. 1, we have coloured the components, black **formal** argument, and grey *informal* work, of mathematics Hal would need to reproduce the proof. In order to verify the results, Hal would need to follow the formal argument, to understand what was written in the proof, but also need to do informal work, to understand the links between concepts for verification.

Hal would come to the problem with a different background and education to Catherine. Although work is necessary for the verification of the results, the reproduction of the reasoning, the work required would be different for Hal and Catherine, based on their respective relevant preparation. However, the language of mathematics carries enough uniformity that Hal can fill in the work he requires to understand the result, from reasoning and mathematical texts. If Catherine were asking a mathematical question computationally, the presentation of the results carries not millennia of development of methodology, as does the noble craft of mathematics, but less than a century of frequently disconnected developments separated by disparate disciplines.

We begin with traditional mathematics and end with answering questions in computational mathematics. To this aim, we adopt, in the manner of Lakatos' *Proofs and Refutations*' conversant conjecture, scenarios between Hal and Catherine, where Hal challenges Catherine over her authorship of the proof. In each scenario, we imagine the challenge would play out for different ways of answering mathematical questions. We argue the thinking work of mathematical science is not as immediately inferable in a computational experimental environment, and that the roots of mathematical science in proof lead to an overconfidence that science is as readily reproducible as a proof.

2.1   Prove It!

Let us suppose Catherine claimed she could demonstrate a property about the order[3] on natural numbers, $\mathbb{N} = \{1, 2, 3, \dots\}$, the counting numbers.

The order on a set of numbers is **dense** if, for any two numbers we can find a number in between. More formally, we say an ordered set P is dense if, for all $x < y$ in P, there exists z in P such that $x < z < y$.

Catherine presents the following argument that the order on $\mathbb{N}$ is not dense. In this case she chooses a type of *indirect* proof, an *existence* proof [4], where she presents a counterexample demonstrating that the density property is not true for all cases for $\mathbb{N}$.

Proof. **The order on $\mathbb{N}$ is not dense.** Let us, in the spirit of Lewis Carroll[4], be contrary and suppose, by way of contradiction, that the order on $\mathbb{N}$, *is* dense. Then for any two numbers, x and y, in $\mathbb{N}$ such that $x < y$, I should be able to find a distinct number z in $\mathbb{N}$ between them, that is $x < z < y$. But, consider the numbers 3 and 4. Let $x = 3$ and $y = 4$, then $x < y$. There is no distinct number, z, that exists between x and y. Since this rule must be true for any two numbers $x < y$ in the order to be dense, we have shown the order on natural numbers $\mathbb{N}$ is not dense.   □

A standard way to prove something is *not* true, is to assume it *is* true, and derive a contradiction [9]. Arguably, this reasoning goes to the heart of the problem of *dichotomania* lamented by 800 scientists in a recent protest paper about the misinterpretation of statistics in *Nature* [1]. A null hypothesis test of a difference between two groups will assume the opposite of what we suspect is true; we believe there to be a difference between two groups and take a sample from each of the groups and perform a test. This test assumes there is no difference, null, between the two groups and that any observed differences in sampling are due to

---

[3] Let P be a set. An *order* on P is a binary relation $\leqslant$ on P such that, for all $x, y, z \in P$: we have $x \leqslant x$; with $x \leqslant y$ and $y \leqslant x$ imply $x = y$; and, finally, $x \leqslant y$ and $y \leqslant z$ imply $x \leqslant z$. We then say $\leqslant$ is reflexive, antisymmetric, and transitive, for each of these properties, respectively [8].

[4] Lewis Carroll, author of *Alice in Wonderland*, is a writing pseudonym used by Charles Lutwidge Dogson, born in 1832, who taught mathematics at Christ Church, Oxford [7].

random chance. The calculation returned, the p-value, is the likelihood we would observe the difference under those null assumptions. Crucially, the calculation returned is probabilistic, a number between 0 and 1, not a TRUE or FALSE, the logic of a proof by contradiction. The logic does not apply to a situation where, within a single group of people, some people might be resistant to treatment, and some might not be, say, and we have estimated a likelihood of the efficacy of the treatment. Dichotomania is the common misinterpretation of a probabilistic response in a dichotomous framework; scientists are unwittingly framing null hypothesis significance testing in terms of a proof by contradiction.

In order to illustrate our central point, we now turn to a direct argument, rather than the indirect approach of contradiction, in order to examine the process of the making of a proof. In both the case of the direct, and indirect proofs, however, Hal could challenge Catherine, as he did in the play.

"Your dad might have written it and explained it to you later. I'm not saying he did, I'm just saying there's no proof that you wrote this" [2].

2.2   The Steps in the Making of a Proof

Let us now suppose Catherine's proof instead demonstrated a density property on the order on the real numbers,

$$\mathbb{R} = \{\ldots, -3, \ldots, -3.3, \ldots, 0, \ldots, 1, \ldots, 100.23, \ldots\},$$

i.e., the whole numbers, and the decimals between them. Catherine claims the order on $\mathbb{R}$ is dense, which is to say, if we choose any two distinct numbers in the real numbers, we can find a distinct number between them.

Catherine would construct her proof in the manner laid out in the introductory monograph *When is a Proof?* [9], in Table 1, provided to undergraduate mathematics majors at La Trobe University. These steps comprise **formal** and *informal* mathematical work, showing that mathematical *work* comprises more than the *advertising*, as it is labelled in the reproducibility spectrum presented in Fig. 1. In the case of pure mathematics, the advertising would be the paper that outlines the proof, the formal mathematical argument, but the informal work is left out.

Catherine presents the following proof to Hal to show the order on real numbers, $\mathbb{R}$, is dense.

Proof. The order on $\mathbb{R}$ is dense. Let $x < y$ in $\mathbb{R}$. Let[5] $z := (x+y)/2$. To see that $x < z < y$, we begin with $x < y$, so, $x + x < x + y$ and $x + y < y + y$, which gives,

---

[5] In mathematics, we read := as 'be defined as', $\implies$ as 'implies', and < as 'less than but not equal to'.

Table 1. The steps in the making of a proof from Brian A. Davey's primer, *When is a Proof?* [9]. The formal steps that contribute to the final proof are in **bold**, the hidden informal work, in *italics*. These steps are summarised in terms of $p \implies q$ in the final column of the table.

Step -1	**Translate the statement to be proved into ordinary English and look up appropriate definitions**	
Step 0	**Write down what you are asked to prove. Where appropriate, isolate the assumptions, p, and the conclusion, q**	$p \implies q$
Step 1	**Write down the assumptions, p: "Let … "**	Assume p
Step 2	**Expand Step 1 by writing out definitions: "i.e., … "**	Define p
Step 3	*Write down the conclusion, q, which is to be proved: "To prove: … "*	*State q*
Step 4	*Expand Step 3 by writing out definitions: "i.e., … "*	*Define q*
Step 5	*Use your head: do some algebraic manipulations, draw a diagram, try to find the relationship between the assumptions and the conclusion*	*Work*
Step 6	**Rewrite your exploration from Steps 3, 4 and 5 into a proof. Justify each statement in your proof**	Formalise work
Step 7	**The last line of the proof**	"Hence q."

$$x + x < x + y < y + y$$

$$\implies \frac{x+x}{2} < \frac{x+y}{2} < \frac{y+y}{2}$$

$$\implies \frac{2x}{2} < \frac{x+y}{2} < \frac{2y}{2}$$

$$\implies x < \frac{x+y}{2} < y$$

$$\implies x < z < y,$$

since $z = (x + y)/2$, as required.

Catherine presents the formal proof, the science that in Fig. 1 is described as the advertising, a subcomponent, of the compendium of research she created in order to arrive at this argument. Hal wishes to verify the results and investigate whether Catherine merely reproduced her father's reasoning. In the case of proof, what is published is the formal argument, but as the steps in Table 1, this is not all of what makes a proof. We could think of the steps presented in Table 1 in terms of a mathematical statement $p \implies q$, which we read as p *implies* q, as given in the final column of the table. We now revisit the proof Catherine offered in terms of these steps.

We begin, step 0; we state what we wish to prove, $p \implies q$, in plain English. We wish to show the real numbers, $\mathbb{R}$, are dense; i.e., for all $x < y$ in $\mathbb{R}$, there exists z such that $x < z < y$.

Step 1, we **assume** p is true. We assume we have two distinct numbers x and y in $\mathbb{R}$ with $x < y$; i.e., x is less than y, and x is not equal to y. Step 2, nothing to define as we are familiar with $<$ and $\mathbb{R}$.

Step 3, we *state* what we wish to prove, q; the order on $\mathbb{R}$ is dense. Step 4, i.e., we need to show there exists z in $\mathbb{R}$ such that $x < z < y$. Now, Catherine has offered a solution $z := (x + y)/2$ that Hal wishes to verify.

Step 5, Suppose Hal asks, what if both x and y are negative numbers? Is it still true that $x < z < y$? Hal might verify his understanding of $+$ by thinking about positive and negative numbers as steps taken to the left or the right. In Fig. 2, Hal considers the case where both numbers are negative, $x, y < 0$. In this case, we have x steps to left, and y steps to the left, which we imagine as arrows of appropriate length. If we lay both arrows end to end, we see the number of combined steps to the left. If we consider the half-way point of x and y laid beside each other, $(x + y)/2$, we see this falls between where the arrow heads of x and y fall.

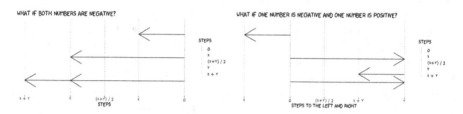

Fig. 2. On the left, Hal might begin to verify his understanding of $+$ by first considering the case where both numbers are negative, $x, y < 0$. In this case, we might think of $+$ as combining x steps to the left with y steps to the left. The halfway point $(x + y)/2$, falls in the middle of the two arrows laid side by side, which also falls between where the two ends of the arrows fall. On the right, Hal considers the case where $x < 0, y > 0$ and $|x| < |y|$. Here $x + y$ can be thought of as y steps to the right and then x steps to the left. Again, the halfway point $(x + y)/2$ falls halfway between the tips of the two arrows above.

Now Hal can flip the arrows in the opposite directions to construct an argument for if both numbers were positive, $x, y > 0$.

But then Hal asks in Fig. 2, what if one number were positive and one number were negative? Is $(x + y)/2$ still halfway between? Let us assume, as mathematicians say, without loss of generality that the magnitude of x is strictly less than y, that is $|x| < |y|$, the number of steps in x is less than the number of steps of y. Hal now considered where one would end up if one took y steps to the right and then x steps to left. He checks that he does not need to consider two cases, as he would end up in the same place if he took x steps to the left and then y steps to the right. Again, $(x + y)/2$ falls between where he would start and where he would end.

Now Hal has verified his understanding of $+$, which may or may not be the way that Catherine arrived at her result, but after this work he is capable of fully reproducing the mathematical result presented. He reads the proof Catherine has provided, and verifies Steps 6, and Step 7. Catherine has proved that the order on $\mathbb{R}$ is dense. With this proof, as with the proof presented in Sect. 2.1, Hal cannot disqualify the possibility that Catherine merely reproduced her father's work.

Even in these relatively simple proofs, Step 5, the informal work of verification and understanding vastly outweighs what goes into the formal proof. But these toy examples belie a process of redefinition and re-examination, as illustrated in the discussion within a hypothetical mathematics classroom that forms the narrative of Lakatos' *Proofs and Refutations* [21]. We now move to a recently published proof to illustrate this process of redefinition.

In the Combat Conditions of New Mathematics. Suppose, now, that Catherine's proof were for the theorem pertaining to quasi-primal algebras, presented in the recent publication 'The homomorphism lattice induced by a finite algebra' [10] in *Order*, a mathematics journal devoted to 'original research on the theory and application of ordered sets'. In addition to the informal work demonstrated by the proof that the order on $\mathbb{R}$ is dense, the making of this proof involved a redefinition of the result proved, through a process writing several proofs. In terms of Table 1, initially a result was considered, $p \implies q$. A proof was written for this result. At this point the mathematicians realised, however, that the converse could be shown, that is, $q \implies p$. And so, a proof was generated for a new result, $p \iff q$. In the case of this proof, the act of writing the proof itself redefined the result in question. In the combat conditions of new mathematics, the process of writing a proof is doing mathematical science, and involves a great deal more work than is presented in the advertising of the science.

Hal may require graduate-level knowledge of abstract algebra to reproduce this proof, but as a professional mathematician, this is not a great leap. More challenging the proof may be, but the process of reproduction would be similar. Even if this were the proof, Hal would not know if Catherine merely reproduced, as he did, her father's proof.

But what if Catherine were posing her mathematical question computationally? Would Hal be able to reproduce her results?

2.3   Is Computational Mathematics Mired in Proof Methodology?

When we are exploring and answering mathematical questions in a computational environment, we consider some aspects of our work to be **formal** and some *informal*. But in omitting the greyed *informal* work in Fig. 1, are we still approaching compendia of research from the perspective of a blackboard mathematician?

Given we use statistics in most science, arguably most scientific questions are posed, to some extent, mathematically. The output format, a published paper, remains similar to mathematics of the pre-computer age. But the informal work of answering mathematical questions has changed significantly. Now that much work is done computationally, there are multiple research outputs that comprise the compendium of science that produces the published paper.

Let us now suppose that Catherine had a statistical estimator for a population parameter of interest. That is, Catherine has an equation that, given some data, she can approximate some value about the population, such as an overall average. Let us further suppose, as is increasingly common, that she does not have a closed-form solution, meaning she cannot write out a mathematical argument in the traditional sense. Instead, she demonstrates the estimator's performance through simulation studies.

Now suppose Hal challenges Catherine to prove that she created the science that produced the paper. Given what is on the piece of paper, how can Hal know that Catherine's code does what she said it does? It is unclear what assumptions were made, about, say, sample size and distribution. How can Hal verify her results? Through adopting research software engineering principles, Catherine can facilitate a process akin to proofs and refutations, the redefinition described in the Sect. 2.2, The combat conditions of new mathematics. The process of redefinition is transcribed by version control, but further to this, the software itself provides a modular framework, such as a theorem in mathematics, for future work to scaffold and extend. New software can be developed that either extends, or redefines the existing software. One analogous way this is occurring is in the rise of metapackages, such as `tidyverse::` [42] and `metaverse::` [39], that collect software to solve particular problems in an opinionated [29] manner, that guide the end-user to what the creators consider to be good enough practice. This is analogous to classes of mathematics, such as group theory or analysis, that collect results, theorems, that rely upon each other, and where certain underlying assumptions, such as the *Axiom of Choice*[6], are made. Indeed, as Martin-Löf proposed a shift in terminology from computer science to computing science, they make the following remark.

It has made programming an activity akin in rigour and beauty to that of proving mathematical theorems [23].

How are contemporary researchers answering mathematical questions? Alex Hayes, current maintainer and one of the many authors of `broom::` [32], an open source R package that amalgamates hundreds of contributions towards providing a suite of tools that *tidily*[7] [15] extract statistical model information from R

---

[6] Turning to the bible of algebra, *Lattices and Order* [8], we learn the *Axiom of Choice* 'asserts that it is possible to find a map which picks one element from each member of a family of non-empty sets'.

[7] From Wickham's *Tidy data* [15], we describe data as *tidy* if

1. Each variable forms a column.
2. Each observation forms a row.
3. Each type of observational unit forms a table.

algorithms, recently noted the underdeveloped nature of the implementation of statistical algorithms [16]:

> In practice, most people end up writing a reference implementation and checking that the reference implementation closely matches the pseudocode of their algorithm. Then they declare this implementation correct. How trustworthy this approach is depends on the clarity of the connection between the algorithm pseudocode and the reference implementation.

This is not to carp upon diligent scientists; we need to do far more to support the software engineering principles we expect from those who answer mathematical questions computationally [28]. Mathematicians are trained to provide enough work such that the hidden steps illustrated in italics in Table 1 can be reproduced by their target audience. The detail of mathematical work shown is tempered for level of the audience, but the same process described in bold in Table 1 is the same. But, does the workflow Alex describes above equip the target audience with enough information such that they can understand all the details of the entire argument put forward?

Code has the appearance of being highly logical, it's easy to assume it's infallible; and whilst the logic of the code is robust, the pipeline that carries the algorithm to implementation may be susceptible to compromising factors, with typos being just one example of inadvertent error.

Because code appears so logical, we assume it is analogous to proof for our intended audience to follow. But we were trained to leave out the informal messy thinking work associated with mathematics; trusting the formal argument provides enough information to verify and reproduce the mathematics. Does our code do what we think it does? In addition to providing the research outputs in the spectrum of reproducibility, Fig. 1, we posit mathematical science should adopt the software development practice of unit testing, to ensure the mathematical results can be verified and reproduced.

3   Testing

Testing is the software engineering tool that is provides a key piece of the correspondence between scientific claim and programming. Just as the Curry-Howard isomorphism expresses proofs-as-programs to link mathematics and programming, we argue that tests are the link between scientific claims more generally and programming. In a test the researcher isolates a scientifically meaningful part of their code, and creates a witness so that others can easily see that the code does what the researcher intends it to do. In this section we consider a 'vital' [40] research output, testing, that it is unlikely the mathematical scientist has been trained in. There are many such under-formalised skills represented in Fig. 1[8]. In 2016, a quarter of packages on R package archives CRAN, Biocon-

---

[8] Indeed, the natural consequence of questioning how we practice mathematical science is how we train the next generation of practitioners. Important, however this may be, this is beyond the scope of this manuscript.

ductor, and rOpenSci, included tests, a repository by repository breakdown of
this is shown in Table 2.

Table 2. Percentage of R packages in repositories that have unit tests included. These
results are from Jim Hester's presentation on covr:: in September 2016 [18].

Repository	Tests	Total	
CRAN	2091	9772	21%
Bioconductor	449	1258	36%
rOpenSci	84	146	58%
	2624	11,176	24%

Now, Hayes advises people against using untested software [16]. It is alarming
that, by this logic, we would be **insane** to use *three quarters* of packages available. But Hayes continues, 'You have two jobs. The first job is to write correct
code. The second job is to convince users that you have written correct code' [16].
The disconnect here suggests a failure to communicate broadly the importance of
testing of algorithms in the dissemination of research. As researchers, we believe
our science is as reproducible as a traditional mathematical proof; however, the
growing literature of the replication crisis demonstrates we have not succeeded
in rendering our science reproducible.

rOpenSci's review system recommends using the `covr::` [19] package to measure how the code behaves with different expected outputs. From the creator of
`covr::`, we obtain the following definition of test coverage.

Test coverage is the proportion of the source code that is executed when
running these tests [19].

3.1   What Is a Test?

Tests demonstrations that a given input produces an expected output. They are
grouped contextually in a file; the context being a certain aspect of the algorithm
that should be tested [40]. An example of a context for a test is the question, does
a given function return the expected result for different inputs? Each test comprises a collection of expectations. Each expectation runs a function or functions
from the package, and checks the returned output is as expected. In this case, we
have a test for the `expect_equal` function: one expectation checks the function
successfully runs when given equal inputs, and another expectation checks that
the function fails when passed two non-equal inputs.

An example test from the `testthat::` [41] contains two expectations.

```
test_that("basically principles of equality hold", {
  expect_success(expect_equal(1, 1))
  expect_failure(expect_equal(1, 2))
})
```

3.2 How Good Are We at *good Enough* testing?

A response to the replication crisis has been to examine *questionable research practices* [13], frequently borne of tradition and convention within different disciplines, deviate from evidence-based best-practice research methodology. We suggest it is a questionable research practice to draw conclusions about the efficacy of statistical estimators from untested code.

Given only a quarter of R packages have unit tests associated with them, we are falling short of best practice in scientific computing [43]. In a recent assessment of what constitutes *good enough* practice in scientific computing [44], unit testing was not included. However, for mathematical science, where the algorithms implemented and the code written is often complex, we suggest that unit testing should be considered good enough practice, in spite of the additional learning curve. With the backdrop of the replication crisis, it is crucial we have confidence in the algorithms we implement.

3.3 Analysis of Testing Code in R Packages

So, what packages have tests? We provide a preliminary analysis of tests in CRAN packages in Fig. 3. The code and data used to generate the results presented here are openly available at https://github.com/softloud/proof.

We provide analysis for packages associated with CRAN task view [46], opinionated [29] collections of R packages that are relevant to a particular type of statistical analysis, maintained voluntarily by experts in their respective fields [46]. CRAN task views provide a convenient taxonomy of R packages for a preliminary exploratory analysis of patterns of test use among R package authors.

Packages listed in a task view are may be interpreted by users as more stable and trustworthy than other packages, because they have passed some kind of inspection by maintainer of the task view who listed the package (however the review and curation process is not open or documented). And yet, even amongst the 4105 packages associated with task views, 1524 packages were without tests; 37 per cent of packages associated with CRAN task view were without tests.

The proportion of task view packages with tests has fallen over the last decade. This does not seem surprising given the uptake of R amongst communities of researchers in applied sciences with little formal programming and computer science training, such as psychology and ecology.

Figure 4 shows that there is wide variation in test coverage. Even the largest and fastest growing CRAN task views have very different proportions of packages with tests (Survival, about 0.23, compared to Web Technologies about 0.66). We find few clear patterns in the presence of tests over time, between different CRAN task views, and with metadata such as the number of authors, the size of the package and the centrality of the package (as measured by the union of the number of reverse dependencies and reverse imports). Based on these data, we suggest there is much work to be done in developing methods and opinionated tools that guide users towards good enough practices.

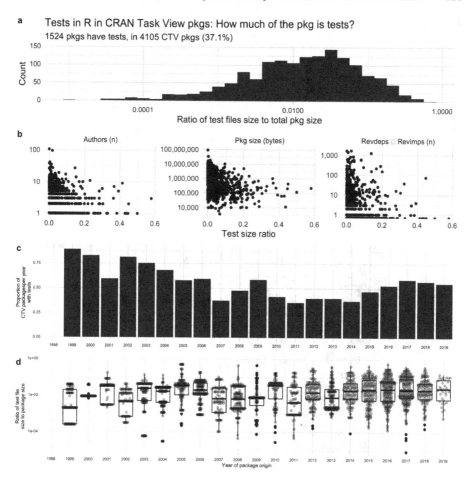

Fig. 3. This panel shows some basic details of tests in R packages listed in CRAN task views [46]. The measure of interest, *test size ratio*, was calculated by dividing the test file size with the overall package source file size from the unofficial CRAN mirror on GitHub. This is a rough indicator of test coverage, future work should consider more precise metrics such as those produced by the covr:: package. (a) the distribution of the ratio of test file size to total package size, test size ratio. (b) scatter plots demonstrate the relationship between test size ratio and number of authors, overall package size, and number of packages imported and calling the package, respectively. (c) the proportion of all task view packages that contain tests over time. (d) boxplot detailing the distribution of file size ratio over time.

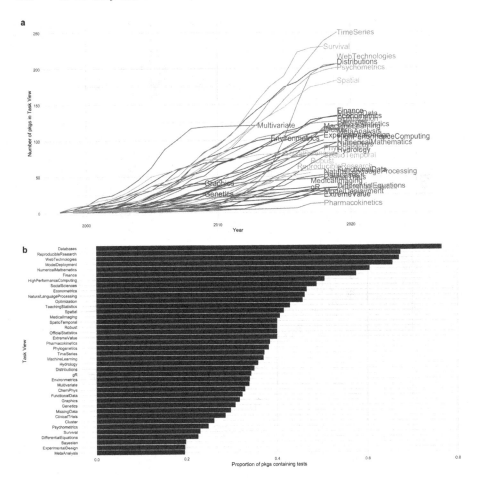

Fig. 4. (a) shows the change in the number of packages in each CRAN task view over time. (b) shows the proportion of packages in each CRAN task view that have tests.

4   Tempered Uncertainty and Computational Proof

It's easy to lie with statistics, but it's even easier without them [27]. In a computational experimental setting, we often cannot achieve the satisfying precision offered by a proof. We can, however, adopt good enough practices in sharing and testing code to increase confidence in our scientific conclusions. Given the prevalence of generalised linear models, we can think of the practice of much science as the interpretation of

$$y \approx bx,$$

where: x represents what we know about the data; y, the observed response of interest that we wish to investigate how it responds to x; and b, the *how* it responds, approximated unknown. It may not be possible to provide the rigour of a closed-form mathematical solution, but we can aim to temper the uncertainty, and bolster confidence, in computational arguments via automated testing, version control, and other computational outputs.

We suggest there is much work to be done in developing good enough practices [44] we can ask mathematical scientists to adopt. For example, we do not have a chance to discuss in this manuscript the role of markdown and html reporting in reproducible science. Indeed, the question of good enough practice can be posed for each research output. Less than offering answers, this manuscript seeks more to suggest there is a rich line of inquiry [28] in the relationship between scientific truth, mathematical proof, and computational reproducibility and rigour.

4.1   Coda

Returning to Catherine and Hal from Auburn's *Proof* [2], we can now imagine her as computational mathematician who provides a compendium of reproducible research. To demonstrate the rigour of her computational work, she would provide unit tests for the algorithms she had implemented. Catherine would share her work openly via her GitHub or similar repository, where the development of her ideas would be timestamped and recorded. The structure of her research compendium of would be automatically standardised via a tool such as `rrtools::` [24]. At publication, her compendium would be deposited on a trustworthy, DOI-issuing repository for others to link to and cite.

And she would feel safe asking questions about good enough practice [44], and how to avoid questionable research practices [13], because there is an understanding in the community that no one is trained in all these things, so we are all always learning.

There would be no struggle, as there was in Auburn's play, to show that the mathematician who created these research outputs was Catherine. But that wouldn't matter - she and Hal would be having far too much fun collaborating on the next question.

References

1. Amrhein, V., Greenland, S., McShane, B.: Scientists rise up against statistical significance. Nature **567**(7748), 305 (2019). https://doi.org/10.1038/d41586-019-00857-9
2. Auburn, D.: Proof: A Play. Farrar, Straus and Giroux, New York (2001). Google-Books-ID: 6AUtQVhrY90C
3. Bertot, Y.: A short presentation of Coq. In: Mohamed, O.A., Muñoz, C., Tahar, S. (eds.) TPHOLs 2008. LNCS, vol. 5170, pp. 12–16. Springer, Heidelberg (2008). https://doi.org/10.1007/978-3-540-71067-7_3
4. Brown, S.: Partial unpacking and indirect proofs: a study of students' productive use of the symbolic proof scheme. In: Proceedings of the 16th Annual Conference on Research in Undergraduate Mathematics Education, vol. 2, pp. 47–54 (2013)
5. Bryan, J.: Excuse me, do you have a moment to talk about version control? Am. Stat. **72**(1), 20–27 (2018). https://doi.org/10.1080/00031305.2017.1399928

6. Camerer, C.F., et al.: Evaluating replicability of laboratory experiments in economics. Science **351**(6280), 1433–1436 (2016). https://doi.org/10.1126/science.aaf0918

7. Carroll, L.: The Annotated Alice: The, Definitive Edition, updated, subsequent edn. W. W. Norton & Company, New York (1999)

8. Davey, B.A., Priestley, H.A.: Introduction to Lattices and Order. Cambridge University Press, Cambridge (2002). Google-Books-ID: vVVTxeuiyvQC

9. Davey, B.A.: When is a Proof?, 2nd edn. La Trobe University, Bundoora (2009)

10. Davey, B.A., Gray, C.T., Pitkethly, J.G.: The homomorphism lattice induced by a finite algebra. Order **35**(2), 193–214 (2018). https://doi.org/10.1007/s11083-017-9426-3

11. Donoho, D.L.: An invitation to reproducible computational research. Biostatistics **11**(3), 385–388 (2010). https://doi.org/10.1093/biostatistics/kxq028

12. Fidler, F., Wilcox, J.: Reproducibility of scientific results. In: Zalta, E.N. (ed.) The Stanford Encyclopedia of Philosophy. Metaphysics Research Lab, Stanford University, winter 2018 edn. (2018)

13. Fraser, H., Parker, T., Nakagawa, S., Barnett, A., Fidler, F.: Questionable research practices in ecology and evolution. PLOS One **13**(7), e0200303 (2018). https://doi.org/10.1371/journal.pone.0200303

14. Haack, S.: Defending Science - within Reason: Between Scientism and Cynicism. Prometheus Books, Buffalo (2011). Google-Books-ID: RhXxaPTc_EYC

15. Wickham, H.: Tidy data. J. Stat. Softw. **59**(1), 1–23 (2014). https://doi.org/10.18637/jss.v059.i10

16. Hayes, A.: testing statistical software - aleatoric, July 2019. https://www.alexpghayes.com/blog/testing-statistical-software/

17. Head, M.L., Holman, L., Lanfear, R., Kahn, A.T., Jennions, M.D.: The extent and consequences of p-hacking in science. PLOS Biol. **13**(3), e1002106 (2015). https://doi.org/10.1371/journal.pbio.1002106

18. Hester, J.: covr: Bringing test coverage to R, January 2016. https://www.rstudio.com/resources/webinars/covr-bringing-test-coverage-to-r/

19. Hester, J.: covr: Test Coverage for Packages (2018). https://CRAN.R-project.org/package=covr

20. Katz, D.S., McHenry, K.: Super RSEs: Combining research and service in three dimensions of Research Software Engineering, July 2019. https://danielskatzblog.wordpress.com/2019/07/12/

21. Lakatos, I.: Proofs and Refutations: The Logic of Mathematical Discovery, reissue edn. Cambridge University Press, Cambridge (2015)

22. LeVeque, R.J., Mitchell, I.M., Stodden, V.: Reproducible research for scientific computing: tools and strategies for changing the culture. Comput. Sci. Eng. **14** (2012). https://doi.org/10.1109/mcse.2012.38

23. Martin-Löf, P.: Constructive mathematics and computer programming. In: Cohen, L.J., Łoś, J., Pfeiffer, H., Podewski, K.P. (eds.) Studies in Logic and the Foundations of Mathematics, Logic, Methodology and Philosophy of Science VI, vol. 104, pp. 153–175. Elsevier (1982). https://doi.org/10.1016/S0049-237X(09)70189-2

24. Marwick, B.: rrtools: Creates a reproducible research compendium (2018). https://github.com/benmarwick/rrtools

25. Marwick, B., Boettiger, C., Mullen, L.: Packaging data analytical work reproducibly using R (and friends). Technical report e3192v2, PeerJ Inc., March 2018. https://doi.org/10.7287/peerj.preprints.3192v2, https://peerj.com/preprints/3192

26. Merton, R.K.: On Social Structure and Science. University of Chicago Press, Chicago (1996). Google-Books-ID: j94XiVDwAZEC

27. Murray, C.: How to accuse the other guy of lying with statistics. Stat. Sci. **20**(3), 239–241 (2005). https://www.jstor.org/stable/20061179
28. Nowogrodzki, A.: How to support open-source software and stay sane. Nature **571**, 133 (2019). https://doi.org/10.1038/d41586-019-02046-0
29. Parker, H.: Opinionated analysis development. preprint (2017). https://doi.org/10.7287/peerj.preprints.3210v1
30. Peng, R.D.: Reproducible research in computational science. Science **334**(6060), 1226–1227 (2011). https://doi.org/10.1126/science.1213847
31. Pickering, A.: The Mangle of Practice: Time, Agency, and Science. University of Chicago Press, Chicago (2010)
32. Robinson, D., Hayes, A.: broom: Convert Statistical Analysis Objects into Tidy Tibbles (2019). https://CRAN.R-project.org/package=broom
33. Rodriguez-Sanchez, F., Pérez-Luque, A.J., Bartomeus, I., Varela, S.: Ciencia reproducible: qué, por qué, cómo. Revista Ecosistemas **25**(2), 83–92-92 (2016). https://doi.org/10.7818/re.2014.25-2.00, https://www.revistaecosistemas.net/index.php/ecosistemas/article/view/1178
34. Shapin, S., Schaffer, S.: Leviathan and the Air-Pump: Hobbes, Boyle, and the Experimental Life (New in paper), vol. 32. Princeton University Press, Princeton (2011)
35. Stodden, V.: What scientific idea is ready for retirement? (2014). https://www.edge.org/response-detail/25340.%202014
36. Stodden, V., Borwein, J., Bailey, D.H.: "Setting the default to reproducible" in computational science research. SIAM News **46**(5), 4–6 (2013)
37. Sørensen, M.H., Urzyczyn, P.: Lectures on the Curry-Howard Isomorphism, vol. 149. Elsevier, Amsterdam (2006)
38. Wallach, J.D., Boyack, K.W., Ioannidis, J.P.A.: Reproducible research practices, transparency, and open access data in the biomedical literature, 2015–2017. PLOS Biol. **16**(11), e2006930 (2018). https://doi.org/10.1371/journal.pbio.2006930
39. Westgate, M., et al.: metaverse: Workflows for evidence synthesis projects (2019). https://github.com/rmetaverse/metaverse, r package version 0.0.1
40. Wickham, H.: R Packages: Organize, Test, Document, and Share Your Code. O'Reilly Media, Sebastopol (2015). https://books.google.com.au/books?id=DqSxBwAAQBAJ
41. Wickham, H.: testthat: Get Started with Testing (2011)
42. Wickham, H.: tidyverse: Easily Install and Load the 'Tidyverse' (2017). https://CRAN.R-project.org/package=tidyverse
43. Wilson, G., et al.: Best practices for scientific computing. PLoS Biol. **12**(1), e1001745 (2014). https://doi.org/10.1371/journal.pbio.1001745
44. Wilson, G., Bryan, J., Cranston, K., Kitzes, J., Nederbragt, L., Teal, T.K.: Good enough practices in scientific computing. PLOS Comput. Biol. **13**(6), e1005510 (2017). https://doi.org/10.1371/journal.pcbi.1005510
45. Wyatt, C.: Research Software Engineers Association (2019). https://rse.ac.uk/
46. Zeileis, A.: CRAN task views. R News **5**(1), 39–40 (2005). https://CRAN.R-project.org/doc/Rnews/

On Adaptive Gauss-Hermite Quadrature for Estimation in GLMM's

Paul Kabaila$^{(\boxtimes)}$(iD) and Nishika Ranathunga

Department of Mathematics and Statistics, La Trobe University,
Melbourne, Australia
{P.Kabaila,n.kapuruge}@latrobe.edu.au

Abstract. Adaptive Gauss-Hermite quadrature is used for the computation of the log-likelihood function for generalized linear mixed models. The basic first step in this method is to multiply and divide the integrand of interest by a carefully chosen probability density function. The same first step is used for the computation of this log-likelihood function using simulations that employ importance sampling. We compare these two methods by considering in detail a single cluster from a well-known teratology data set that is modelled using a logistic regression with random intercept. We show that while importance sampling fails for this computation, adaptive Gauss-Hermite quadrature does not. We derive a new upper bound on the error of approximation of adaptive Gauss-Hermite quadrature. Using this new upper bound, we show that the feature of this problem that makes importance sampling fail is useful in disclosing why adaptive Gauss-Hermite quadrature succeeds.

Keywords: Adaptive Gauss-Hermite quadrature · Generalized linear mixed models · Importance sampling · Maximum likelihood estimation

1 Introduction

For Gauss-Hermite quadrature, Liu and Pierce [1] present a method of transforming the variable of integration so that the integrand is sampled at relatively important values. This method has found application in the computation of the log-likelihood function for generalized linear mixed models (GLMM's), see e.g. [2–10]. This computation is needed for the evaluation of the maximum likelihood estimate and the observed information matrix. Pinheiro and Bates [2] refer to the method put forward by Liu and Pierce [1] as 'adaptive' Gauss-Hermite quadrature. The basic first step in this method is to multiply and divide the integrand of interest by a carefully chosen probability density function (pdf). The same first step is used for the computation of this log-likelihood function using simulations that employ importance sampling. As is well-known, see e.g. p. 102 of [11] and [12], importance sampling needs to be applied with extreme care to be successful. This raises the question:

© Springer Nature Singapore Pte Ltd. 2019
H. Nguyen (Ed.): RSSDS 2019, CCIS 1150, pp. 130–139, 2019.
https://doi.org/10.1007/978-981-15-1960-4_9

How similar are adaptive Gauss-Hermite quadrature and importance sampling in the context of the computation of the log-likelihood function for GLMM's?

If the answer to this question is that they are very similar then we would expect adaptive Gauss-Hermite quadrature to display the same potential fragility as importance sampling.

To answer this question, we compare the properties of importance sampling and adaptive Gauss-Hermite quadrature in the particular context of the computation of the log-likelihood function for a logistic regression with random intercept. In Sect. 2 we describe this model, the log-likelihood function for this model and adaptive Gauss-Hermite quadrature in this context. In Sect. 3, we introduce the teratology data set of Weil [13]. We also show that importance sampling fails spectacularly, in the sense of leading to an estimator with infinite variance, for the computation of the log-likelihood function. In Sect. 4 we show that, by contrast, adaptive Gauss-Hermite quadrature works well for this data set. In that section we also derive a new upper bound on the error of approximation of adaptive Gauss-Hermite quadrature (Theorem 1). Using this upper bound, we provide a very detailed explanation for why adaptive Gauss-Hermite quadrature works well in the context of cluster number 29. This explanation relies on the fact that a particular function increases faster than any polynomial as its argument diverges to ∞ or $-\infty$. Interestingly, this property is precisely what causes the importance sampling method to fail spectacularly (see Remark 1 for details).

2   The Logistic Regression with Random Intercept Model, Its Log-Likelihood and Adaptive Gauss-Hermite Quadrature

We consider a logistic regression model with random intercept. Let y_i and x_i denote the response and covariate, respectively, for the i'th cluster ($i = 1, \ldots, N$). Let $\boldsymbol{\eta} = (\eta_1, \ldots, \eta_N)$, where the η_i's are independent and identically $N(0, \sigma^2)$ distributed. Also let $\boldsymbol{z} = (z_1, \ldots, z_N)$. Suppose that, conditional on $\boldsymbol{\eta} = \boldsymbol{z}$, the y_i's are independent and $y_i \sim \text{Binomial}(J_i, \pi_i)$, where J_i denotes the size of the i'th cluster and

$$\log\left(\frac{\pi_i}{1 - \pi_i}\right) = \beta_1 + \beta_2\, x_i + z_i$$

for $i = 1, \ldots, N$. Let $\widehat{\beta}_1$, $\widehat{\beta}_2$ and $\widehat{\sigma}$ denote the maximum likelihood estimates obtained from all of the data.

Let $\phi(t; \mu, \sigma^2)$ denote the $N(\mu, \sigma^2)$ pdf, evaluated at t. The additive contribution of a given cluster of size J to the log-likelihood function is, to within an additive constant, the logarithm of

$$\int_{-\infty}^{\infty} \frac{\exp\left[(\beta_1 + \beta_2\, x + t)y\right]}{[1 + \exp(\beta_1 + \beta_2\, x + t)]^J}\, \phi(t; 0, \sigma^2)\, dt, \tag{1}$$

where y and x denote the observed response and covariate, respectively, for this cluster. Obviously, (1) is equal to $c(\theta, \sigma)$ which we define to be

$$\int_{-\infty}^{\infty} g(t; \theta, \sigma) \, \phi(t; 0, \sigma^2) \, dt, \tag{2}$$

where

$$g(t; \theta, \sigma) = \frac{\exp\left[(\theta + t)y\right]}{[1 + \exp(\theta + t)]^J},$$

with $\theta = \beta_1 + \beta_2 \, x$.

We now describe the adaptive Gauss-Hermite quadrature method for the computation of $c(\theta, \sigma)$. Let

$\mu = $ posterior mode of $g(t; \theta, \sigma) \, \phi(t; 0, \sigma^2)$, considered as a function of t

$$\tau = \left[-\frac{\partial^2}{\partial t^2} \left(\log \left[g(t; \theta, \sigma) \, \phi(t; 0, \sigma^2) \right] \right) \right]^{-1}.$$

This method is based on employing $\phi(t; \mu, \tau^2)$ as an approximation to $g(t; \theta, \sigma) \, \phi(t; 0, \sigma^2)$. The first basic step in the description of adaptive Gauss-Hermite quadrature is to write

$$c(\theta, \sigma) = \int_{-\infty}^{\infty} g(t; \theta, \sigma) \, \phi(t; 0, \sigma^2) \, dt = \int_{-\infty}^{\infty} \frac{g(t; \theta, \sigma) \, \phi(t; 0, \sigma^2)}{\phi(t; \mu, \tau^2)} \, \phi(t; \mu, \tau^2) \, dt. \tag{3}$$

This step of multiplying and dividing the integrand by a pdf is common to both importance sampling and adaptive Gauss-Hermite quadrature. We re-express (3) as

$$c(\theta, \sigma) = \int_{-\infty}^{\infty} h(t; \theta, \sigma) \, \phi(t; \mu, \tau^2) \, dt, \tag{4}$$

where

$$h(t; \theta, \sigma) = \frac{g(t; \theta, \sigma) \, \phi(t; 0, \sigma^2)}{\phi(t; \mu, \tau^2)}.$$

We now change the variable of integration in (4) to $z = (t - \mu)/(\sqrt{2}\tau)$ to obtain

$$c(\theta, \sigma) = \frac{1}{\sqrt{\pi}} \int_{-\infty}^{\infty} h(\mu + \sqrt{2}\tau z; \theta, \sigma) \, \exp(-z^2) \, dz. \tag{5}$$

Let

$$\sum_{i=1}^{m} f(z_i) \, w_i$$

be the m-node Gauss-Hermite quadrature approximation to

$$\int_{-\infty}^{\infty} f(z) \, \exp(-z^2) \, dz.$$

The m-node adaptive Gauss-Hermite quadrature approximation to $c(\theta, \sigma)$ is

$$c_m(\theta, \sigma) = \frac{1}{\sqrt{\pi}} \sum_{i=1}^{m} h(\mu + \sqrt{2}\tau z_i; \theta, \sigma) \, w_i. \tag{6}$$

3   The Teratology Data and Importance Sampling

We will compare the adaptive Gauss-Hermite quadrature approximation (6) with importance sampling, using the importance pdf $\phi(t; \mu, \tau^2)$ in (4), in the particular context of the teratology data described by Weil [13]. This dataset lists the number of rat pups in 16 control litters that survived and the number of rat pups in 16 treated litters that survived. Each litter is treated as a cluster, so that the total number of clusters $N = 32$. The covariate x_i takes the value 1 for a litter i that is treated and the value 0 for a litter i that is a control. For this dataset, J_i and y_i denote the number of pups and the number of surviving pups, respectively, in litter i. This data is shown in Table 1. The litters are numbered from 1 up to 32 with the litter in row j and column k allocated the number $i = 8(j - 1) + k$.

Table 1. Teratology data set of [13]. This data lists the number of rat pups in 16 control litters that survived and the number of rat pups in 16 treated litters that survived.

	(Number survived, number dead)							
Control	(13, 0)	(12, 0)	(9, 0)	(9, 0)	(8, 0)	(8, 0)	(12, 1)	(11, 1)
	(9, 1)	(9, 1)	(8, 1)	(11, 2)	(4, 1)	(5, 2)	(7, 3)	(7, 3)
Treatment	(12, 0)	(11, 0)	(10, 0)	(9, 0)	(10, 1)	(9, 1)	(9, 1)	(8, 1)
	(8, 1)	(4, 1)	(7, 2)	(4, 3)	(5, 5)	(3, 3)	(3, 7)	(0, 7)

We now describe a simulation method that employs importance sampling for the computation of (6), using the importance pdf $\phi(t; \mu, \tau^2)$ in (4). We suppose that this simulation consists of M independent simulation runs. Let v_i denote the observation obtained in the i'th simulation run of a random variable with pdf $\phi(t; \mu, \tau^2)$. The importance sampling estimator of $c(\theta, \sigma)$ is

$$\widetilde{c}_M(\theta, \sigma) = \frac{1}{M} \sum_{i=1}^{M} h(v_i; \theta, \sigma).$$

This is an unbiased estimator of $c(\theta, \sigma)$ and its variance is $\widetilde{\sigma}^2/M$, where

$$\widetilde{\sigma}^2 = \int_{-\infty}^{\infty} \frac{[g(t; \theta, \sigma)\, \phi(t; 0, \sigma^2)]^2}{\phi(t; \mu, \tau^2)}\, dt - c^2(\theta, \sigma),$$

see e.g. [12]. Hence

$$\widetilde{\sigma}^2 = \int_{-\infty}^{\infty} r^2(t; \theta, \sigma)\, dt - c^2(\theta, \sigma),$$

where

$$r(t; \theta, \sigma) = h(t; \theta, \sigma) \left[\phi(t; \mu, \tau^2) \right]^{1/2}. \tag{7}$$

It may be shown that

$$r(t;\theta,\sigma) = c_0 \frac{\exp\left(c_1 + c_2 t + c_3\, t^2\right)}{\left[1 + \exp(\theta + t)\right]^J},$$

where $c_0 = (\tau/\sqrt{2\pi}\sigma^2)^{1/2}, c_1 = \theta y + \mu^2/(4\tau^2), c_2 = y - \mu/(2\tau^2), c_3 = 1/(4\tau^2) - 1/(2\sigma^2)$. It follows from the definition of τ that $c_3 > -1/(4\sigma^2)$. If $c_3 > 0$ then $r(t;\theta,\sigma) \to \infty$ as $t \to \infty$ and as $t \to -\infty$. Thus, if $c_3 > 0$ then $\tilde{\sigma}^2 = \infty$ and importance sampling fails spectacularly. If, however, $c_3 < 0$ then (a) $r(t;\theta,\sigma) \to 0$ as $t \to \infty$ and as $t \to -\infty$ and (b) $\tilde{\sigma}^2$ is finite. Of course, even if $c_3 < 0$, importance sampling may still fail to improve on simple Monte Carlo simulation.

   Consider the particular values $\beta_1 = 2.6$, $\beta_2 = -1.1$ and $\sigma = 1.3$. These are examples of values that might be encountered during the computation of the maximum likelihood estimates, which are $\hat{\beta}_1 = 2.625651357$, $\hat{\beta}_2 = -1.082405923$ and $\hat{\sigma} = 1.345703142$. For the particular values of β_1, β_2 and σ that we have chosen, $\theta = 2.6$ for each litter in the control group and $\theta = 1.5$ for each litter in the treatment group. For litters 7–12, 14–19 and 21–32, we find that $c_3 > 0$, so that $\tilde{\sigma}^2 = \infty$ and importance sampling fails spectacularly for these litter numbers. Figure 1 presents graphs of $\log r(t;\theta,\sigma)$, considered as a function of t, for control litters 12 and 15 and treatment litters 29 and 32. These graphs confirm that $r(t;\theta,\sigma) \to \infty$ as $t \to \infty$ and as $t \to -\infty$. In other words, these graphs confirm that $\tilde{\sigma}^2 = \infty$ for these litters.

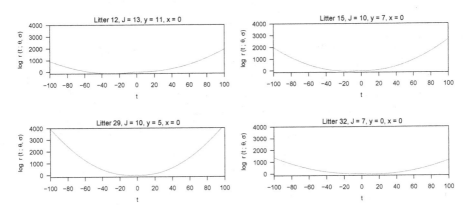

Fig. 1. Graphs of $\log r(t;\theta,\sigma)$ as a function of t for $\sigma = 1.3$, $\theta = 2.6$ for a control litter and $\theta = 1.5$ for a treatment litter. The top two graphs are for control litters 12 and 15. The bottom two graphs are for treatment litters 29 and 32.

4 The Performance of Adaptive Gauss-Hermite Quadrature for Cluster 29 of the Teratology Data

The following theorem, which uses a well-known type of argument for bounding the error of Gauss-Legendre quadrature, see e.g. Atkinson [14] (pp. 277–278), describes a new upper bound on $\left|c(\theta,\sigma) - c_m(\theta,\sigma)\right|$.

Theorem 1. *Suppose that m, z_ℓ and z_u, where $z_\ell \leq z_1$ and $z_u \geq z_m$, are given. Also suppose that J, y, θ and σ are given. For notational convenience let $k(z) = h\left(\mu + \sqrt{2}\tau z; \theta, \sigma\right)$. Define $q_m(z)$ to be the polynomial $p_m(z)$ of degree $2m-1$ that minimizes (either exactly or numerically)*

$$\max_{z \in [z_\ell, z_u]} \left|k(z) - p_m(z)\right|.$$

Let e_m denote this minimized value. Then

$$\left|c(\theta,\sigma) - c_m(\theta,\sigma)\right| \leq 2\,e_m + |a_\ell| + |a_u|, \tag{8}$$

where

$$a_\ell = \frac{1}{\sqrt{\pi}} \int_{-\infty}^{z_\ell} (k(z) - q_m(z))\, \exp(-z^2)\, dz \quad \text{and} \quad a_u = \frac{1}{\sqrt{\pi}} \int_{z_u}^{\infty} (k(z) - q_m(z))\, \exp(-z^2)\, dz.$$

Furthermore:

If $k(z) \geq q_m(z) \geq 0$ for all $z \leq z_\ell$ then $0 \leq a_\ell \leq \displaystyle\int_{-\infty}^{\mu + \sqrt{2}\tau z_\ell} g(t; \theta, \sigma)\, \phi(t; 0, \sigma^2)\, dt.$ (9)

and

If $k(z) \geq q_m(z) \geq 0$ for all $z \geq z_u$ then $0 \leq a_u \leq \displaystyle\int_{\mu + \sqrt{2}\tau z_u}^{\infty} g(t; \theta, \sigma)\, \phi(t; 0, \sigma^2)\, dt.$ (10)

Proof. It follows from

$$\int_{-\infty}^{\infty} q_m(z)\, \exp(-z^2)\, dz = \sum_{i=1}^{m} q_m(z_i)\, w_i$$

that

$$c(\theta,\sigma) - c_m(\theta,\sigma) = \frac{1}{\sqrt{\pi}} \int_{-\infty}^{\infty} (k(z) - q_m(z))\, \exp(-z^2)\, dz + \frac{1}{\sqrt{\pi}} \sum_{i=1}^{m} (q_m(z_i) - k(z_i))\, w_i$$

$$= a_\ell + a_u + \frac{1}{\sqrt{\pi}} \int_{z_\ell}^{z_u} (k(z) - q_m(z))\, \exp(-z^2)\, dz$$

$$+ \frac{1}{\sqrt{\pi}} \sum_{i=1}^{m} (q_m(z_i) - k(z_i))\, w_i.$$

Now

$$\left| \frac{1}{\sqrt{\pi}} \int_{z_\ell}^{z_u} (k(z) - q_m(z)) \exp(-z^2) \, dz \right| \leq \frac{1}{\sqrt{\pi}} \int_{z_\ell}^{z_u} |k(z) - q_m(z)| \, \exp(-z^2) \, dz$$

$$\leq e_m \frac{1}{\sqrt{\pi}} \int_{-\infty}^{\infty} \exp(-z^2) \, dz = e_m$$

and

$$\left| \frac{1}{\sqrt{\pi}} \sum_{i=1}^{m} \{q_m(z_i) - k(z_i)\} w_i \right| \leq \frac{1}{\sqrt{\pi}} \sum_{i=1}^{m} |q_m(z_i) - k(z_i)| \, w_i \leq e_m \frac{1}{\sqrt{\pi}} \sum_{i=1}^{m} w_i = e_m.$$

The proofs of (9) and (10) are very similar. For the sake of brevity, we present only the proof of (9). If $k(z) \geq q_m(z)$ for all $z \leq z_\ell$ then

$$0 \leq \frac{1}{\sqrt{\pi}} \int_{-\infty}^{z_\ell} (k(z) - q_m(z)) \exp(-z^2) \, dz \leq \frac{1}{\sqrt{\pi}} \int_{-\infty}^{z_\ell} k(z) \exp(-z^2) \, dz$$

$$\leq \int_{-\infty}^{\mu + \sqrt{2}\tau z_\ell} g(t; \theta, \sigma) \, \phi(t; 0, \sigma^2) \, dt,$$

by changing the variable of integration from z to $t = \mu + \sqrt{2}\,\tau\,z$.

This theorem shows that if $k(z) \geq q_m(z)$ for all $z \leq z_\ell = z_1$ and for all $z \geq z_u = z_m$ for every m in an increasing sequence of values of m then $|a_\ell| + |a_u| \to 0$ as m increases through these values, since $z_m \to \infty$ as $m \to \infty$ (see e.g. Szegö [15], p. 130).

Figure 2 presents a graph of $k(z) - q_m(z) = h(\mu + \sqrt{2}\tau z; \theta, \sigma) - q_m(z)$, considered as a function of z, for litter 29 for adaptive Gauss-Hermite quadrature with $m = 5$ nodes. The smallest and largest nodes z_1 and z_m, respectively, are shown. This graph shows that, in this case, $k(z) \geq q_m(z)$ for all $z \leq z_\ell = z_1$ and for all $z \geq z_u = z_m$.

Table 2 presents the values of e_m for cluster 29, where z_1 and z_m are the smallest and largest nodes for Gaussian-Hermite with m nodes, where $m = 3, 5, 7, 9$ and 11. Here z_ℓ and z_u are the lower and upper limits such that $z_\ell \leq z_1$ and $z_u \geq z_m$. For every value of m in this table, $k(z) \geq q_m(z) \geq 0$ for all $z \leq z_\ell$ and for all $z \geq z_u$. In other words, the upper bound on $|c(\theta, \sigma) - c_m(\theta, \sigma)|$ that results from (8), (9) and (10) applies. Note that the values of e_m in this table decrease as m increases. Also, since z_ℓ decreases and z_u increases as m increases, $|a_\ell|$ and $|a_u|$ decrease as m increases through the values $m = 3, 5, 7, 9$ and 11. Our conclusion from this table is that the upper bound on $|c(\theta, \sigma) - c_m(\theta, \sigma)|$ decreases as m increases through these values. Figure 2 and Table 2 were obtained using the "minimax" command in Maple.

Remark 1. As noted in Sect. 3, if $c_3 > 0$ then $r(t; \theta, \sigma) \to \infty$ as $t \to \infty$ and as $t \to -\infty$, so that $\widetilde{\sigma}^2 = \infty$ and importance sampling fails spectacularly. However, (7) implies that if $c_3 > 0$ then $h(t; \theta, \sigma) \to \infty$, as $t \to \infty$ and as $t \to -\infty$, faster

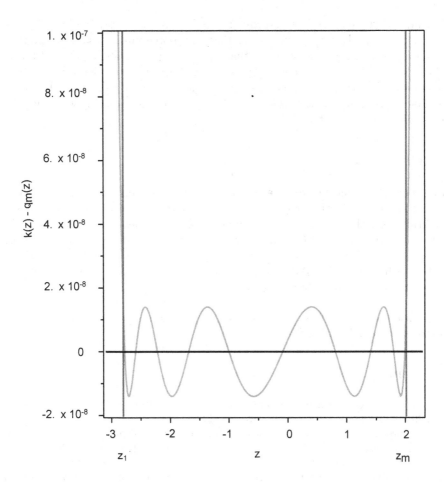

Fig. 2. Graph of the function $k(z) - q_m(z) = h(\mu + \sqrt{2}\tau z; \theta, \sigma) - q_m(z)$, considered as a function of z, for litter 29 for adaptive Gauss-Hermite quadrature with $m = 5$ nodes. The smallest and largest nodes z_1 and z_m, respectively, are shown.

Table 2. Values of e_m for cluster 29, where z_1 and z_m are the smallest and largest nodes for Gaussian-Hermite with m nodes. Here z_ℓ and z_u are the lower and upper limits such that $z_\ell \leq z_1$ and $z_u \geq z_m$.

m	z_ℓ	z_1	z_m	z_u	e_m
3	-2.100000000	-1.224744871	1.224744871	1.280000000	$3.945318530 \times 10^{-8}$
5	-2.800000000	-2.020182870	2.020182870	2.020182870	$1.406266160 \times 10^{-8}$
7	-3.400000000	-2.651961357	2.651961357	2.651961357	$7.927392539 \times 10^{-9}$
9	-3.930000000	-3.190993202	3.190993202	3.190993202	$5.198304537 \times 10^{-9}$
11	-4.400000000	-3.668470847	3.668470847	3.668470847	$3.976752578 \times 10^{-9}$

than any polynomial. In other words, if $c_3 > 0$ then $k(z) = h(\mu + \sqrt{2}\tau z; \theta, \sigma) \to \infty$, as $z \to \infty$ and as $z \to -\infty$, faster than any polynomial. It is this fact than makes it possible for the condition that $k(z) \geq q_m(z)$ for all $z \leq z_\ell$ and for all $z \geq z_u$ to be satisfied. This condition is used in Theorem 1 to bound $|a_\ell|$ and $|a_u|$ from above. In other words, the property of the function $h(t; \theta, \sigma)$ of t that leads to importance sampling failing spectacularly is useful in disclosing why adaptive Gauss-Hermite quadrature succeeds.

5 Discussion

Adaptive Gauss-Hermite quadrature and importance sampling share the same basic first step of multiplying and dividing the integrand of interest by a chosen pdf. However, this is where the similarity between these two methods ends. Extreme care is required to apply importance sampling effectively. Fortunately for users of adaptive Gauss-Hermite quadrature for the computation of the log-likelihood function of GLMM's this method can be applied effectively without anywhere near the same level of care.

Acknowledgements. This work was supported by an Australian Government Research Training Program Scholarship.

References

1. Liu, Q., Pierce, D.A.: A note on Gauss-Hermite quadrature. Biometrika **81**, 624–629 (1994). https://doi.org/10.2307/2337136
2. Pinheiro, J.C., Bates, D.M.: Approximations to the log-likelihood function in nonlinear mixed-effects models. J. Comput. Graph. Stat. **4**, 12–35 (1995). https://doi.org/10.2307/1390625
3. Lesaffre, E., Spiessens, B.: On the effect of the number of quadrature points in logistic random-effects model: an example. Appl. Stat. **50**, 325–335 (2001). https://doi.org/10.1111/1467-9876.00237
4. Rabe-Hesketh, S., Skrondal, A., Pickles, A.: Reliable estimation of generalized linear mixed models using adaptive quadrature. Stata J. **2**, 1–21 (2002)
5. Demidenko, E.: Mixed Models : Theory and Applications. Wiley, Hoboken (2004)
6. Hedeker, D., Gibbons, R.D.: Longitudinal Data Analysis. Wiley, Hoboken (2006)
7. Tuerlinckx, F., Rijmen, F., Verbeke, G., De Boeck, P.: Statistical inference in generalized linear mixed models: a review. Br. J. Math. Stat. Psychol. **59**(2), 225–255 (2006). https://doi.org/10.1348/000711005X79857
8. Rabe-Hesketh, S., Skrondal, A.: Multilevel and Longitudinal Modeling Using Stata, 2nd edn. Stata Press, College Station (2008)
9. Kim, Y., Choi, Y., Emery, S.: Logistic regression with multiple random effects: a simulation study of estimation methods and statistical packages. Am. Stat. **67**(3), 171–182 (2013). https://doi.org/10.1080/00031305.2013.817357
10. Chang, B., Hoaglin, D.: Meta-analysis of odds ratios: current good practices. Med. Care **55**(4), 328–335 (2017). https://doi.org/10.1097/MLR.0000000000000696
11. Robert, C.P., Casella, G.: Monte Carlo Statistical Methods, 2nd edn. Springer, New York (2004). https://doi.org/10.1007/978-1-4757-4145-2

12. Owen, A.B.: Monte Carlo Theory, Methods and Examples (Chap. 9) (2013). http://statweb.stanford.edu/~owen/mc/. Accessed 01 Aug 2018
13. Weil, C.: Selection of the valid number of sampling units and a consideration of their combination in toxicological studies involving reproduction, teratogenisis or carcinogenisis. Food Cosmet. Toxicol. **8**, 177–182 (1970)
14. Atkinson, K.E.: An Introduction to Numerical Analysis, 2nd edn. Wiley, New York (1989)
15. Szegö, G.: Orthogonal Polynomials, vol. 23, 3rd edn. American Mathematical Society, Providence (1967)

Deep Learning with Periodic Features and Applications in Particle Physics

Steffen Maeland[1]([✉])[iD] and Inga Strümke[2][iD]

[1] NORSAR, Gunnar Randers vei 15, 2007 Kjeller, Norway
`steffen.maeland@norsar.no`
[2] PricewaterhouseCoopers, Dronning Eufemias gate 71, 0194 Oslo, Norway
`inga.struemke@pwc.com`

Abstract. We introduce a periodic loss function and corresponding activation function, to be used for neural network regression and autoencoding task involving periodic targets. Such target features, typically represented in non-Cartesian coordinates, arise mainly from angular distributions, but also include repeating time series, e.g. 24-h cycles or seasonal intervals. To demonstrate the use of this loss function, two different usecases within the context of high-energy physics are presented. The first is a simple regression network, trained to predict the angle between particles emerging from the decay of a heavier, unstable particle. Next, we look at the same particle decay, but train an autoencoder to reproduce all inputs, which include both cyclic and noncyclic features. All examples show that failing to incorporate the cyclic property of the targets into the loss and activation function significantly degrades the performance of the model predictions.

Keywords: Particle physics · Machine learning · Loss function · Autoencoder · Regression

1 Introduction

Observables repeating over a given interval appear everywhere within the physical sciences; these may be angles defined on the interval $[0, 2\pi)$, or measurements repeating in a daily or a yearly cycle. A model predicting the value of a periodic observable necessarily needs to incorporate the bounds defined by the nature of the problem at hand. Particularly in the field of high-energy particle physics, angular observables are ubiquitous, as particle trajectories are typically given in cylindrical coordinates. This is due to the design of modern-day particle detectors, which are built as concentric cylinders incorporating different detection technologies. The current leading detectors are located at CERN, Geneva, observing collisions of atomic nuclei which are accelerated by the Large Hadron Collider (LHC). The energy density of the collisions resemble those from moments after the Big Bang, and cause multitudes of energetic particles to emerge in all directions from the collision point.

© Springer Nature Singapore Pte Ltd. 2019
H. Nguyen (Ed.): RSSDS 2019, CCIS 1150, pp. 140–147, 2019.
https://doi.org/10.1007/978-981-15-1960-4_10

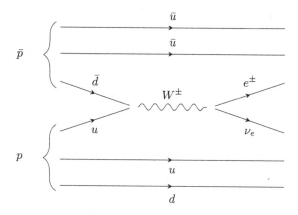

Fig. 1. Creation of an electroweak boson $W^\pm$ in a proton-proton collision, and subsequent decay into an electron or positron $e^\pm$ with its associated electron neutrino.

1.1 Data: Physics Observables from Colliders

Upon successful reconstruction of a collision event, the basic observables available are the directions of the outgoing particles, along with their energies and electric charges. Unstable particles may decay before reaching the detector; hence their existence can only be inferred through their decay products. The resolution of the reconstructed energy and direction of such an intermediate particle necessarily depends on how well the decay products are measured. While the larger part of a collision can be measured with high precision, this is typically not the case for electrically neutral particles, particularly so for the weakly interacting neutrino, which will always traverse the sensor systems undetected. Although it cannot be observed directly, the initial conditions of the collision allows us to constrain the possibility space of its momentum. As the colliding particles approach each other along the z axis, they have zero energy in the transverse $(x\text{-}y)$ plane, and this property is also conserved after the collision. Hence, if we observe a net energy imbalance $\mathbf{E}^{\text{miss}}_{xy}$ in the $x\text{-}y$ plane after the collision, we can attribute the 'missing' energy to neutrinos travelling in the opposite direction. This case is investigated further in the next section.

In all following demonstrations, we use realistic, simulated data, for which the true regression targets are precisely known. The data are generated using Pythia [7] version 8.2.40, configured to match the collision energy of the LHC at present. For computational efficiency we simulate only certain physical processes; the specifics of the processes considered are given in the description of each case below.

2 Periodic Loss and Activation Function

The crucial part for setting up a loss function $L(y_{\text{pred}}, y_{\text{true}})$ to be applied to a periodic target y_{true}, where $y_{\text{true}} \in [a, b)$, is that $\lim_{\epsilon \to 0} L(a, b - \epsilon)$ should

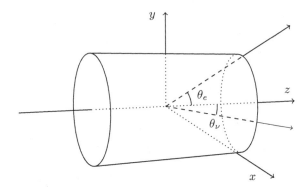

Fig. 2. A sketch illustrating the angles of the electron and the neutrino (θ_e and θ_ν, respectively), with respect to the z axis.

evaluate to 0. For 'standard' mean squared error loss, this is obviously not the case, since one would rather obtain the maximum possible difference. To solve this, we convert y_{pred} and y_{true} to points on the unit circle, and take, as a loss value, the norm of their vector difference. The absolute-value norm corresponds to mean absolute error (MAE), while the Euclidean norm corresponds to mean squared error (MSE). The following shows an implementation in **python** using the **TensorFlow** [1] library:

```python
from math import pi
import tensorflow as tf

def to_unit_circle(yvals):
    """
    Convert points in range [a, b) to point on full unit circle
    """
    c = 2*pi/(b-a)
    siny, cosy = tf.sin(a+c*yvals), tf.cos(a+c*yvals)

    return tf.stack(values=[siny, cosy], axis=-1)


def periodic_mae(y_true, y_pred):
    """
    Absolute mean error function for periodic target
    """
    y_t = to_unit_circle(y_true)
    y_p = to_unit_circle(y_pred)

    return tf.mean(tf.norm(y_t-y_p, axis=1))
```

For a regression task, one would typically use a linear activation function in the output layer, affinely combining the outputs of the nodes in the previous layer.

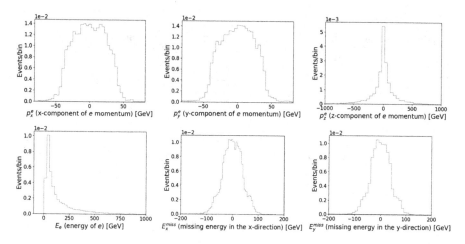

Fig. 3. Histograms of the input features. Descriptions are given inside the parentheses and units are provided in the brackets.

We need a final modification, however, to ensure that the range of network output coincides with the interval on which y_{true} is defined. This can be done by introducing the modulus operator in the activation function of the output layer:

```
from math import pi

def bounded_linear_activation(y):
    """
    Map y to the range [a, b]
    """

    ans = a+tf.mod(y, b)

    if ans == 0:
        return tf.keras.backend.epsilon()

    return ans
```

The following examples illustrate cases where the combination of these loss and activation functions is required, and show how 'common' loss functions without the periodic property fail to properly reconstruct the targets.

3   Example 1: Predicting the Angle of an Invisible Particle

As a first example, we consider the decay of an electroweak boson W^- into an electron e^- and an electron neutrino ν_e. This can for instance take place in a proton-proton collision at the LHC, via the process depicted in Fig. 1. The W decays immediately, leaving behind only an electron, which is accurately

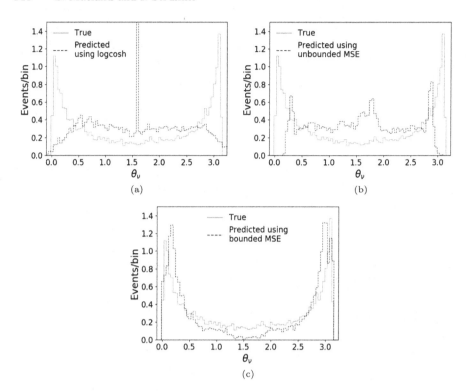

Fig. 4. Three different loss functions are used in a regression network to predict the angle θ; (a) the logcosh, (b) the mean squared error and (c) a periodic mean squared error function.

measured by the detector, and a neutrino. As mentioned, the neutrino is unde-tectable, leaving us only with an energy imbalance $\mathbf{E}_{xy}^{\mathrm{miss}}$, and no knowledge of its energy in the z direction. In this example we wish to predict the z-component of the neutrino momentum, or more specifically, its angle θ_ν relative to the z axis, as shown in Fig. 2.

A regression network with $(150, 100, 50, 1)$ nodes is trained on one million W decay events. The input features are the three components of the electron momentum vector, denoted p_x^e, p_y^e and p_z^e, the electron energy E^e, and the energy imbalance components E_x^{miss} and E_y^{miss}. The distributions of the features are shown in Figure 3. The network is implemented in Keras [3], using ReLU acti-vation in all except the last layer, with the default setting for random weight initialisation, and is optimised using Adam [5], again with default settings. Three different loss functions are tested; unbounded ('standard') logcosh, unbounded MSE, and bounded MSE. We train these networks until an early stopping cri-terion is met, which is when the loss on the validation dataset has shown no improvement for 25 consecutive epochs. The distributions of the network pre-dictions, superimposed on the true target values, are shown in Fig. 4. As can

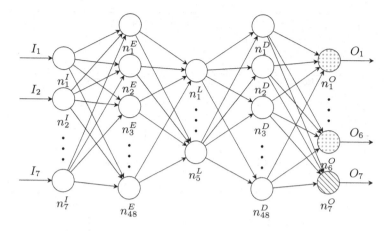

Fig. 5. Autoencoder architecture. The dotted circles in the last layer represent nodes with unbounded linear activation functions, while the shaded circle represents the node with bounded activation function.

be seen, when using the unbounded loss functions (Fig. 4a and b) the network fails to properly reconstruct the targets close to the boundaries; this problem is ameliorated by using the bounded loss and final-layer activation functions (Fig. 4c).

4   Example 2: Autoencoding Periodic Features

Turning now to autoencoders [2,4,6], the task of these deep neural networks is unsupervised representation learning, performed in two steps; the first part of the network, referred to as the encoder, encodes the features into latent representations. These latent representations are then used by the second part, the decoder, to reconstruct the original features. Autoencoders are commonly used, e.g. for dimensionality reduction in image recognition, which works when the autoencoder is able to ignore noise while learning relevant structure in the images [8].

We set up a simple autoencoder to reconstruct the features of the data in Sect. 3, including θ_ν, meaning it operates on in total seven features, of which one is periodic whilst the others are not. The encoder consists of an input and two encoding layers with $(7, 48, 5)$ nodes. The decoder consists of a layer with 48 nodes, followed by two parallel layers; one with six nodes corresponding to the nonperiodic features, and one with a single node corresponding to the periodic feature, see Fig. 5. The former output layer is given the standard MSE loss and linear activation, while the latter has the bounded MSE loss and periodic linear activation. For comparison, we also train an autoencoder using unbounded MSE loss for all features. The results are shown in Fig. 6. On the left (Fig. 6a) we see the reconstructed features using only unbounded MSE, which near-perfectly recover the nonperiodic features, but fails completely to reconstruct the angle

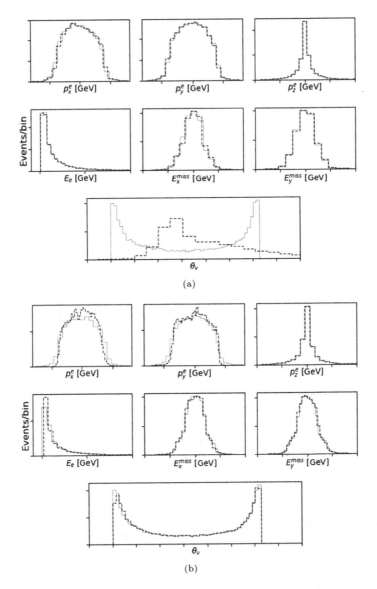

Fig. 6. Original and decoded features, using (a) standard unbounded loss and activation function for all features, and (b) unbounded loss and activation for features 1 through 6, and bounded loss and activation for the periodic θ_ν (lowermost panel). The axis ranges are not indicated, but are the same as in Fig. 3.

θ_ν. On the right (Fig. 6b) we see the adapted network, which recovers the correct distribution for θ_ν. We note, however, that this implementation of splitting the decoder output into parallel layers with different loss functions, introduces a new hyperparameter, which is the relative weighting to apply to the loss values of the two output layers. This requires some case-dependent tuning, since the maximum value of the bounded loss function is limited by its domain, while the unbounded one can in principle yield arbitrarily large values, hence 'overwhelming' the other.

5 Conclusion

We have shown that when applying neural networks to regression tasks involving periodic targets, such tasks including also autoencoders, the periodicity of the targets need to be specifically incorporated into the loss and final-layer activation functions of the network, in order to accurately reconstruct target values close to the boundary. An example implementation of these functions is given in Sect. 2. As a use-case we have investigated a scenario in high-energy physics, where angular and hence periodic features are frequently encountered.

References

1. Abadi, M., et al.: TensorFlow: large-scale machine learning on heterogeneous systems (2015). tensorflow.org
2. Bourlard, H., Kamp, Y.: Auto-association by multilayer perceptrons and singular value decomposition. Biol. Cybern. **59**(4), 291–294 (1988). https://doi.org/10.1007/BF00332918
3. Chollet, F., et al.: Keras (2015). https://github.com/fchollet/keras
4. Hinton, G.E., Zemel, R.S.: Autoencoders, minimum description length and Belmholtz free energy. In: Proceedings of the 6th International Conference on Neural Information Processing Systems, NIPS 1993, pp. 3–10. Morgan Kaufmann Publishers Inc., San Francisco (1993). http://dl.acm.org/citation.cfm?id=2987189.2987190
5. Kingma, D.P., Ba, J.: Adam: a method for stochastic optimization. CoRR abs/1412.6980 (2014)
6. Lecun, Y., Fogelman Soulie, F.: Modeles connexionnistes de l'apprentissage. Ph.D. thesis, March 1987
7. Sjöstrand, T., et al.: An Introduction to PYTHIA 8.2. Comput. Phys. Commun. **191**, 159–177 (2015). https://doi.org/10.1016/j.cpc.2015.01.024
8. Vincent, P., Larochelle, H., Lajoie, I., Bengio, Y., Manzagol, P.A.: Stacked denoising autoencoders: learning useful representations in a deep network with a local denoising criterion. J. Mach. Learn. Res. **11**, 3371–3408 (2010). http://dl.acm.org/citation.cfm?id=1756006.1953039

Copula Modelling of Nurses' Agitation-Sedation Rating of ICU Patients

Ainura Tursunalieva[1]([✉])(iD), Irene Hudson[2](iD), and Geoff Chase[3](iD)

[1] Swinburne University of Technology, Hawthorn 3122, Australia
atursunalieva@swin.edu.au
[2] Royal Melbourne Institute of Technology, Melbourne, VIC 3000, Australia
irene.hudson@rmit.edu.au
[3] University of Canterbury, Christchurch 8140, New Zealand
geoff.chase@canterbury.ac.nz

Abstract. Inadequate assessment of the agitation associated with clinical outcomes has an adverse impact on a patient's wellbeing including under or oversedation. Earlier research found that the majority of nurses under-estimate more severe pain and over-estimate mild pain.

Empirical distributions of the nurses' ratings of a patient's agitation levels and the administered dose of a sedative are often positively skewed so that their joint distributions are non-elliptical. Therefore, the high nurses' ratings of a patient's agitation levels may not correspond to the cases with large doses of sedative.

Copulas measure nonlinear dependencies capturing the dependence between skewed distributions. Therefore, we propose to use a copula-based dependence measure between the nurses' rating of patients' agitation level and the automated sedation dose to identify the patient-specific thresholds that separate the regions of mild and severe agitation. Delineating the regions with different agitation intensities allows us to establish the regions where nurses are more likely to over or under-estimate the patient's agitation levels.

This study uses agitation-sedation profiles of two patients collected at Christchurch Hospital, Christchurch School of Medicine and Health Sciences, NZ. Classification of patients into poor and good trackers based on Wavelet Probability Band. The best-fitting copula shows that the dependency structure between the nurses' rating of a patient's agitation level and the administered dose of sedative for both patients has an upper tail. Specifically, the value of the tail threshold is lower and the average magnitude of the bias in the nurses' rating of a patient's agitation level is smaller for a good tracker compared with a poor tracker.

Establishing the presence of tail dependence and patient-specific thresholds for areas with different agitation intensities has significant implications for the effective administration of sedatives. Better management of agitation-sedation states will allow clinicians to improve the efficacy of care and reduce healthcare costs.

© Springer Nature Singapore Pte Ltd. 2019
H. Nguyen (Ed.): RSSDS 2019, CCIS 1150, pp. 148–161, 2019.
https://doi.org/10.1007/978-981-15-1960-4_11

Keywords: Copula · Kendall plot · Tail dependence · Pain assessment · Nurses' Rating

1   Introduction

1.1   Background

Earlier research found that the majority of nurses under-estimate more severe pain and over-estimate mild pain [1]. Optimised sedation management in the Intensive Care Unit (ICU) is crucial for better pain management, reduced agitation, and survival of patients in ICU. A trend towards lighter sedation has been evident in many ICUs in Australia and Europe [2]. In Australia, the number of critically ill patients managed in the emergency department (ED) is increasing [3]. Between 2011 and 2016, the number of critically ill patients presenting to the ED increased by nearly 60% [4] with over a third of patients (39%) needing intubation and mechanical ventilation. Although the care of critically ill patients traditionally occurs in ICUs, emergency staff are increasingly having to manage critically ill mechanically ventilated patients for extended periods of time [5]. Pain management is an essential component of quality care delivery for the critically ill patient. However, as many as 79% of patients experience moderate to severe pain, whilst intubated and mechanically ventilated from both their initial reason for presentation (e.g. trauma) and required treatments. Intravenous analgesia is commonly administered to alleviate pain, suffering, adverse physiological and psychological effects [6], unplanned self-extubation, accidental removal of invasive monitoring devices, or injury to staff [7].

Several recent studies have emphasised the cost and health-care advantages of drug delivery protocols based on assessment scales of agitation and sedation. Typically agitation-sedation cycling in critically ill patients involves oscillations between states of agitation and over-sedation, which is detrimental to patient health and increases hospital length of stay (LoS) [8–11]. The goal of the research specifically in reference [11] was to develop a physiologically representative model that captures the fundamental dynamics of the agitation-sedation system. The resulting model serves as a platform to develop and test semi-automated sedation management controllers that offer the potential of improved agitation management and reduce LoS in ICU.

The work of [8,11,12] developed a physiological agitation model which can be used in feedback protocols for medical decision support systems and eventually automated sedation administration. A minimal differential equation model to predict or simulate each patient's agitation-sedation status over time was presented in [11] for 37 ICU patients and was shown to capture patient agitation-sedation (A-S) dynamics (see Table 7 of [13]). Current agitation management methods rely on subjective agitation assessment and an appropriate sedation input response from recorded-at-bedside agitation scales [14,15]. The carers then select an appropriate infusion rate based upon their evaluation of these scales, their experience, and intuition [16]. This approach usually leads to largely continuous infusions which lack a bolus-focused approach, commonly resulting in over- or under-sedation. Further details of this process can be found in [11]. A more refined A-S model, which utilised kernel regression with an Epanechnikov kernel and better captured the fundamental A-S dynamics was formulated later [10,11].

A-S cycling in critically ill ICU patients is damaging to health. Therefore, the use of quantitative modelling to enhance understanding of A-S system and the provision of an A-S simulation platform are key tools in the area of patient critical care. A Bayesian approach using densities and wavelet shrinkage methods is suggested as a means to assess a previously derived deterministic, parametric A-S model [13]. Research on A-S pharmacodynamics by [8,9,12,13,17] has helped enhance the understanding of the underlying A-S dynamics and enabled development of advanced protocols for semi-automated sedation administration technology. This work has successfully challenged the practice of sedating ICU patients using continuous infusions. Specifically, [13,17] has shown that wavelets provide a diagnostic and visualization tool to assess A-S models, and provided alternative numerical metrics of A-S control, using a density estimation approach via wavelet smoothing and discrete wavelet transform in assessing the validity of the earlier developed A-S deterministic DE dynamic models (see Table 7 of [13]).

The Bayesian approach of using densities and wavelet shrinkage methods suggested in [13] as a means to assess a previously derived deterministic, parametric A-S model incorporated two steps to construct a so-called wavelet probability band (WPB) for the model and use of this as a basis to evaluate whether the nonparametric regression curve lies within the band. This WPB approach yielded graphical assessments along with numerical metrics at a patient by patient level.

The WPB was constructed for each of 37 ICU patients, and the time and duration of any deviations from the WPB were recorded for each patient [13]. A 70% value for a WPB implies that for at least 70% of the time, the estimated mean value of the given patient's administered dose of sedative lies within the band. A density profile was also successfully used to define two alternative metrics, the average normalized wavelet density (ANWD) and the relative average normalized wavelet density (RANWD), as estimates of comparability between the patient's simulated and nurses' recorded rates. The WPB and related statistics were shown to be excellent tools for detecting regions where the nurses' rating and the automated (modelled) infusion rate do not track, thus providing ways distinguish between good versus poor trackers [13,17] to help improve and distil the deterministic A-S model. Therefore, the difference between good and poor trackers is as follows: a good tracker is a patient whose simulated profiles was "close" to the mean profile (a majority of the time profile), a poor tracker is a patient for whom this was not the case [13].

Nurses play a major role in rating a patient's agitation levels. Assessing the severity of agitation is a challenging clinical problem as variability related to drug metabolism for each individual is often subjective. A multitude of previous studies suggests that nurses tend to underestimate more severe pain and overestimate mild pain [18,19].

The empirical distributions of the nurses' ratings of the patient's agitation level and the administered dose of sedative estimated for the full set of 37 patients shows that both variables have positively skewed distributions [20]. As a consequence, the joint distribution between nurses' rating and the automated sedation doses is non-elliptical. In a non-elliptical joint distribution the high nurses' ratings of a patient's agitation levels may not necessarily correspond to the cases with large doses of sedative.

The aim of this paper is to build on an earlier work [20] to address the gap in the methodology by integrating the non-elliptical dependency structure between nurses' rating of a patient's agitation level and the automated sedation dose thereby accounting for possible nonlinear relationships between two variables. A similar approach was used by [21] to capture the non-elliptical dependence between exchange rates. In an earlier work [20] the tail thresholds were determined visually, whereas in this paper we employ the dynamic programming algorithm [22] to establish the tail thresholds.

Copula [23,24] is a multivariate functional form for the joint distribution of random variables derived purely from pre-specified parametric marginal distributions of each random variable. The copula-based approach retains a parametric specification for the bivariate dependency but allows testing of several parametric structures to characterise the dependency including one between skewed distributions. The empirical context in the current paper is a model of nurses' rating of agitation level with respect to the automated simulated sedation dose. The copula decomposes the multivariate distribution function of nurses rating of agitation level and the patient automated dose profile into univariate marginals and into a function that quantifies their statistical dependency. Linear correlation coefficients are suitable for measuring dependence of elliptical joint distributions. If conventional correlation measures were applied outside the class of elliptical distributions and linear relationships, a possibility of pitfalls and erroneous results could have occurred [24,25].

Copulas allow the strength of dependence to vary in different quantiles. The Kendall plot is used to determine the quantiles with significantly different strength of dependence. Establishing the regions with different agitation intensities between nurses' rating of patients' agitation level and the automated sedation dose allow us to identify regions where nurses are more likely to over and under-estimate the patient-specific agitation severity levels.

2   Methodology

This study uses two intensive care unit patients' agitation-sedation profiles collected at Christchurch Hospital, Christchurch School of Medicine and Health Sciences, NZ [12,13,17]. Two variables were recorded for each patient: (1) the nurses' ratings of a patient's agitation level and (2) an automated sedation dose. Infusion data were recorded using an electronic drug infusion device for all admitted ICU patients during a nine-month observation period and requiring more than 24 h of sedation. Infusion data containing less than 48 h of continuous data, or data from patients whose sedation requirements were extreme, such as those with severe head injuries, were excluded. A total of 37 ICU patients met these requirements and were enrolled in this study. Approval for this research was obtained from the Canterbury Ethics Board.

We will use the Kendall plot (K-plot) to determine the bivariate patient-specific thresholds which split the data into two regions with significantly different strength of dependence between nurses' rating of a patient's agitation level and the automated sedation dose: (1) the main region with an approximately linear relationship and (2) the tail region with a non-linear relationship.

The K-plot [26] adopts the familiar probability plot (Q-Q plot) to detect dependence. A lack of linearity of the standard Q-Q plot is an indication of non-normality of the distribution of a random variable. Similarly, in the absence of association between two variables, the K-plot is close to a straight line, while the amount of curvature in the plot is characteristic of the degree of dependence in the data, and is related in a definite way, to the underlying copula. This method is closely related to the Kendall's τ-statistic [27], from which it takes the name.

The construction of a K-plot requires ordering H_i, as $H_{(1)} \leq \dots \leq H_{(n)}$. For a given pair (X_i, Y_i) with $1 \leq i \leq n$, H_i is defined as follows:

$$H_i = \frac{1}{n-1} \#\{j \neq i : X_j \leq X_i, Y_j \leq Y_i\}. \tag{1}$$

Now, using the definition of the density of an order statistic, we have the expected value $W_{(i:n)}$ for the ith order statistic $H_{(i)}$ under the null hypothesis of independence for all $1 \leq i \leq n$:

$$W_{i:n} = n \binom{n-1}{i-1} \int_0^1 \omega K_0(\omega)^{i-1} \times \{1 - K_0(\omega)\}^{n-1} dK_0(\omega). \tag{2}$$

Now to obtain a K-plot, we plot the pairs $(W_{i:n}, H_{(i)})$, for $1 \leq i \leq n$ where $W_{i:n}$ is the expectation of the ith-order statistic in a random sample of size n, drawn from the distribution K_0 of the H_i, under the null of independence. The form of the bivariate distribution K_0 is given as follows:

$$K(\omega) = K_0(\omega) = Pr\{UV \leq \omega\} = \omega - \omega \log(\omega), 0 \leq \omega \leq 1 \tag{3}$$

where U and V are independent uniform random variables on the interval [0,1]. This choice of $K_0(\omega)$ is then used to compute the $W_{i:n}$ required for the plot.

To identify a patient-specific threshold, we employ the dynamic programming algorithm discussed in Sect. 3.3 of [22] to estimate threshold in the dependence measure H_i between two variables. The dynamic programming algorithm [22] captures multiple thresholds, however, to be consistent with the objective of this paper, we focus on determining the tail threshold. In our context, the tail threshold corresponds to the highest threshold.

3   Results

In this section, we consider two patients from the pool of 37 patients. The first patient is a poor tracker (patient 27) and the second patient is a good tracker (patient 8). Classification of patients into poor and good trackers based on both Wavelet Probability Bands (WPB) discussed in [13,17]. While patients 27 has WPB of 47.27% and is considered a poor tracker, patient 8 has WPB of 87.5% and is considered a good tracker (see Table 3 in [17]).

First, we provide results for patient 27. The line plot of nurses' rating of a patient's agitation and the automated sedation dose in Fig. 1(a) indicates that two variables tend to move together most of the time.

The scatter plot with the line of the best fit in Fig. 1(c) shows that most of the observations are close to the regression line, however, some observations lie further away from the main cluster of the data. The simple linear regression for predicting nurses' rating of a patient's agitation for patient 27:

$$\hat{\text{Score}} = 1.06\text{Dose} \tag{4}$$

However, the estimated Eq. (4) is not valid due to the fact that the variance is not constant as evident from the scatterplot in Fig. 1(c). To identify a patient-specific threshold, first, we construct the K-plot shown in Fig. 2(a). We use the K-plot to assess if the dependence between two variables varies in different quantiles. The K-plot shows strong dependence between the automated sedation dose and the nurses' rating. The strength of the dependence is lower in the upper quantiles as the observations are closer to the diagonal line of independence. The dynamic algorithm applied to the dependence measure H_i between nurses' rating of a patient's agitation level and the automated sedation dose identifies the tail threshold (178th order statistic) shown as a broken vertical line, see Fig. 2(c).

The identified bivariate threshold corresponds to 6.8 mg/ml for the automated sedation dose and a score of 5.7 for the nurses' rating. The histograms for both variables with thresholds are shown in Fig. 3(a) and (b). Both variables have positively skewed distributions, therefore the contour plot in Fig. 4(a) shows that the bivariate relationship between nurses' rating of patient's agitation level and the automated sedation dose for patient 27 is non-elliptical. The long right tails of both distributions stretch the shape of the joint distribution along the vertical and horizontal axes. The *BiCopSelect* function from VineCopula R package selects survival BB8 copula as the best fitting copula [29]. This copula has an upper tail dependence, therefore we introduce a tail dummy variable into the regression model:

$$\hat{\text{Score}} = 0.79\text{Dose} + 4.28(\text{tail} == \text{"Yes"}) \tag{5}$$

The slope of 0.79 indicates that, on average when a patient is experiencing a mild or moderate agitation, nurses tend to under-estimate a patient's agitation level. However, when a patient is experiencing a severe agitation, nurses tend to assign a score that is on average 4.28 points higher than expected for the patient's agitation level. This over-estimation occurs one in every seven ratings, as there are 32 out of a total of 223 occurrences of severe agitation. For patient 27 incorporating the tail dummy variable improved predictions of the nurses rating by increasing the adjusted R^2 values by 17%. The patient-specific thresholds

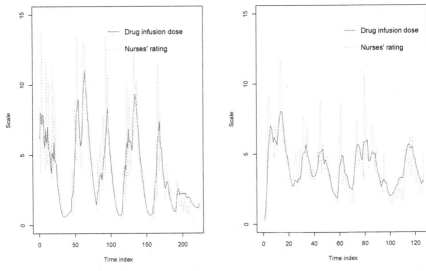

(a) The line plot of nurses' rating and the drug infusion dose for patient 27.

(b) The line plot of nurses' rating and the drug infusion for patient 8.

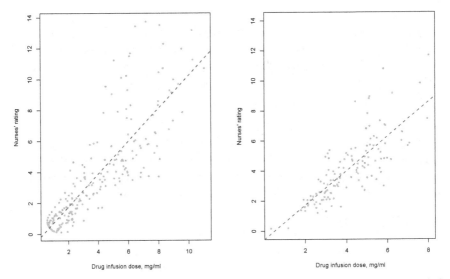

(c) The scatter plot with the line of the best fit (dashed line) for patient 27.

(d) The scatter plot with the line of the best fit (dashed line) for patient 8.

Fig. 1. Line plots and scatterplots.

split the data into four quadrants with the top two quadrants representing the region where nurses tend to over-estimate patients' agitation level, see Fig. 4(c).

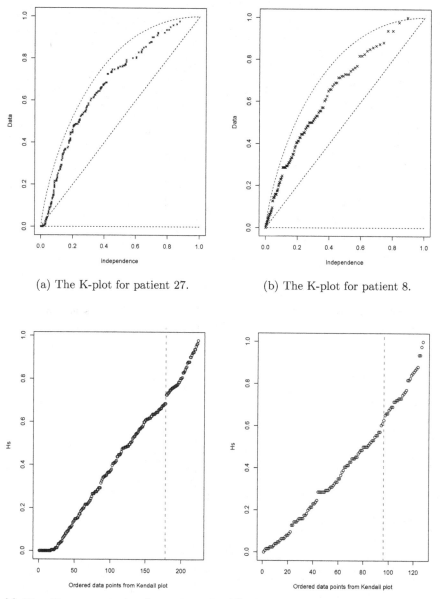

(a) The K-plot for patient 27.

(b) The K-plot for patient 8.

(c) The H_i measure plot for patient 27 with the bivariate threshold (dashed line).

(d) The H_i measure plot for patient 8 with the bivariate threshold (dashed line).

Fig. 2. K-plots and the H_i measure plot

Now, we provide the results for patient 8. The line plot of nurses' rating of a patient's agitation and the automated sedation dose in Fig. 1(b) indicates that both variables tend to move together most of the time.

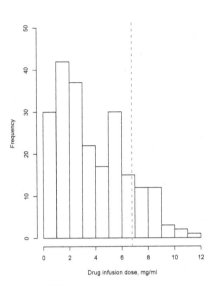

(a) The histogram for nurses' rating
for patient 27.

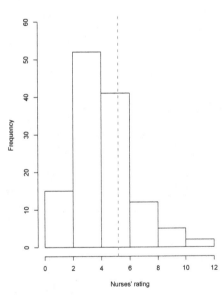

(b) The histogram for the sedation dose
for patient 27.

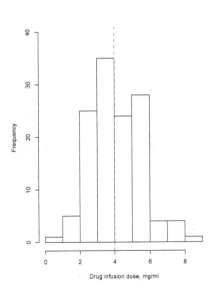

(c) The histogram for nurses' rating
for patient 8.

(d) The histogram for the sedation dose
for patient 8.

Fig. 3. Histograms with the threshold (dashed line).

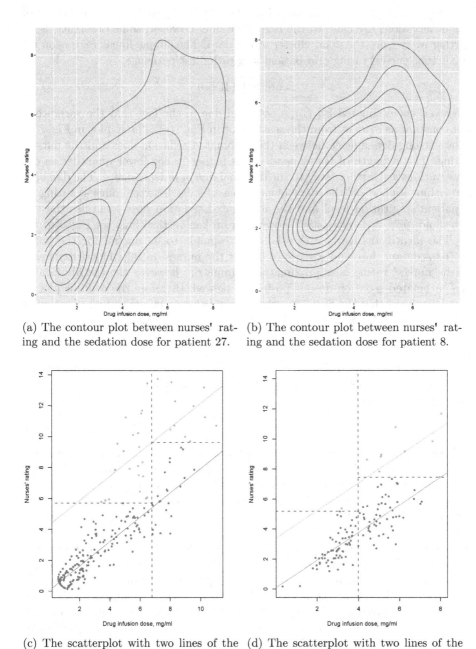

(a) The contour plot between nurses' rating and the sedation dose for patient 27.

(b) The contour plot between nurses' rating and the sedation dose for patient 8.

(c) The scatterplot with two lines of the best fit and tail thresholds (dashed lines) for patient 27.

(d) The scatterplot with two lines of the best fit and tail thresholds (dashed lines) for patient 8.

Fig. 4. Contour plots and scatterplots with thresholds

The scatter plot with the line of the best fit in Fig. 1(d) shows that most of the observations are close to the regression line, however, some observations lie further away from the main cluster of the data. The simple linear regression for predicting nurses' rating of a patient's agitation for patient 8:

$$\hat{Score} = 1.14Dose \tag{6}$$

However, the estimated Eq. (6) is not valid due to the fact that the variance is not constant as evident from the scatterplot in Fig. 1(c). The K-plot in Fig. 2(b) shows strong dependence between the automated sedation dose and the nurses' rating. The strength of the dependence is higher in the lower quantiles as the observations are farther away to the diagonal line of independence. The dynamic algorithm applied to the dependence measure H_i between nurses' rating of a patient's agitation level and the automated sedation dose identifies the tail threshold (96th order statistic) shown as a broken vertical line, see Fig. 2(d).

The identified bivariate threshold corresponds to 4.0 mg/ml for the automated sedation dose and a score of 5.2 for the nurses' rating. The distribution of the nurses' rating is approximately symmetric, however the distribution of the automated sedation dose is positively skewed, therefore the contour plot in Fig. 4(b) shows that the bivariate relationship between nurses' rating of patient's agitation level and the automated sedation dose for patient 8 is non-elliptical. The long right tail of the distribution for the nurses' rating stretched the shape of the joint distribution along the vertical axis. The *BiCopSelect* function from VineCopula R package selects survival Clayton copula as the best fitting copula [29]. This copula has an upper tail dependence, therefore we introduce a tail dummy variable into the regression model:

$$\hat{Score} = 0.92Dose + 3.38(tail == \text{``Yes''}) \tag{7}$$

The slope of 0.92 indicates that, on average when a patient is experiencing a mild or moderate agitation, nurses tend to slightly under-estimate the patient's agitation level. Since this estimated coefficient is closer to one, the magnitude of the systematic under-estimation is slightly lower for patient 8 compared with the patient 27. However, when a patient is experiencing a severe agitation, nurses tend to assign a rating that is, on average, 3.38 points higher than expected for the patient's given agitation level. This over-estimation occurs one in every thirteen ratings, as there are 10 out of a total of 127 occurrences of severe agitation for this patient. For patient 8 incorporating the tail dummy variable improved predictions of the nurses rating by increasing the adjusted R^2 values by 16%. The patient-specific thresholds split the data into four quadrants with the top two quadrants representing the region where nurses tend to overestimate patients' agitation level, see Fig. 4(d).

4   Conclusion

The aim of the paper was to integrate the non-elliptical dependency structure between nurses' rating of a patient's agitation level and the automated sedation dose thereby accounting for possible nonlinear relationships between two variables. The best-fitting copula shows that the dependency structure between the nurses' rating of a patient's agitation level and the administered dose of sedative for both patients has an upper tail. Specifically, the value of the tail threshold is lower for a good tracker compared with a poor tracker. Moreover, the frequency of the cases with severe agitation is lower for a good tracker compared with a poor tracker. And finally, the average magnitude of the bias in the nurses' rating of a patient's agitation level is smaller for a good tracker. Overall, the results show that the estimated copula provides valuable information regarding whether a tail dummy variable has to be included in the regression model. Incorporating the tail dummy variable improved predictions of the nurses' rating by increasing the adjusted R^2 values by a minimum of 16%.

A great variety of copula functions allows for modelling different dependency structures including dependence between skewed distributions and measuring nonlinear dependencies. Incorporating the copula-based dependence structure thresholds into prediction models improves their predictive accuracy and helps clinicians to make more informed decisions. Moreover, the copula-based dependence modelling can be used for segmenting patients into homogeneous groups for the purposes of using the most effective treatment plans. Better predictions and quality of care will have consequent policy implications for improving hospital performance by achieving better health outcomes, improved access to health services, and less waste for the same level of healthcare costs.

The copula approach suggested in this paper is generalisable to any study which investigates the similarity or closeness of bivariate time series of, for example, a large number of units (patients, households) and of time series of disparate lengths and of possibly long length. Hence the aim of future work on the full set of 37 patients is to generalise the result that the copula approach can accurately capture the dynamic (temporal) dependence between the nurses' rating and the automated infusion dose. Future work will investigate the dependence structure between the nurses' rating and the automated infusion dose by identifying both lower and upper thresholds [30]. Further developments could include investigation of commonality of thresholds, copula types, and complexity of a dependence structure for good and bad trackers with the view of segmenting patients into homogeneous groups for the purpose of using the most effective treatment plans.

References

1. Zalon, M.L.: Nurses' assessment of postoperative patients' pain. Pain **54**(3), 329–334 (1993)
2. Egerod, I., Albarran, J.W., Ring, M., Blackwood, B.: Sedation practice in Nordic and non-Nordic ICUs: a European survey. Nurs. Crit. Care **18**(4), 166–175 (2013)
3. Varndell, W., Elliott, D., Fry, M.: Emergency nurses' practices in assessing and administering continuous intravenous sedation for critically ill adult patients: a retrospective record review. Int. Emerg. Nurs. **23**(2), 81–88 (2015)
4. Australian Institute of Health and Welfare: Emergency department care: Australian hospital statistics 2015–16. In: Health Services Series, vol. 72 (2016)

5. O'Connor, G., Geary, U., Moriarty, J.: Critical care in the emergency department. Eur. J. Emerg. Med. **126**(6), 296–300 (2009)
6. Schug, S., Palmer, G., Scott, D., Halliwell, R., Trinca, J.: Acute pain. Management scientific evidence. In: Australian and New Zealand College of Anaesthetists and Faculty of Pain Medicine, 4th edn (2015)
7. Weir, S., O'Neill, A.: Experiences of intensive care nurses assessing sedation/agitation in critically ill patients. Nurs. Crit. Care **13**(4), 185–194 (2008)
8. Rudge, A.D., Chase, J.G., Shaw, G.M., Lee, D., Wake, G.C., Hudson, I.L.: Impact of control on agitation-sedation dynamics. Control Eng. Pract. **13**(9), 1139–1149 (2005)
9. Rudge, A.D., Chase, J.G., Shaw, G.M., Lee, D.: Physiological modelling of agitation-sedation dynamics. Med. Eng. Phys. **28**(1), 49–59 (2006)
10. Rudge, A.D., Chase, J.G., Shaw, G.M., Lee, D.: Physiological modelling of agitation-sedation dynamics including endogenous agitation reduction. Med. Eng. Phys. **28**(7), 629–638 (2006)
11. Chase, J.G., Rudge, A.D., Shaw, G.M., Wake, G.C., Lee, D., Hudson, I.L.: Modeling and control of the agitation-sedation cycle for critical care patients. Med. Eng. Phys. **26**(6), 459–471 (2004)
12. Rudge, A.D., Chase, J.G., Shaw, G.M., Lee, D., Hann, C.E.: Parameter identification and sedative sensitivity analysis of an agitation-sedation model. Comput. Methods Programs Biomed. **83**(3), 211–221 (2006)
13. Kang, I., Hudson, I., Rudge, A., Chase, J.G.: Wavelet signatures and diagnostics for the assessment of ICU Agitation-Sedation protocols. In: Olkkonen, H. (ed.) Discrete Wavelet Transforms, pp. 321–348. InTechOpen (2011)
14. Fraser, G.L., Riker, R.R.: Monitoring sedation, agitation, analgesia, and delirium in critically ill adult patients. Crit. Care Clin. **17**(4), 967–987 (2001)
15. Sessler, C.N., Gosnell, M.S., Grap, M.J., Brophy, G.M., O'Neal, P.V., Keane, K.A.: The Richmond agitation-sedation scale: validity and reliability in adult intensive care unit patients. Am. J. Respir. Crit. Care Med. **166**(10), 1338–1344 (2002)
16. Kress, J.P., Pohlman, A.S., Hall, J.B.: Sedation and analgesia in the intensive care unit. Am. J. Respir. Crit. Care Med. **166**(8), 1024–1028 (2002)
17. Kang, I., Hudson, I.L., Rudge, A., Chase, J.G.: Density estimation and wavelet thresholding via Bayesian methods: a Wavelet Probability Band and related metrics to assess agitation and sedation in ICU patients. In: A-Asmari, A. (ed.) Discrete Wavelet Transforms - A Compendium of New Approaches and Recent Applications, pp. 127–162. InTechOpen (2013)
18. Bondestam, E., Hovgren, K., Johansson, F.G., Jern, S., Herlitz, J., Holmberg, S.: Pain assessment by patients and nurses in the early phase of acute myocardial infarction. J. Adv. Nurs. (Wiley-Blackwell) **12**(6), 677–682 (1987)
19. Puntillo, K., Neighbor, M., Nixon, R.: Accuracy of emergency nurses in assessment of patients' pain. Pain Manag. Nurs. **4**(4), 171–175 (2003)
20. Tursunalieva, A., Hudson, I., Chase, G.: Improved prediction of a nurse's agitation-sedation rating in relation to automated sedation infusion levels in intensive care: a copula approach. In: Joint International Society for Clinical Biostatistics and Australian Statistical Conference (2018)
21. Boero, G., Silvapulle, P., Tursunalieva, A.: Modelling the bivariate dependence structure of exchange rates before and after the introduction of the euro: a semiparametric approach. Int. J. Finance Econ. **16**(4), 357–374 (2010)
22. Bai, J., Perron, P.: Computation and analysis of multiple structural change models. J. Appl. Econ. **18**(1), 1–22 (2003)

23. Nelsen, R.B.: An Introduction to Copulas, 2nd edn. Springer, New York (2006). https://doi.org/10.1007/0-387-28678-0

24. Embrechts, P., Lindskog, F., McNeil, A.: Modelling dependence with copulas and applications to risk management. In: Rachev, S. (ed.) Handbook of Heavy Tailed Distributions in Finance, pp. 331–385. Elsevier (2003)

25. McNeil, R.: Quantitative Risk Management: Concepts, Techniques and Tools. Princeton Series in Finance Princeton University Press Princeton, NJ (2005)

26. Genest, C., Boies, J.-C.: Detecting dependence with Kendall plots. Am. Stat. **57**(4), 275–284 (2003)

27. Kendall, M.G.: A new measure of rank correlation. Biometrika **30**(1/2), 81–93 (1938)

28. Embrechts, P., Hofert, M.: A note on generalized inverses. Math. Methods Oper. Res. **77**(3), 423–432 (2013)

29. Brechmann, E.C., Schepsmeier, U.: Modeling dependence with C- and D-Vine Copulas: the R Package CDVine. J. Stat. Softw. **52**(3), 1–27 (2013)

30. Tursunalieva, A., Hudson, I., Chase, G.: Copula modelling of agitation-sedation rating of ICU patients: towards monitoring and alerting tools. Under review for Proceedings of the 23rd International Congress on Modelling and Simulation (MODSIM2019), Canberra, ACT (2019)

Predicting the Whole Distribution with Methods for Depth Data Analysis Demonstrated on a Colorectal Cancer Treatment Study

D. Vicendese[1,2(✉)], L. Te Marvelde[1,2], P. D. McNair[2], K. Whitfield[2],
D. R. English[1,3], S. Ben Taieb[4], R. J. Hyndman[4], and R. Thomas[5]

[1] Cancer Epidemiology and Intelligence Division, Cancer Council Victoria,
Melbourne, VIC, Australia
D.Vicendese@latrobe.edu.au
[2] Cancer Strategy and Development, Department of Health and Human Services,
Melbourne, VIC, Australia
[3] Centre for Epidemiology and Biostatistics, The University of Melbourne,
Carlton, VIC, Australia
[4] Department of Econometrics and Business Statistics, Monash University,
Clayton, VIC, Australia
[5] Faculty of Medicine, Dentistry and Health Sciences,
The University of Melbourne, Melbourne, VIC, Australia

Abstract. We demonstrate the utility of predicting the whole distribution of an outcome rather than a marginal change. We overcome inconsistent data modelling techniques in a real world problem. A model based on additive quantile regression and boosting was used to predict the whole distribution of length of hospital stay (LOS) following colorectal cancer surgery. The model also assessed the association of hospital and patient characteristics over the whole distribution of LOS. The model recovered the empirical LOS distribution. A counterfactual simulation quantified change in LOS over the whole distribution if an important associated predictor were to be varied. The model showed that important hospital and patient characteristics were differentially associated across the distribution of LOS. Model insights were much richer than just focusing on a marginal change. This method is novel for public health and epidemiological studies and could be applied in other fields of research.

Keywords: Additive quantile regression · Machine learning · Boosting · Density forecast

1 Introduction

Evidence of an association between hospital or surgeon cancer surgery volume and better patient outcomes has been mixed [1–7], however, previous analyses have had important limitations. A 2002 review that examined seven statistical modelling techniques used to assess association between patient factors and length of hospital stay (LOS) in a cardiovascular setting found that choice of model influenced the conclusion

© Springer Nature Singapore Pte Ltd. 2019
H. Nguyen (Ed.): RSSDS 2019, CCIS 1150, pp. 162–182, 2019.
https://doi.org/10.1007/978-981-15-1960-4_12

of the analyses. In some instances the conclusions were reversed. Model results were inconsistent due to LOS being a complex phenomenon and unmet assumptions regarding distributional fit. A small proportion of patients with very long hospital stays made it inherently difficult for simpler parametric methods to effectively model the data [8]. These models have been employed also in colorectal cancer (CRC) studies of association between LOS and provider volume and hence some doubt is cast on both negative and positive conclusions [1, 9–12]. Some studies have used arbitrary thresholds for the categorization of LOS and volumes (low, medium and high) [1, 2, 11, 13, 14] which may reduce statistical power [15]. If the skewness of LOS changes as provider volume changes then arbitrary categorizations of LOS may be problematic [15–17]. Additionally, heterogeneity due to arbitrary categorizations has prevented synthesis of findings [3, 18].

A weakness of some hospital patient outcome studies is bias due to unaccounted for correlations between outcomes within a hospital - sometimes referred to as random effects [19]. Furthermore, surgeries are performed in the context of a hospital with a distinct infrastructure and management that affect their outcome [2]. Modelling this contextual effect may yield important information regarding its association with patient outcomes such as LOS [20–22]. The systematic differences in patients' outcomes across hospitals that persist after differences in patients' risk profiles have been accounted for reflect differences in hospitals' quality of care [23]. In this study we examined the relationship between LOS following CRC surgery and provider volume by using a quantile regression model which makes no distributional assumption about LOS or error terms [24, 25] and avoids arbitrarily categorizing LOS or provider volume [15]. The model formulation we specified took the individual patient as the unit of analysis and used the clustering of patients within a hospital to analyse the association between the hospital context and LOS [20–22]. Furthermore, we did not focus on just a marginal change, such as a mean or median or some other quantile, but instead modelled the whole distribution so as to give a more in depth understanding of the interplay between LOS and hospital and patient characteristics over the whole distribution of LOS [26].

2   Methods

2.1   Data Details

The Victorian Admitted Episode Dataset (VAED) includes all separations (discharges and transfers) undertaken within all Victorian hospitals. All separations between 1 Jul 2005 and 30 Jun 2015 recorded in the VAED, that included one of 30 ICD-10-AM Australian Classification of Health Interventions procedure codes for colorectal surgery, as the primary reason for admission, were identified. There were 62,774 admissions for 57,446 patients. Analysis was restricted to admissions whose principal diagnosis was for CRC, ICD-10-AM codes C18.x to C21.x which resulted in a final data set of 28,343 admissions for 27,633 patients. Provider volume was defined as the number of colorectal surgical procedures performed by a hospital within a fiscal year (1 July to 30 June), whether patients had a principal diagnosis of CRC or not. That is, annual volume (AV). Length of stay was defined as the number of days from admission to discharge for the episode of care including transfers to other hospitals and geriatric and rehabilitation centres.

2.2   Modelling Details

As it was conceivable that LOS and provider volume were not necessarily linearly related, we used an additive quantile regression (AQR) model which does not require a predetermined functional fit but instead determines the best fit from the data [15, 25, 27, 28]. The specification we used was based on a formulation by Mundlak [29] and is in a class commonly referred to as a 'within and between' effects model [20, 22, 30]. It required that we enter both AV and mean annual volume (MAV). Mean annual volume (MAV) was defined for each hospital as the mean of all AV over the number of years the hospital operated within the 10 year study period. Not all hospitals performed colorectal surgical procedures in every study year [2]. The within effect was modelled by AV. It estimated the effect on LOS within hospitals as AV varied and its interpretation is equivalent to any fixed effect estimator [30]. The between effect was modelled by MAV. Due to the model formulation used, it estimated the effect on LOS if a patient were to attend another hospital with a different MAV, that is, the hospital contextual effect [20, 21, 23]. This method draws comparisons across hospitals and estimates the effect of hospital choice on patient LOS, or in other words, hospitals' quality of care or efficiency regarding LOS [23]. The model was adjusted for various patient and hospital factors that may confound the association between provider volume and LOS [19, 31–33].

We tested how well the model represented the data by predicting the empirical cumulative distribution (CDF) of LOS and assessed its fit. We refer to this as the recovered distribution. As the model formulation estimated the effect on LOS if a patient were to attend another hospital [20, 21, 23], we used the model to simulate the change in LOS if CRC patients were to have counterfactually attended a hospital that the model indicated to be more LOS efficient. We predicted each percentile 1 to 99 which were combined to obtain the predicted CDF of LOS [26]. This was termed the counterfactual CDF. The area under the counterfactual CDF was calculated and compared to the area under the recovered CDF. As the area under each CDF directly related to the total sum of LOS days, the difference in areas estimated the change in total sum of LOS due to this hypothetical experiment. This enabled us to calculate a dollar value for the difference by allowing $1000 per LOS day [34]. A 95% confidence interval (CI) was computed for the estimated change. Statistical significance was set at the 0.05 level.

2.2.1   Boosting to Assess Variable Selection and Functional Fit

To aid model building and variable selection we used boosting [15, 26, 35–37]. Boosting is a statistical algorithm in the class of machine learning methods. It determines the most appropriate model by optimizing the fit to the data while limiting over fitting. It will discard variables, along with their proposed functional fit that do not aid optimization where optimization is defined as the largest reduction in the model fit error at each iterative step. For the actual form of the loss function to determine model fit error, please see [38]. Hence it is a variable and functional form selection method that does not resort to heuristic techniques such as ad hoc stepwise variable selection [24]. It also works well in the setting of many predictors with possibly high correlation between them [38]. We used component-wise gradient boosting [38]. Continuous variables may be entered simultaneously as linear and non-linear components into the model and the boosting process is able to determine which variable and which functional fit aids model fit to the signal in the data. Categorical variables are entered linearly. This process is thoroughly described in the following references [15, 24, 26, 37]. Intrinsic to this method is the choice of step length and optimal number of iterations. We used the default of 0.1 for step length. General cross validation (GCV) was used to choose 5000 as an optimal number of iterations. We carried this out with the freeware R version 3.3.3 [39], using the package mboost [24, 40]. All continuous variables were mean centered as recommended for the boosting algorithm [24, 25]. We assessed possible random effects [14] by including a random intercept for hospital into the boosting process where hospital was represented by a dummy variable.

2.2.2   Model Specification for Additive Quantile Regression with Boosting
The following model was implemented using the boosting algorithm:

$Q_t(\text{LOS})$ modelled with *intercept*
$+ f_{nl}(\text{mean annual surgery volume}) + f_l(\text{mean annual surgery volume})$
$+ f_{nl}(\text{annual surgery volume}) + f_l(\text{annual surgery volume})$
$+ f_{nl}(\text{age}) + f_l(\text{age})$
$+ f_{nl}(\text{daily surgery admissions}) + f_l(\text{daily surgery admissions})$
$+ f_{nl}(\text{Elixhauser comorbidity score}) + f_l(\text{Elixhauser comorbidity score})$
$+ f_{nl}(\text{year of discharge}) + f_l(\text{year of discharge})$ (1)
$+ f_{nl\,cyclic}(\text{month of discharge}) + f_l(\text{month of discharge})$
$+ f_l(\text{sex}) + f_l(\text{ASA}) + f_l(\text{cancer site}) + f_l(\text{metastatic cancer})$
$+ f_l(\text{laparoscope}) + f_l(\text{admission type}) + f_l(\text{separation mode})$
$+ f_l(\text{hospital type}) + f_l(\text{collocated}) + + f_l(\text{surgical procedure})$
$+ f_{ri}(\text{hospital campus}),$

where: Q_t = modelled quantile, $t = 1$ to 99;

f_{nl} = non linear function;

f_l = ordinary linear squares function for continuous or categorical data;

$f_{nl\,cyclic}$ = non linear function using cyclic splines;

f_{ri} = random intercept to adjust for correlations in LOS outcomes within a hospital.

2.2.3   Recovering the Unconditional Predicted Quantile
As with ordinary linear regression that predicts a mean of a dependent variable conditional on independent covariates, quantile regression also predicts quantiles of the distribution conditional on independent covariates. That is, if we refer to the quantiles

we are modelling as $Q(z)$ for the zth quantile and the vector of covariates as x, then we are modelling $\widehat{Q}(z|x)$, where $\widehat{Q}$ indicates an estimate of Q. To obtain the unconditional estimate $Q(z)$ we need to average out the effect of x over the distribution that the sample is drawn from. That is,

$$Q(z) = \int_{-\infty}^{+\infty} f(x)\hat{Q}(z|x)dx \tag{2}$$

where $f(x)$ is the joint probability distribution of the covariates. We don't know the mathematical formulation for $f(x)$, however, the above expression is equivalent to $E\left[\widehat{Q}(z|x)\right]$, that is, the mean of $\widehat{Q}(z|x)$. Since our sample is clearly representative of $f(x)$ and the number of estimates is very large (28,343 = the sample size), taking the mean will converge to $Q(z)$ by the law of large numbers.

2.2.4   Smoothing Count Data – a Technicality

Due to the combination of LOS being a count variable and the estimation of the conditional quantile involves a non-smooth objective function, a certain amount of smoothness needs to be imposed on the data. This is done by adding a specific form of random noise, referred to as jittering, which preserves the one to one relationship between the jittered and un-jittered data. Hence the model estimates based on the jittered data are readily converted to their un-jittered values. This is well described by Machado and Silva [41]. To accomplish this, we used the dither function in the R package quantreg [42]. The counts were recovered from the modelled jittered LOS by applying Theorem 2 by Machado and Silva [41],

$$\widehat{Q}(z|x) = \text{ceiling}\left[\text{jittered}\left\{\widehat{Q}(z|x)\right\} - 1\right] \tag{3}$$

where $\widehat{Q}(z|x)$ **is** the estimated quantile conditional on x, the vector of covariates, and ceiling[n] denotes the smallest integer greater than or equal to n.

2.2.5   Building the Second Additive Quantile Regression Model Without Boosting

We used the results of the boosting to build a second AQR model. The boosting indicated that random effects for hospital was not important for predicting LOS and that all entered variables may be important for predicting LOS except for Elixhauser (co-morbidity) score, co-location status and sex – see Fig. 1. Month and separation mode were the strongest predictors in the sense of reduction of model fit error by up to approximately 3 and 1 respectively for some of the percentiles. Other variable contributions were below about 0.5 for all percentiles. Although boosting suggested that sex was not an important variable, we still included it due to its generally important relevance in epidemiological studies. This decision was eventually justified, see Sect. 3.3.4 below. Boosting also suggested that non-linear fits for MAV, AV, age and month better predicted LOS than linear but a linear fit for year and number of daily colorectal surgery admissions better predicted LOS. We still entered the number of daily admissions and year as a non-linear fit as model regularization would likely result in a linear fit without taxing degrees of freedom.

The variables that were indicated as important for predicting LOS were then entered into the second AQR. This was done without including random effects for hospital as suggested by the boosting. The AQR was applied with a regularization of the model fit to the data to reduce over fitting and to increase prediction accuracy. This is done by the calculation of smoothing parameters for the continuous variables and the use of a least absolute shrinkage and selection operator (lasso) for the categorical variables [27, 35, 36]. The selection of the smoothing parameter for each continuous variable is first initiated on a univariate basis. These initial values are then passed onto a function, along with a starting value for the lasso, which then recalibrates them over the whole parameter space for each model fit to each percentile. These functions and the R computer code for their implementation are well described by Koenker [27].

This is the specification of the second AQR model that we implemented following variable selection by the boosting process.

$Q_t(\text{LOS})$ modelled with *intercept*

$+ f_{nl}(\text{mean annual surgery volume}) + f_{nl}(\text{annual surgery volume})$

$+ f_{nl}(\text{age}) + f_{nl}(\text{daily surgery admissions}) + f_{nl}(\text{year of discharge})$

$+ f_l(\text{month of discharge}) + f_l(\text{sex}) + f_l(\text{ASA}) + f_l(\text{cancer site})$ (4)

$+ f_l(\text{metastatic cancer}) + f_l(\text{laparoscope}) + f_l(\text{admission type})$

$+ f_l(\text{separation mode}) + f_l(\text{hospital type}) + f_l(\text{surgical procedure}).$

The second AQR was carried out with the R package quantreg [42] and was compared to the boosted model.

2.2.6 Comparing the AQR Models - with Boosting to Without Boosting

The models were compared in four ways. Firstly we used the continuous ranked probability score (CRPS) [26, 43, 44] as formulated by Gneiting and Ranjan to compare density forecasts [43] - see Eq. 6 in their paper. Secondly, we calculated the Akaike information criterion (AIC) for each percentile and added the results. We defined the AIC as 2 * (number of variables - logarithm of the likelihood). Thirdly we compared by eye how well the graph of the recovered distribution represented the actual empirical distribution of LOS. We termed this graphical fit. Fourthly, we computed the areas under the graphs of the empirical and recovered distribution, between the 1st and 99th percentiles and compared them. A recovered distribution should give an area close to the area under the empirical distribution. Computing the area under the graph was done using the R package flux using the auc function [45].

Table 1 compares model capability of both models in recovering the empirical distribution of LOS. Although the boosted model had overall lower AIC, due to the importance of the continuous ranked probability score (CRPS) for comparing density forecasts, the superior approximation of the area under the graph and better graphical fit (Fig. 2) we proceeded with the second AQR model in assessing the association between provider volume and LOS.

Table 1. Model performance indicators. See Sect. 2.2.6.

	Boosted model	2nd AQR
CRPS Lower is better	4.27	3.95
Total AIC over 1–99% Lower is better	12,173,948	19,758,132
Area under graph ratio to empirical LOS Close to 1 is better	0.95	1.02

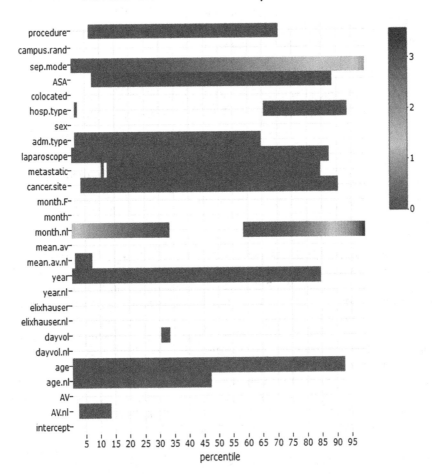

Fig. 1. Variable and functional fit importance indicated by boosting. The scale is total reduction in model fit error over all iterations. See Sects. 2.2.1 and 2.2.5.

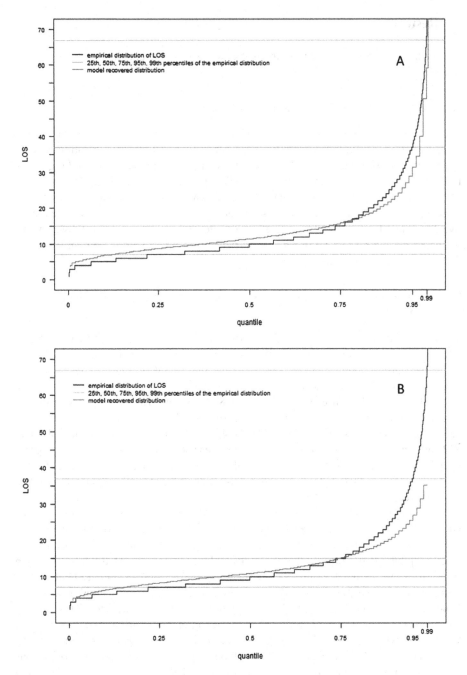

Fig. 2. The empirical distribution of LOS in black compared to the recovered distribution by the AQR models in red: **A** without boosting; **B**: with boosting. (Color figure online)

3 Results

For the sake of brevity we present the important results that illustrate the utility of our model.

3.1 Mean Annual Volume Association with LOS

The model graphs (Fig. 3) indicate that hospitals' performances (contextual effect) regarding patient LOS varied greatly for all percentiles and that this variation was not systematically associated with MAV. At almost any point in the graph, looking left or right displays both lower and higher LOS. The graphs display initial falls in LOS for MAV up to 33 approximately which are then followed by marked variation, with low points in LOS at MAV of 122.1 and 245.8 and a high point at 105.6 MAV. The former two MAV had the lowest LOS over all percentiles 40 and 20 times respectively while the latter had the highest 89 times. For all percentiles, the p values from an F statistic for model fit were less than 1×10^{-6}.

3.2 Counterfactual Prediction of Change in LOS Contingent on Change in MAV

To carry out the counterfactual prediction, we selected the hospital with MAV of 122.1 as an LOS efficient hospital due to its consistent association with reduced LOS over many percentiles. There were 68 hospitals that had MAV of 122.1 or less and they generated 106,488 LOS days (27.5%) from 7,979 episodes of care (28.1%). The counterfactual prediction estimated a fall of 8.5% in total LOS over all patients and hospitals, p < 0.007, with 95% CI (2.9%, 14.6%). This equated to a reduction of 32,842 total LOS days, 95% CI (11,225, 56,434 days) or a predicted saving of about $32 million in present day terms over the ten year study period by allowing approximately $1000 per LOS day [34]. Figure 4 demonstrates that, for the counterfactual experiment, the predicted savings mainly came from reduced LOS for patients who had LOS between percentiles 14 and 94.

By using the model's counterfactual prediction capacity, an index of performance with a 95% CI can be obtained for any hospital. The prediction would immediately indicate if the hospital was performing better, worse or at about the same level as all other hospitals. If performing the counterfactual experiment, using the hospital being assessed as the basis of comparison, resulted in a 5% drop in total LOS then that hospital would have an index of .95 with an associated confidence interval. Lower than 1 indicates superior efficiency, higher than 1 inferior and 1 no difference. This index is independent of any distributional or model fit assumptions or arbitrary categorization. For the analysis of LOS, the Victorian Auditor General's report resorted to using trimmed data in a linear regression. This method may be subject to statistical objections if a mean is not representative of the whole data and because information in the tail(s) of the distribution, that may represent patients who may not fit the profile of a mean patient, is discarded [25, 34, 46, 47].

Our model could be used in national or international settings as it can allow for nesting in those levels, and so help assess hospital efficiency in regard to LOS in broader contexts. This would assist with synthesis of future international studies when in the past, diverse categorization methods had impeded synthesis. It can be extended to analyse variation between hospitals regarding other outcomes such as mortality and readmission following CRC surgery.

3.3   Further Results for Patient and Hospital Factors

3.3.1   Laparoscope Use – See Fig. 5

Use of laparoscope was associated with reduced LOS. There were growing reductions of between 0.5 to 4 days with increasing LOS percentile which indicated the importance of laparoscope use, where possible, for helping to reduce unnecessarily long LOS. Use of laparoscope was the main modifiable feature of our analysis. This suggested where possible, use of laparoscope is important for patient outcomes and resource allocation, as observed and recommended by others [11]. This is an example of where, if solely a marginal change had been predicted, the result would have been just one coefficient comparing laparoscope use to non – use. The one coefficient would have represented an overall average effect and would have been graphically illustrated as a flat line across all percentiles. Instead, with quantile regression, we see how the effect due to laparoscope use varied over across all percentiles of LOS. This observation holds for all the presented variable results. All p values for all coefficients over all percentiles were less than 1×10^{-4}.

3.3.2   Separation Mode – See Fig. 6

It is evident from Fig. 6 that discharge to transition care, an aged care residential facility, other acute hospital or a statistical separation contributed heavily to protracted LOS with between 2–50 days greater than patients who were transferred to home. This difference increased with increasing LOS percentile. Left against medical advice shows no association with LOS as the coefficients for nearly all percentiles are close to zero and non-significant. It seems that patients who died in hospital after colorectal cancer surgery did so either early in their hospital stay or at the end of a protracted stay. The p values for all statistically significant coefficients, were mainly less than 0.01.

There may be some scope for improvements in LOS for patients who are transferred for extended care to other acute hospitals, rehabilitation or geriatric care centres. These patients had between 2–50 days more LOS compared to patients who were transferred to home which accounted for 104,497 days (27%) of total LOS. If with vigilant follow up and management, a modest 10% of these days were to be saved, that would have amounted to a saving of about 10,400 days and a potential saving of about $10 million in present terms over the 10 year study period [34].

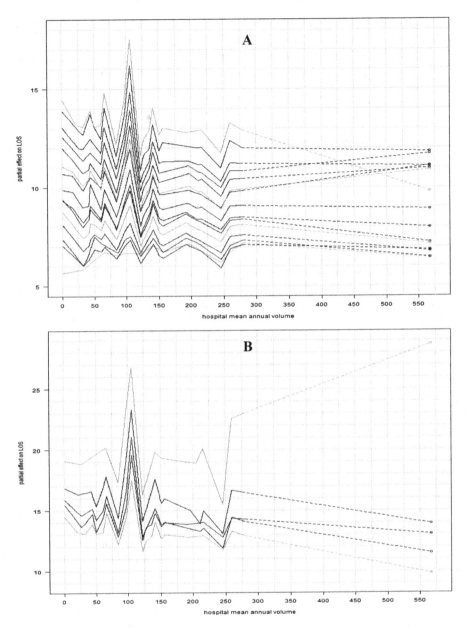

Fig. 3. Model estimates for between hospital differences (contextual effect) in the association between annual volumes and LOS; **A**-percentiles 5–75, **B**-percentiles 75–95, both in intervals of 5. Red lines are percentiles 5, 25, 50, 75 and 95. The lines are dotted between 279.6 and 566.7 as there were no hospitals with MAV between these values. A few of the percentile fits display quantile crossing (see Sect. 3.4). All p values, from an F statistic that assessed model fit for percentiles 1–95, were less than 1×10^{-6}. (Color figure online)

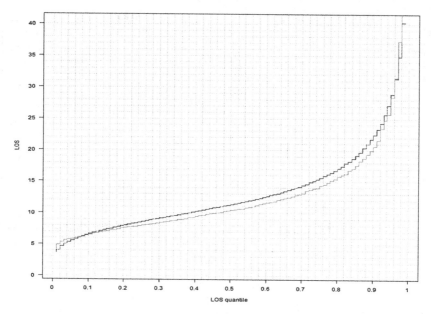

Fig. 4. In black we have the recovered distribution. Red is the counterfactual distribution obtained by setting all MAV to 122.1 and all AV to the same AV in each year as generated by the hospital with MAV of 122.1. (Color figure online)

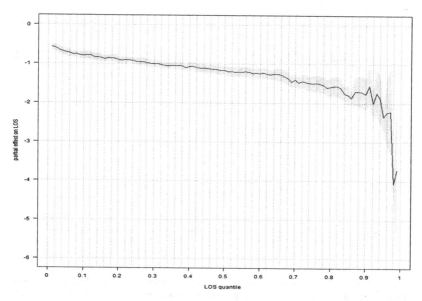

Fig. 5. The association between laparoscope use and LOS. Use of laparoscope is compared to non-use of laparoscope. The shading represents the coefficient 95% CI for each percentile.

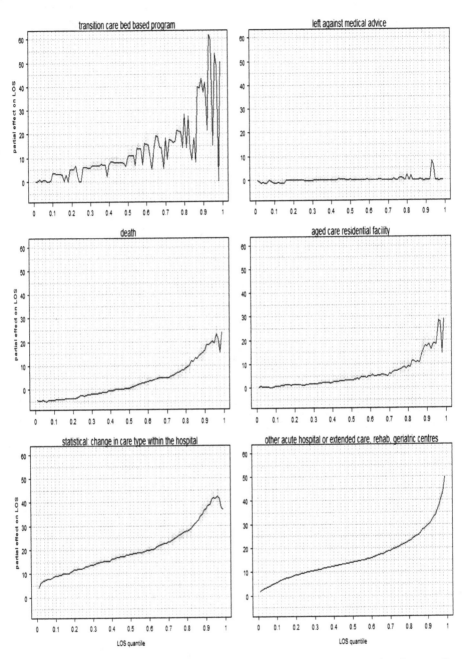

Fig. 6. Model estimates for the association between separation mode and LOS for all percentiles 1–99. These coefficients use separation to private residence or accommodation as the reference level. The shading represents the coefficient 95% CI for each percentile.

3.3.3 Month of Year – See Fig. 7

Please note the months are within a fiscal year and are ordered from July to June. Compared to July, month of year differences in LOS ranged between −4.6 to 3.9 days over all percentiles but the bulk of differences were between −0.5 to 0.5 days. The larger magnitudes were for the higher LOS percentiles of 90 or more. Except for January, all months mostly had lower LOS than July but only May, June, October and December showed consistency (Fig. 7C) and strong statistical evidence (Fig. 7B) for this difference. January tended to have higher LOS however, zero was not extreme for this relative difference compared to July (Fig. 7C) as can also be seen by lack of statistical significance (Fig. 7B). Most likely these monthly effects were due to seasonally adjusted hospital administration factors rather than an environmental seasonal effect.

The association between month of year and LOS was a surprise finding. This is more likely to reflect seasonally modified discharge practices rather than seasonal environmental factors [48] and so may potentially be another modifiable feature. If the same efficiencies that seem to have applied to discharges in May, June, December and October were to be applied to discharges in all other months of the year, then potentially there could have been reductions of approximately 1000–2000 days of LOS which would translate to savings of between $1 million to $2 million dollars approximately over the 10 year study period.

Adding these savings and savings due to better management of separation mode (see Sect. 3.3.2) to savings reaped from increased hospital efficiency indicated in the main results (see Sect. 3.2), potential total savings due to improved LOS efficiency could be more than $4 million per year for CRC surgery.

3.3.4 Sex - See Fig. 8

For percentiles up to about 20 there were no statistical differences between LOS for women and men. For higher percentiles we see a gradual decrease for women to about 3 days less LOS compared to men. This may be related to being in better general health at surgery due to health promoting behaviours that are more likely to be exhibited by women than men.

3.4 Quantile Crossing

When the predicted unconditional quantiles are combined to recover the full distribution, the monotonicity of the cumulative distribution function (CDF) may not be retained. That is, at times a predicted value of LOS that does not respect the strictly increasing property of quantiles may be produced by a model. This has been referred to as quantile crossing [25, 28]. We used the process of monotone rearranging to restore the required monotonicity property to the CDF for the recovered distribution of LOS [26, 49]. The rearrangement process was based on mathematical results by Hardy, Littlewood and Polya [50, 51] and was implemented with the rearrangement function in the R package, quantreg [42].

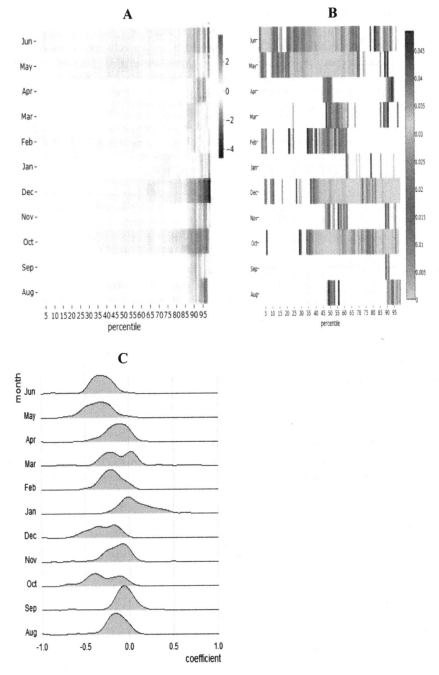

Fig. 7. A - Coefficients for month of year compared to July for percentiles 1–99. **B** - p values for month coefficients for percentiles 1–99. White indicates a p value > 0.05. **C** - Distribution of coefficients for each month.

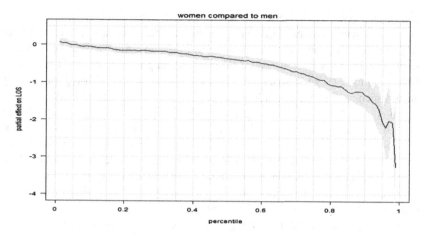

Fig. 8. Model estimates for the association between sex and LOS for all percentiles 1–99. These coefficients are based on comparison of women to men as the reference level. The shading represents the coefficient 95% CI for each percentile.

Quantile crossing may also occur when generally modelling the association between different quantiles of a dependent variable as a *function* of an independent variable. This occurred in this study a small number of times in modelling the association between quantiles of LOS and mean annual volume, annual volume and age of patient – only the results for mean annual volume, Fig. 3, are displayed for brevity purposes. It is evident that fitted lines (functions) for a few of the percentiles cross the lines of other percentiles. The advantage of quantile regression in not requiring global distributional assumptions but instead using the data near the specified quantile, sometimes has the disadvantage of producing quantile crossing. That is, in modelling one quantile, quantile regression is ignorant of other quantiles [25]. However, quantile crossing usually occurs in sparse areas of the data and is also sensitive to outlier values for the *independent* covariate (quantile regression is robust to *dependent* variable outlier values) [25, 52]. Linear regression may also be affected by these conditions [25]. The 99th LOS percentile for age was affected by both conditions as the 99th percentile is a sparse region of the data and there was a patient of 103 years of age. For mean annual volume and annual volume, there was a clear outlier with a relatively large distance from the next lowest values and no intervening values (dotted lines in those graphs). The fit determined by data to the left of the dotted line cannot anticipate lack of data to the right. If quantile crossing occurs substantially often, then this may indicate model misspecification [25]. Quantile crossing is an area of ongoing research [25, 49].

4   Discussion

Our model presented itself as a useful method for assessing provider performance regarding LOS, adjusted for pertinent patient and hospital factors [19]. The additive quantile regression approach proved a suitable method to cope with the difficulties inherent in analysing LOS due to its greatly skewed distribution and non-linear association with provider volume. Assuming a distributional fit or arbitrary categorization of LOS and focusing only on the mean would not have allowed the assessment of associations with LOS over its entire distribution. Imposing a linear fit causes non-linear associations to go undetected, if present in the data. Focusing on only one parameter, such as a mean, would have assumed that change in mean LOS would have translated equally to all patients or that the "average" patient was representative of all patients. It is quite possible that there may be important changes in the tails of a distribution yet, the mean is unchanged [25]. A linear, Cox or logistic regression would have resulted in only one (average) coefficient for all percentiles and would not have given any insight into differential associations across LOS percentiles. That is, focusing only on mean outcomes may cause important information about patients who may not fit well into the frame of average to go unobserved.

General linear models focus on marginal change in the outcome variable dependent on predictive factors, however it is quite possible that there is a multifaceted interchange between the outcome and predictive factors which may vary for different levels of the outcome. If we aim to understand the full impact of possible predictors on an outcome, predicting a marginal change is not sufficiently meaningful [26, 53–55]. To model a possible multifaceted interchange between predictors and outcomes, we need to examine change in the outcome, conditional on the predictor, across the outcome distribution in its entirety or at least a sufficiently representative number of percentiles. Our model obtained a more realistic idea of the change in the outcome variable, conditioned on important covariates, by predicting all percentiles. This has been likened to the difference between calculating only "what changes" compared to "who changes" and by how much [46] where, in our case, surgery patients were the who. These are ongoing and important issues for epidemiological studies and data analysis generally [47].

Our model did not assume any distributional fit which may bias estimates or reduce sensitivity to detect associations if the assumption is an over simplification and therefore not justified [56]. John Tukey referred to this as having more honest foundations for data analysis [47]. Furthermore, we did not use any arbitrary categorical definitions of the outcome, or continuous predictors which has hindered synthesis of past studies. John Ioannidis in his seminal work, Why Most Published Research Papers are False, stated in his 4th Corollary that the greater the flexibility in designs, definitions, outcomes, and analytical modes in a scientific field, the less likely the research findings are to be true [57].

Although we used a Mundlak formulation in our example, it is not necessary to do so. The basic idea of predicting the outcome distribution can be used in any typical regression formulation for a continuous or count outcome. Our model was further complemented by shrinkage and penalization to obtain more accurate estimates and reduce statistical error [19, 27].

Our model did have the limitation of being data hungry. A large sample size is required to reliably employ our model especially if many covariates are entered into the analysis. However, this age of big data somewhat mitigates this limitation. Our model is also computer resource intensive. To predict percentiles 1–99 and run the counterfactual simulation took approximately 70 min on a 64-bit PC with 64 GB RAM and a

Xeon® 3.5 GHz CPU but 5.5 h on a 64-bit PC with 16 GB of RAM and an i5 2.5 GHz CPU. The recalibration of the lasso and smoothing parameters over the whole parameter space for each model fit for percentiles 1–99, took approximately 36 h on a 64-bit PC with 16 GB of RAM and an i5 2.5 GHz CPU.

This model can only be applied to continuous or count outcomes but there has been some recent work that has applied quantile regression to time to event data and work that seeks to apply it to binary response data [25].

5 Conclusion

Predicting the whole outcome distribution was useful in providing an in depth description of the complexity of the associations between hospital and patient factors across the whole distribution of LOS. The facility of the model to indicate change over the whole distribution is useful where predicting a change in mean or median outcome is an oversimplification of the data and does not provide insight that is sufficiently indicative and real-world. With sufficient data, our model can be applied to any continuous or count outcome. Improvements in optimizing the time to predict the percentiles and calculate lasso and smoothing parameters are required. As far as we can tell, this method is novel for public health and epidemiological studies and may have further uses in these areas as well as other fields of scientific research.

References

1. Borowski, D.W., Bradburn, D.M., Mills, S.J., Bharathan, B., Wilson, R.G., Ratcliffe, A.A., et al.: Volume-outcome analysis of colorectal cancer-related outcomes. Br. J. Surg. 97(9), 1416–1430 (2010)
2. Burns, E.M., Bottle, A., Almoudaris, A.M., Mamidanna, R., Aylin, P., Darzi, A., et al.: Hierarchical multilevel analysis of increased caseload volume and postoperative outcome after elective colorectal surgery. Br. J. Surg. 100(11), 1531–1538 (2013)
3. Chowdhury, M.M., Dagash, H., Pierro, A.: A systematic review of the impact of volume of surgery and specialization on patient outcome. Br. J. Surg. 94(2), 145–161 (2007)
4. Faiz, O.: The volume–outcome relationship in colorectal surgery. Tech. Coloproctol. 18(10), 961–962 (2014). Official Journal of SICCR, MSCP, ISCRS, ECTA, Colorectal Anal Group of Surgical Section of Chinese Medical Association, MSPFD
5. Killeen, S.D., O'Sullivan, M.J., Coffey, J.C., Kirwan, W.O., Redmond, H.P.: Provider volume and outcomes for oncological procedures. Br. J. Surg. 92(4), 389–402 (2005)
6. Kizer, K.W.: The volume-outcome conundrum. New Engl. J. Med. 349(22), 2159–2161 (2003)
7. McGrath, D.R., Leong, D.C., Gibberd, R., Armstrong, B., Spigelman, A.D.: Surgeon and hospital volume and the management of colorectal cancer patients in Australia. ANZ J. Surg. 75(10), 901–910 (2005)
8. Austin, P.C., Rothwell, D.M., Tu, J.V.: A comparison of statistical modeling strategies for analyzing length of stay after CABG surgery. Health Serv. Outcomes Res. Method. 3(2), 107–133 (2002)

9. Gatt, M., Anderson, A.D., Reddy, B.S., Hayward-Sampson, P., Tring, I.C., MacFie, J.: Randomized clinical trial of multimodal optimization of surgical care in patients undergoing major colonic resection. Br. J. Surg. **92**(11), 1354–1362 (2005)

10. Huebner, M., Hubner, M., Cima, R.R., Larson, D.W.: Timing of complications and length of stay after rectal cancer surgery. J. Am. Coll. Surg. **218**(5), 914–919 (2014)

11. Thompson, B.S., Coory, M.D., Gordon, L.G., Lumley, J.W.: Cost savings for elective laparoscopic resection compared with open resection for colorectal cancer in a region of high uptake. Surg. Endosc. **28**(5), 1515–1521 (2014)

12. Zheng, Z., Hanna, N., Onukwugha, E., Bikov, K.A., Mullins, C.D.: Hospital center effect for laparoscopic colectomy among elderly stage I-III colon cancer patients. Ann. Surg. **259**(5), 924–929 (2014)

13. Faiz, O., Haji, A., Burns, E., Bottle, A., Kennedy, R., Aylin, P.: Hospital stay amongst patients undergoing major elective colorectal surgery: predicting prolonged stay and readmissions in NHS hospitals. Colorectal Dis.: Off. J. Assoc. Coloproctol. Great Br. Irel. **13**(7), 816–822 (2011)

14. Gruen, R.L., Pitt, V., Green, S., Parkhill, A., Campbell, D., Jolley, D.: The effect of provider case volume on cancer mortality: systematic review and meta-analysis. CA: Cancer J. Clin. **59**(3), 192–211 (2009)

15. Fenske, N., Fahrmeir, L., Hothorn, T., Rzehak, P., Hohle, M.: Boosting structured additive quantile regression for longitudinal childhood obesity data. Int. J. Biostat. **9**(1), 1–8 (2013)

16. Borghi, E., de Onis, M., Garza, C., Van den Broeck, J., Frongillo, E.A., Grummer-Strawn, L., et al.: Construction of the World Health Organization child growth standards: selection of methods for attained growth curves. Stat. Med. **25**(2), 247–265 (2006)

17. Wang, X., Dey, D.K.: Generalized extreme value regression for binary response data: an application to B2B electronic payments system adoption. Ann. Appl. Stat. **4**(4), 2000–2023 (2010)

18. Archampong, D., Borowski, D., Willejrgensen, P., Iversen, L.H.: Workload and surgeons specialty for outcome after colorectal cancer surgery. Cochrane Colorectal Cancer Group **3**(3) (2012)

19. Ash, A.S., Fienberg, S.F., Louis, T.A., Norm, S.T., Stukel, T.A., Utts, J.: Statistical issues in assessing hospital performance. Committee of Presidents of Statistical Societies The COPSS-CMS White Paper Committee (2011). http://citeseerx.ist.psu.edu/viewdoc/summary?doi=10.1.1.352.6798

20. Bell, A., Fairbrother, M., Jones, K.: Fixed and random effects models: making an informed choice 2017 March 2018. https://www.researchgate.net/publication/299604336_Fixed_and_Random_effects_models_making_an_informed_choice

21. Feaster, D., Brincks, A., Robbins, M., Szapocznik, J.: Multilevel models to identify contextual effects on individual group member outcomes: a family example (Report). Fam. Process **50**(2), 167 (2011)

22. Dieleman, J.L., Templin, T.: Random-effects, fixed-effects and the within-between specification for clustered data in observational health studies: a simulation study. PLoS ONE **9**(10), e110257 (2014)

23. Danks, L., Duckett, S.: All complications should count: using our data to make hospitals safer (Methodological supplement) (2018). https://grattan.edu.au/wp-content/uploads/2018/02/897-All-complications-should-count-methodological-supplement.pdf

24. Hofner, B., Mayr, A., Robinzonov, N., Schmid, M.: Model-based boosting in R: a hands-on tutorial using the R package mboost. Comput. Stat. **29**(1), 3–35 (2014)

25. Koenker, R.: Quantile Regression. Cambridge University Press, Cambridge (2005)

26. Taieb, S.B., Huser, R., Hyndman, R.J., Genton, M.G.: Forecasting uncertainty in electricity smart meter data by boosting additive quantile regression. IEEE Trans. Smart Grid **7**(5), 2448–2455 (2016)
27. Koenker, R.: Additive models for quantile regression: model selection and confidence bandaids. Braz. J. Probab. Stat. **25**(3), 239–262 (2011)
28. Koenker, R.: Quantile Regression in R: A Vignette, 6 March 2018 (2018). https://cran.r-project.org/web/packages/quantreg/vignettes/rq.pdf
29. Mundlak, Y.: On the pooling of time series and cross section data. Econometrica **46**(1), 69 (1978)
30. van de Pol, M., Wright, J.: A simple method for distinguishing within - versus between-subject effects using mixed models. Anim. Behav. **77**(3), 753–758 (2009)
31. Aravani, A., Samy, E.F., Thomas, J.D., Quirke, P., Morris, E.J., Finan, P.J.: A retrospective observational study of length of stay in hospital after colorectal cancer surgery in England (1998–2010). Medicine **95**(47), e5064 (2016)
32. Cologne, K.G., Byers, S., Rosen, D.R., Hwang, G.S., Ortega, A.E., Ault, G.T., et al.: Factors associated with a short (<2 Days) or Long (>10 Days) length of stay after colectomy: a multivariate analysis of over 400 patients. Am. Surg. **82**(10), 960–963 (2016)
33. Field, K., Shapiro, J., Wong, H.L., Tacey, M., Nott, L., Tran, B., et al.: Treatment and outcomes of metastatic colorectal cancer in Australia: defining differences between public and private practice. Intern. Med. J. **45**(3), 267–274 (2015)
34. Frost, P.: Victorian auditor-general's report: hospital performance: length of stay. In: Victorian, Auditor-General's, Office (eds.) (2016)
35. Efron, B., Hastie, T.: Computer Age Statistical Inference: Algorithms, Evidence, and Data Science. Cambridge University Press, Cambridge (2016) https://web.stanford.edu/~hastie/CASI_files/PDF/casi.pdf
36. Hastie, T., Tibshirani, R., Friedman, J.: The Elements of Statistical Learning: Data Mining, Inference, and Prediction, 2nd edn. Springer, New York (2009) https://web.stanford.edu/~hastie/Papers/ESLII.pdf. https://doi.org/10.1007/978-0-387-84858-7
37. Hofner, B., Hothorn, T., Kneib, T., Schmid, M.: A framework for unbiased model selection based on boosting. J. Comput. Graph. Stat. **20**(4), 956–971 (2011)
38. Bühlmann, P., Hothorn, T.: Boosting algorithms: regularization, prediction and model fitting. Stat. Sci. **22**(4), 477–505 (2007)
39. R Core Team: R: a language and environment for statistical computing Vienna, Austria: R Foundation for Statistical Computing (2017) http://www.R-project.org/
40. Hothorn, T., Buehlmann, P., Kneib, T., Schmid, M., Hofner, B.: mboost: Model-Based Boosting (2017) https://CRAN.R-project.org/package=mboost
41. Machado, J.A.F., Silva, J.M.C.S.: Quantiles for counts. J. Am. Stat. Assoc. **100**(472), 1226–1237 (2005)
42. Koenker, R.: Quantreg: Quantile Regression (2017) https://CRAN.R-project.org/package=quantreg
43. Gneiting, T., Ranjan, R.: comparing density forecasts using threshold - and quantile-weighted scoring rules. J. Bus. Econ. Stat. **29**(3), 411–422 (2011)
44. Hersbach, H.: Decomposition of the continuous ranked probability score for ensemble prediction systems. Weather Forecast. **15**(5) (2000) https://journals.ametsoc.org/action/doSearch?AllField=Decomposition+of+the+Continuous+Ranked+Probability+Score+for+Ensemble+Prediction+Systems
45. Jurasinski, G., Koebsch, F., Guenther, A., Beetz, S.: Flux: flux rate calculation from dynamic closed chamber measurements (2014). https://CRAN.R-project.org/package=flux

46. Hohl, K.: Beyond the average case: the mean focus fallacy of standard linear regression and the use of quantile regression for the social sciences. SSRN, Elsevier (2009). https://ssrn.com/abstract=1434418

47. Tukey, J.: More honest foundations for data analysis. J. Stat. Plan. Inference 57(1), 21–28 (1997)

48. Kc, D.S., Terwiesch, C.: Impact of workload on service time and patient safety: an econometric analysis of hospital operations. Manag. Sci. 55(9), 1486–1498 (2009)

49. Chernozhukov, V., Fernández-Val, I., Galichon, A.: Quantile and probability curves without crossing. Econometrica 78(3), 1093–1125 (2010)

50. Burchard, A.: A short course on rearrangement inequalities (2018). https://www.math.toronto.edu/almut/rearrange.pdf

51. Hardy, H.G., Littlewood, J.E., Pólya, G.: Inequalities. Acta Applicandae Mathematica 23(1), 95 (1991)

52. John, O.O.: Robustness of quantile regression to outliers. Am. J. Appl. Math. Stat. 3(2), 86–88 (2015)

53. Kneib, T.: Beyond mean regression. Stat. Model. 13(4), 275–303 (2013)

54. Harvey, A.: Discussion of 'Beyond mean regression'. Stat. Model. 13(4), 363–372 (2013)

55. Le Cook, B., Manning, W.G.: Thinking beyond the mean: a practical guide for using quantile regression methods for health services research. Shanghai Arch. Psychiatry 25(1), 55–59 (2013)

56. Fenske, N., Burns, J., Hothorn, T., Rehfuess, E.A.: Understanding child stunting in India: a comprehensive analysis of socio-economic, nutritional and environmental determinants using additive quantile regression. PLoS ONE 8(11), e78692 (2013)

57. Ioannidis, J.P.: Why most published research findings are false. PLoS Med. 2(8), e124 (2005)

Resilient and Deep Network for Internet of Things (IoT) Malware Detection

Nazanin Bakhshinejad[(⊠)] and Ali Hamzeh

Department of Computer Science and Engineering and IT, Shiraz University,
Shiraz, Iran
Nazanin.bakhshinejad@shirazu.ac.ir,
ali@cse.shirazu.ac.ir

Abstract. Nowadays, digital devices and the internet make our life remarkably easy since a massive number of daily activities can be carried out simply through the internet. Internet of Things (IoT) devices are increasingly employed in diverse industries with a wide range of purposes such as sensing or collecting environmental data. The development of IoT brings many opportunities but also many security challenges. Recently, the presence of IoT in a numerous number of applications and their improving computing and processing abilities make them a vulnerable attack target. Hence, developing a method that is capable of proactively detect and prevent malware in IoT is a perpetual demand. In the recent years, machine learning techniques have been applied in the field of malware detection and achieved acceptable results, however; these approaches have inherently a challenging step called feature extraction. Therefore, we need a method that has the ability to automatically extract features which is significantly time-consuming and error-prone process. The introduction of deep learning, a new area of artificial intelligence, helps the malware detection by automating the feature extraction due to its multi-layer training. This paper proposes a novel architecture of Convolutional Neural Network (CNN) that utilizes raw bytes as input and eliminates the need to extract high-level features manually. In addition, we benefit from the reputed embedding techniques to generate numerical vectors of bytes since deep networks only accept numerical vectors as input. Our results indicate that the proposed approach can achieve high detection rate of malware among IoT devices, outperforming traditional machine learning based methods which reveals the merit of deep learning techniques in IoT malware detection.

Keywords: Malware detection · Internet of Things · Deep learning · CNN · Embedding · Word2vec

1 Introduction

Electronic devices and the Internet are rapidly becoming a vital part of our daily life. It is essential to make digital devices significantly secure because they work with sensitive data that needs high level of privacy. In fact, this gain in acceptance has not come without its costs. A huge number of computer attacks occur yearly and diverse types of

© Springer Nature Singapore Pte Ltd. 2019
H. Nguyen (Ed.): RSSDS 2019, CCIS 1150, pp. 183–197, 2019.
https://doi.org/10.1007/978-981-15-1960-4_13

malware are designed with different destructive intentions to infiltrate IoT devices and threaten the users' security and privacy [1].

A typical deployment of Internet of Things (IoT) consists of a wide pervasive networks of Internet-connected devices, Internet-connected vehicles, sensors, embedded systems and other devices or systems that autonomously sense, store, transfer and process collected data [2]. Owing to the growing popularity of Internet of Things (IoT) devices and the lack of approaches capable of providing security for these devices, they can be highly vulnerable to malware attacks [3]. According to the Kaspersky Lab, the majority of IoT devices assessed in 2016 were insecure, most of these devices had either default password or unpatched vulnerabilities. This means that IoT devices can be easily targeted and compromised by malware. Therefore, we need specialized tools, techniques and procedures for making IoT networks secure and collecting, preserving and analyzing residual evidences of IoT environments.

Malware (short for malicious software) is a software created to deliberately actualize the detrimental intent of an attacker [4]. Malware can be categorized in several groups in terms of their specifications and purposes such as viruses, Trojans, botnets, worms, spyware and backdoors [5]. Gaining access to private systems, stealing user's confidential information or infecting files on system are merely some of security attacks that can be carried out by malware and caused drastic damages as well as financial loss to users.

Signature-based methods, which are based on the unparalleled binary patterns of each malicious sample, have been employed by commercial anti-malware [6–8]. The complexity and diversity of malware samples are swiftly increased each year which makes the efficiency of signature-based techniques remarkably blur since they are only able to detect pre-known malware and they fail to detect obfuscated and unknown samples. According to these drawbacks of traditional anti-malware approaches, researchers are persuaded to find a series of intelligent methods using machine learning techniques to deal with samples [9, 10].

In general, special models are built in machine learning techniques in order to learn from input data and then to process unseen samples. These input data should be in the form of features such as integers or floating numbers, Boolean and categorical values. Owing to the fact that the efficiency of machine learning approaches is highly dependent on the input features, selecting and processing features are the most challenging steps. In fact, desirable results cannot be achieved if proper features not extracted, even if we have a robust classifier. Two methods are available to analyze and extract features from files: dynamic [11] and static analysis [9, 12].

Accordingly, a group of machine learning techniques called deep learning is selected in this article to automatically perform the complicated feature extraction process. It has motivated a large number of prosperous applications in image classification, speech recognition and natural language processing. A multi-layer deep learning can come up with a high-level representation of data by associating features. Hence, it can provide pre-training multiple layers of feature detectors from the lowest to the highest level to build the final classification model [13]. Moreover, using the deep architecture leads to the simultaneous execution of feature extraction and dimension reduction.

This paper introduces a new architecture of CNN that applies raw bytes of files as input. Since the length of byte sequences is different for each file, we employ the header (usually the first 1024 bytes) of files as the input of CNN. Furthermore, the input of deep network should be vectors with identical size, thus embedding techniques are implemented in this article to create fixed size and unique numerical vectors for each byte. Since embedding techniques are often used in text classification problems so we treat each byte as a word and apply word2vec embedding method.

The remainder of this paper is organized as follows. In Sect. 2, some related works conducted in this field will be discussed. Our proposed method and its required background is described step by step in Sect. 3. In Sect. 4, experimental results are presented to evaluate our method and compare it to other relevant approaches. Finally, conclusion of our work and a discussion of possible future work is considered in Sect. 5.

2 Related Works

In this section some of the similar recent works done in the field of malware detection will be explained.

Farrokhmanesh et al. [14] proposed a static byte-level approach which is implemented by voice processing technique. In this method, malware detection problem is treated as music classification due to the same content of music and malware files that both of them are an ordinal sequence of components (malware is a sequence of bytes and music is sequence of notes). In fact, bytes of an executable file are converted to MIDI instructions by the goal of generating a real Audio file. To acquire features of audio files, algorithms for feature extracting in music classification are utilized. The most challenging and the main flaw of this work is ascertaining the strategies for converting bytes to music notes. Since conversion strategy has direct impact on the final results, it should be able to convert a byte into simultaneous notes in such a way that they can be distinguishable by feature extraction algorithms.

Hashemi et al. [15] provided an image-based approach to detect unknown malware. They used the idea that different behaviors of benign and malware files create different micro-patterns that can be utilized to distinguish between files of each class. First, each executable file is converted to gray scale image by two techniques named row representation and Markov chain. Then, Local Binary Pattern (LBP) method is employed to extract micro-patterns from created images. They used KNN machine learning classifier to classify samples.

Xiao et al. [16] introduced a detection method based in deep learning to distinguish android malware from trusted applications. As regards to the fact that there is some semantic information in system call sequences of android as the natural language, they treat one system call sequence as a sentence in the language and construct a classifier based on the Long Short-Term Memory (LSTM) language model. In the classifier, at first two LSTM models are trained respectively by the system call sequences extracted from malware and those from benign applications. Then, two similarity scores are calculated respecting to the benign and malware models. Finally, the classifier defines whether the application under analysis is malicious or trusted by the greater score.

A heterogeneous deep learning framework for intelligent malware detection is proposed by Ye et al. [17] which is based on the Windows Application Programming interface calls (API) extracted from the portable executable (PE) files. Their model composed of an AutoEncoder stacked up with multi-layer restricted Boltzmann machines (RBMs) [18] and a layer of associative memory to detect newly unknown malware.

Gibert et al. [19] developed a file agnostic deep learning system that learns visual features from executable files for classifying malware into different families. Their idea was motivated by the visual similarity between malware samples of the same family. Visualization techniques are beneficial for detecting variations in samples since they have the ability to capture minor changes while keeping the global structure. Hence, malware executables were visualized as gray scale images and then a convolutional neural network was trained on the representation of malware's binary content as gray scale images. Their proposed neural network consists of an input layer and three 4-stage feature extractors learning hierarchical features through convolution, activation, pooling and normalization layers and classify malware samples intro various families at the end.

In another research, a deep learning based method is introduced by Cui et al. [20] for detecting malware variants. First, malicious codes were converted to gray scale images which were then delivered to the convolutional neural network as input to classify malware samples into their corresponding families. Usually, the number of malicious code variations differs greatly among various code families which is called data imbalance problem. To resolve this issue, they designed a data equilibrium approach based on the bat algorithm that can deal with a huge volume of variations and work well across malware families.

3   Proposed Method

We design a novel architecture for malware detection problem by the help of CNN [21], one of the robust deep learning networks employed in various areas. Our CNN has an effectual layout of layers demonstrating remarkably more acceptable and preferable results in comparison with simple architecture of CNN. As stated before, our method applies raw bytes of files directly as dealing with bytes is easier and faster than using the source code [18]. Also, raw bytes are always available while high-level features may not be possible to be obtained properly in some cases like when anti-disassembling techniques are applied [9]. Indeed, in this paper each file is considered as a sequence of bytes.

Deep learning algorithms only accept numerical vectors as input, hence it is required to convert bytes to vectors before delivering them to the CNN. Due to this, embedding techniques are applied in this paper with respect to their remarkable performance in generating numerical vectors. Generally, embedding is the name of a group of approaches used in text processing to create vectors for each word of a sentence. In fact, the identical structure of executable files and sentence (both of them consist of a sequence of something, sentence is the sequence of words and file is the sequence of bytes) motivates us to apply word embedding techniques and benefit from them to generate more intelligent numerical vectors for bytes. Therefore, each byte in the byte sequence is treated as a word in the sentence and a unique numerical vector is generated for each byte value by the help of embedding approaches.

Different embedding approaches are provided with a wide range of capabilities that are applied in various fields in accordance to their application. One-hot encoding is one of the most rudimentary embedding techniques created vectors in an uncomplicated way [22]. In this method, a list is firstly created containing all of the distinct words of the text and consecutive indexes are assigned to these words. Then, to represent each word in vector form, an embedding vector with the specified size equivalent to the length of the list is generated. All of the value of this vector is zero except where associated to the index of the target word which its value is one.

One-hot encoding produces vector for each word independently regardless the other words, thus these vectors are not effectual enough as they cannot provide any helpful information related to the meaning of the word, its relation with others that are semantically close to this word, its position in the text, etc. In other words, one-hot encoding vectors are not informative and that's the reason we have employed the other group of embedding methods in this paper called word2vec.

3.1 Word2vec Model

Word2vec names a group of embedding models which recently gained notable popularity in the analysis of natural language text [22]. It is interesting to know that, these models are two-layered shallow networks employed by deep learning models. They take a large corpus of text as input and create a vector space of words. In comparison to the dimensionality of one-hot embedding space which is the size of the list of distinguished words, the dimensionality of the word2vec embedding space is usually lower and also its vectors are denser. The models designed based on the word2vec are considerably effective and achieved reasonable results. Continuous Bag of Words (CBOW) [22] and Skip-Gram [22] are the two eminent architectures of word2vec.

Our proposed embedding network is based on the Skip-Gram model that converts the byte values of each file to numerical vectors. Skip-gram architecture attempts to predict the surrounding words, given the center word. In word2vec models, after the completion of learning process, each word in the text has a distinctive embedding vector with predefined size. Regarding the fact that these vectors are generated by considering their position in the sentence, they have the ability to prevail over the momentous drawback of one-hot encoding. Therefore, by utilizing this superb property of word2vec, the distance between the vectors of semantically close words is marginal. One of the important novelties of this paper is the implementation of embedding techniques in the field of malware detection.

The overall structure of our introduced word2vec network is depicted in Fig. 1. As stated before, we treat each byte as a word and create embedding vectors for all of the byte values. It is essential to remark that, only the header part of each executable file (which is usually the first 1024 bytes) is chosen in this work as input since they contain valuable information about the file. Also, despite of the better results that may obtained using all bytes of each file, the computational complexity and processing time are increased considerably. In the first step, the header bytes of all training samples are extracted and the frequency of each byte in the dataset is calculated. Then, bytes are arranged in descending order with respect to their frequencies to build a dictionary which maps each byte to its index. Consequently, index 1 of the dictionary refers to the byte with the highest frequency and the last index that in our case is 256 (the total number of distinct bytes), is assigned to the byte with the lowest frequency.

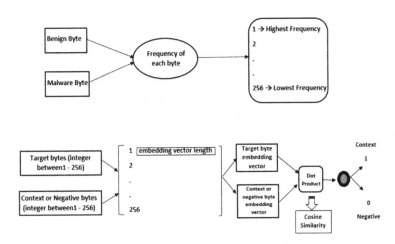

Fig. 1. Overall structure of the proposed word2vec model

It can be seen from Fig. 1 that our byte2vec network takes two one-hot vectors as input and produces a value between 0–1 in the output. This value implies that how much the first input byte is similar to the second corresponding input byte. Since each byte has 256 different modes, the length of the input one-hot vectors should be 256. In our network, each sample is read byte by byte and when a byte is selected as a target, it is delivered to the network as the first input byte. Then, n random bytes from its context and n random bytes from out of its context called negative bytes are chosen that each time one of them is opted as the second input byte. In Skip-Gram, a window with predefined size is considered which the context, negative and target bytes are defined based on this window. Accordingly, bytes surrounded the target byte are determined as context and those that are outside the window are negatives. Therefore, byte2vec network is trained in a way that each time for a target byte, if the second input is a context byte, the output value is set to be 1, otherwise the output value is 0.

The most momentous part of this network is the embedding matrix which is located in the next layer and provided the final embedding vectors of bytes supplied to the deep learning network as the output of word2vec model. The size of this matrix is considered to be 256 * d, where d is the length of embedding vectors. In this article, the value of d is specified to be 50, therefore the size of our embedding matrix is 256 * 50. After the training of word2vec model, each row of this matrix represents the embedding vector generated for the associated byte value.

This matrix is initialized randomly and updated during the training procedure regarding the values of input bytes to generate effectual numeric vectors. The element-wise multiplication is applied between the input one-hot vectors and the embedding matrix; as a result, the value of all the rows of matrix became 0 except two rows that are relevant to the input indexes. It is significant to note that, these two rows are the produced embedding vectors of the target byte and its negative or context bytes utilized as the numerical vectors of byte values after the completion of training process. Then, these rows are retrieved from the matrix and a dot product operation is performed between them to evaluate the similarity. Ultimately, the achieved similarity is passed to a sigmoid layer to assign label 0 or 1. Label 1 implies that the input byte pair has been (target and context) and 0 is and indicative of the (target and negative) pair.

To train our byte2vec model, the window size in the Skip-Gram is defined to be 5. Indeed, each training sample is scanned to pick a target byte along with the 5 context bytes from within the windows of bytes around the target and 5 negative bytes outside the window are opted randomly. Also, it is required to set a label of 0 or 1 depending on whether the supplied byte is context or negative. To ensure that truly similar bytes have vectors with high similarity score, the embedding matrix is updated regularly through the back-propagation of errors. As mentioned before, we consider only the header part of each executable file, so that these steps are carried out for the total bytes of each header. Once training is completed, each row of the embedding matrix is considered as the embedding vector for the associated byte.

3.2 The Proposed CNN Architecture

As mentioned before, our introduced deep network considers each executable file as a sequence of raw bytes. One of the most outstanding characteristics of this network is that, it is endeavored to train a neural network which has the ability of extracting n-gram patterns simultaneously with different size (2-gram, 3-gram, 4-gram, etc.) from bytes. These patterns are employed to make a distinction between malware and benign files. To reach our goal, we modified the layout of basic CNN model in such a way that its filters can be able to exploit n-gram features from the byte sequences.

In typical approaches, it is very difficult to find a salutary algorithm for extracting diverse features of n-grams from input bytes; whereas in our method, various n-grams (ex. 2-gram) are obtained automatically and concurrently by taking the advantages of the properties of CNN. Additionally, we generalized the initial CNN to a parallel network in order to make the possibility of simultaneous extraction of n-gram features with several values of n and then use them as the feature set. Indeed, the implementation of these procedures in commonplace networks and methods is severely hard and time-consuming. Thus, our proposed approach can exploit varied set of features for different n-grams with diverse values of n in a straightforward way and significantly short time. Moreover, several features can be considered for each size of n which leads to attain more acceptable results. Owing to the utilization of embedding vectors, bytes which frequently appear close to each other in the dataset, have more similar vectors and this has a remarkable influence on the quality of produced n-gram features compared to the other methods that directly consider the value of each byte.

The overall structure of our proposed network is illustrated in Fig. 2. Due to the fact that this network employs filters in parallel, we called it Parallel-CNN. The input of our Parallel-CNN is a two-dimensional matrix (n, d); the first dimension (n) is the number of byte sequences extracted from each file that in our case is equal to the number of header bytes (usually the first 1024 bytes). The second dimension (d) is always fixed and equal to the size of the embedding vectors obtained in the previous step which is defined to be 50 in this paper. In the convolution layer, 3 filters with different size and number are utilized in parallel and simultaneously. As shown in Fig. 2, filters are two-dimensional (f, d) similar to the input matrix and their second dimension (d) is exactly identical to the second dimension of input. It is essential to note that, these parallel filters give us the ability to extract n-gram features from the input and in fact this is one of the most valuable properties of the deep learning algorithms and peculiarly our Parallel-CNN which can exploit wealthy features automatically.

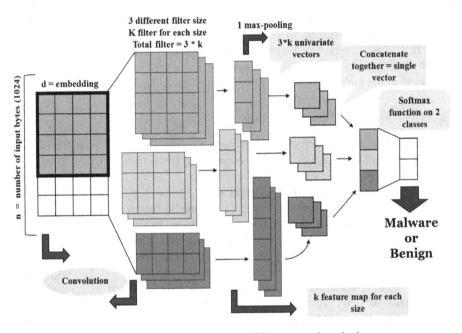

Fig. 2. The overall structure of the proposed method

Next, filters are placed concurrently on the input to carry out the convolutional operation; an element-wise multiply between filter and part of the input (n) which is equal to the size of the considered filter. Notice that the weights of these filters are initialized randomly at first then updated by back-propagation during training along with the rest of the network's parameters. We choose 3 parallel filters (f) with different size and number (k). Consequently, the output of the convolution layer will be multiple scalar lists (3 * k) with the size of $n - f + 1$. For example, if we have 20 of each filter, the output of this layer is 60 (3 * 20) scalar lists.

In the max pooling layer, the maximum value of each list is selected, therefore, the output of this layer will be 3 * k number of scalars. It is significant to remark that, we assessed the two other functions of pooling, average and minimum pooling, but the results of max pooling are much more desirable.

In the next layer, all the achieved scalars are concatenated to form a single and fixed-size vector as a final feature that is independent from the size of filters. Finally, to allocate a label for each input data, a two-class Softmax function is applied to the last layer which is a linear layer to determine the possibility distribution of labels (malware and benign). Hence, each label that has the highest possibility is opted as the true label of the input file.

It should be noted that, the more the number of parallel filters, the more acceptable the final results; nevertheless, computational complexity and the processing time of our model will be strikingly escalated. To assess the performance of our method, different number and size of filters are specified and the results of our conducted various experiments are provided in the next section.

4   Experiments and Results

To evaluate our method, various experiments are conducted and their results are provided in this section. Since the purpose of our method is malware detection of IoT, two datasets with different type of files are used to evaluate the performance of our CNN. This section considers the experimental results on each dataset and the impact of different parameters on the final accuracy of our method. Also, we compare our introduced CNN model with other malware detection techniques.

4.1   Dataset

Our experiments are carried out on two datasets. Dataset 1 is an android dataset comprises of the program file format of android files called apk. DREBIN [23, 24] is an android malware dataset contains 5560 samples from 179 different malware families. In this article, a subset of DREBIN includes 2000 files is selected as malware dataset. Also, for benign dataset, 2000 apk files are randomly downloaded from Google Play Store which are opted from various categories such as game, entertainments, learning programs, etc. To be insure that these files are not contain malicious content, we scanned them with ESET NOD32 and Kaspersky.

Dataset 2 is created from 280 malwares and 270 benign samples for ARM-based IoT applications [25]. All of our malware samples were selected utilizing VirusTotal Threat Intelligence platform. Moreover, all goodware were collected form a number of official IoT App stores like Pi Store.

4.2   Experimental Environment and Evaluation Metrics

All the experiments of this paper are accomplished on the machine with the following specifications: Hardware: Intel Core i7 8700k with 12 threads, 32.0 GB of RAM, GeForce GTX 950 with 2 GB of VRAM as GPU. Software: Ubuntu 16.04 and Python 3.0. Also, we have used Tensorflow with Keras [26] library of Python for creating deep learning network with different layers. As deep learning algorithm can run both on the CPU and GPU, we choose GPU due to the high rapidity.

Common machine-learning evaluation metrics are applied to assess the results of this paper. These are True Positive Ratio (TPR), False Positive Ratio (FPR), Precision, Recall, Accuracy and F-measure (F1) [27].

4.3   Experiments

In this section, first, the experiments on different parameters of the proposed model are considered. Then, our Parallel-CNN is compared to other malware detection methods and the achieved results are discussed in details.

Experiments on Different Parameters of the Proposed Network. This section provides the results of experiments carried out with various values of parameters of our model. As mentioned before, three parallel filter set are used simultaneously by our Parallel-CNN, hence we have two different parameters which affect the detection results: the number and the size of each parallel filter.

As regards to the structure of filters and input matrix, our network is capable of extracting n-gram features from input. Therefore, by considering different filter sizes, various n-grams are achieved concurrently. Obviously, this represents the effectiveness of employing deep learning techniques and peculiarly the proposed layout of CNN which can easily and automatically extract features. Moreover, the number of each parallel filter have a remarkable impact on appropriately detecting malware. This parameter is equal to the number of features exploited from input data, thus the higher the number of filters, the more the information obtained from data.

Due to the fact that, by considering the whole content (bytes) of a file, size of the input matrix will be huge and processing will take more time and memory, we choose only the header of each file as it is the most informative part of files. To determine the sensitiveness of our model to these parameters, experiments are accomplished on both datasets. Since the essence of files in these datasets is completely different, our parameters are tuned for each of them separately to better comprehend the effectiveness of them. The results of experiments on Dataset 1 and Dataset 2 are represented in Tables 1 and 2, respectively. In all of our experiments we consider 80% of data for training and the rest for the testing phase.

From the results, it is clear that by increasing the number of each parallel filter, results are growth slightly. Indeed, more features can be extracted from files by rising the number of filters and this leads to the more accurate malware detection. On the other hand, a large number of filters may enhance the computations and time of processing. Also, the size of filters has a considerable impact on the final accuracy. It is essential to note that, by performing various experiments with different filter sizes, we figured out that beyond a certain value, an increment in size does not have significant effect on accuracy and even in some cases it reduces the results marginally. Furthermore, it should be said that we examined various number of filters and understood that beyond a certain number of filters (which is in our case is 64) the accuracy stay constant or decrease which is caused by overfitting. Hence, the larger number of filters will not always make better results since it leads to the higher computation time as well as lower accuracy.

Table 1. Result of different number and size of filters on Dataset 1 (FN = Filter Number in each parallel filter set & FS = Filter size, ACC = Accuracy)

FN	FS = (2, 3, 4)				FS = (3, 4, 5)				FS = (4, 5, 6)			
	ACC	FPR	TPR	F1	ACC	FPR	TPR	F1	ACC	FPR	TPR	F1
8	0.9245	0.1120	0.9376	0.9201	0.9367	0.0845	0.9416	0.9328	0.9486	0.0763	0.9499	0.9418
16	0.9410	0.0923	0.9491	0.9398	0.9541	0.0812	0.9562	0.9491	0.9673	0.0701	0.9702	0.9635
64	0.9626	0.0532	0.9745	0.9583	0.9695	0.0501	0.9736	0.9623	**0.9710**	**0.0745**	**0.9571**	**0.9694**

Table 2. Result of different number and size of filters on Dataset 2 (FN = Filter Number in each parallel filter set & FS = Filter Size, ACC = Accuracy)

FN	FS = (2, 3, 4)				FS = (3, 4, 5)				FS = (4, 5, 6)			
	ACC	FPR	TPR	F1	ACC	FPR	TPR	F1	ACC	FPR	TPR	F1
8	0.9804	0.090	0.9378	0.9498	0.9855	0.118	0.9743	0.9654	0.9862	0.091	0.9667	0.9651
16	0.9877	0.054	0.9596	0.9688	0.9862	0.082	0.9637	0.9651	0.9899	0.062	0.9742	0.9744
64	0.9935	0.027	0.9778	0.9832	0.9971	0.009	0.9891	0.9925	**0.9987**	**0.017**	**0.1**	**0.9965**

By comparing the results on both Datasets, it can be seen that the experiments on Dataset 2 (IoT Dataset) achieve significantly better results. As mentioned before, Dataset 1 consists of android apk files. It should be noted that apk is the format of android's executable files and it is identical to a zip file which contain all of the files of android program. In this apk package, dex file is most important among others that includes the most valuable information about the application. Accordingly, we extract the byte of dex to perform our experiments. It seems that, the difference between the structure of android dex and other types of executable files causing the results to be lower on this dataset.

According to the results, it is obvious that by considering only the header of the file, the accuracy of the proposed method is reasonable. It is not astounding, because header is a short-sized piece of data that is strikingly informative and is the only part of the file that is not encrypted. It is possible that by employing high-level features like opcodes, we can achieve better results, but our goal is to demonstrate that malware detection is prone to increasing the speed and simplicity without disassembling.

Experiments on Comparison with Other Malware Detection Methods. In this section, the proficiency of our method is compared with Farrokhmanesh et al. [14],

Hashemi et al. [15], Gibert et al. [19], Cui et al. [20] and Xiao et al. [16]. To have a better analysis, all of the malware detection approaches that are selected for comparison utilize raw bytes similar to our Parallel-CNN model. We ran all of these methods on the same environment explained in Sect. 4.2. It should be mentioned that the best parameter values of our technique achieved in previous section are employed in these experiments in order to compare them with other malware detection methods. Results on Dataset 1 and Dataset 2 are provided in Tables 3 and 4, respectively. Also, Table 5 represents the comparison of running time with other approaches. Running times are obtained by getting average among 10 times of executing the network.

Table 3. Results of comparison with other malware detection methods on Dataset 1

Classifier	TPR	FPR	Accuracy	F-measure
Proposed method	0.9571	0.0745	0.9710	0.9694
Hashemi et al. [15]	0.8993	0.0843	0.9087	0.8995
Farrokhmanesh et al. [14]	0.9560	0.0340	0.96	0.9608
Gibert et al. [19]	0.9387	0.0735	0.9406	0.9317
Cui et al. [20]	0.9255	0.1120	0.9346	0.9221
Xiao et al. [16]	0.9358	0.099	0.9094	0.7110

In spite of the fact that the accuracy of our method in these experiments is higher than others, it cannot be argued that it performs always better than other approaches. It is obvious that the f-measure of our Parallel-CNN is marginally greater than Farrokhmanesh et al. [14] method while Hashemi et al. [15] obtained less f-measure in detecting malware on Dataset 1 compared to our technique. Our goal is to depict that with the help of deep learning algorithm, appropriate feature extraction is accomplished and accuracy is enhanced sensibly. Also, it can be concluded that although Gibert et al. [19] and Cui et al. [20] are capable of detecting malware with reasonable accuracy, they consume a huge amount of time. This issue is caused by the visualization of samples which needs much more time for processing compared with the raw bytes.

Table 4. Results of comparison with other malware detection methods on Dataset 2

Classifier	TPR	FPR	Accuracy	F-measure
Proposed method	0.9571	0.0745	0.9710	0.9694
Hashemi et al. [15]	0.8974	0.181	0.9652	0.9099
Farrokhmanesh et al. [14]	0.9816	0.072	0.9906	0.9764
Gibert et al. [19]	0.9410	0.081	0.9819	0.9531
Cui et al. [20]	0.8392	0.099	0.9602	0.8912
Xiao et al. [16]	0.9720	0.054	0.9325	0.7959

Table 5. Running time (in seconds) comparison with other malware detection methods

Classifier	Running time on (Dataset 1)	Running time on (Dataset 2)
Proposed method	**3,820**	**2,700**
Hashemi et al. [15]	5,241	4,200
Farrokhmanesh et al. [14]	4,850	3,670
Gibert et al. [19]	32,953	25,200
Cui et al. [20]	45,391	31,487
Xiao et al. [16]	7,200	4,550

5 Summary and Conclusion

The fast pace of development and nature of IoT environments causes a lot of security and forensics challenges. In this paper, a parallel and novel architecture of CNN is designed which works by raw bytes with the purpose of IoT malware detection. The layout of filters in our network are considered in a way that give us the ability to extract n-gram features from the input data. To generate numerical vectors for bytes, we applied word2vec, one of the most powerful word embedding methods which are gained a noticeable acceptance among text classification researches. In fact, we choose word2vec because its produced vectors are meaningful and demonstrates the similarity between bytes. It is essential to note that, to work with embedding technique, each byte in a sequence is considered as a word in sentence and a scalar vector is created for each byte. Due to the byte-level essence of our method, it does not require any disassembling compared to other machine learning methods which are wasted a great amount of time for disassembling and exploiting features. Indeed, we directly utilized files to create models with raw bytes and classify unseen instances. Additionally, as deep learning methods are capable of automatically extracting features, the need of finding an effective feature extraction algorithm which is one of the most challenging steps among other machine learning methods is eliminated.

References

1. Varsha, M.V., Vinod, P., Dhanya, K.A.: Identification of malicious android app using manifest and opcode features. J. Comput. Virol. Hacking Tech. **13**(2), 125–138 (2016)
2. Conti, M., Dehghantanha, A., Franke, K., Watson, S.: Internet of Things security and forensics: challenges and opportunities. Future Gener. Comput. Syst. **78**, 544–546 (2018)
3. Mosenia, A., Jha, N.: A comprehensive study of security of Internet-of-Things. IEEE Trans. Emerg. Top. Comput. **5**(4), 586–602 (2017)
4. Egele, M., Scholte, T., Kirda, E., Kruegel, C.: A survey on automated dynamic malware-analysis techniques and tools. ACM Comput. Surv. **44**(2), 1–42 (2012)

5. Bazrafshan, Z., Hashemi, H., Fard, S.M.H., Hamzeh, A.: A survey on heuristic malware detection techniques. In: The 5th Conference on Information and Knowledge Technology (2013)
6. Filiol, E.: Malware pattern scanning schemes secure against blackbox analysis. J. Comput. Virol. **2**(1), 35–50 (2006)
7. Filiol, E., Jacob, G., Liard, M.L.: Evaluation methodology and theoretical model for antiviral behavioural detection strategies. J. Comput. Virol. **3**(1), 27–37 (2007)
8. Abou-assaleh, T., Cercone, N., Keselj, V., Sweidan, R.: N-gram based detection of new malicious code. In: 2004 Proceedings of the 28th Annual International Conference on Computer Software and Applications, vol. 2, no. 1, pp. 41–42 (2004)
9. Vinod, P., Jaipur, R., Laxmi, V., Gaur, M.: Survey on malware detection methods. In: Proceedings of the 3rd Hackers' Workshop on Computer and Internet Security (IITKHACK 2009), pp. 74–79 (2009)
10. Zolotukhin, M., Hamalainen, T.: Detection of zero-day malware based on the analysis of opcode sequences. In: 2014 IEEE 11th Consumer Communications and Networking Conference (CCNC) (2014)
11. Shabtai, A., Kanonov, U., Elovici, Y., Glezer, C., Weiss, Y.: "Andromaly": a behavioral malware detection framework for android devices. J. Intell. Inf. Syst. **38**(1), 161–190 (2011)
12. Sanz, B., Santos, I., Laorden, C., Ugarte-Pedrero, X., Bringas, P.G., Álvarez, G.: PUMA: permission usage to detect malware in android. In: Herrero, Á., et al. (eds.) International Joint Conference CISIS 2012-ICEUTE´12-SOCO´12 Special Sessions. Advances in Intelligent Systems and Computing, vol. 189, pp. 289–298. Springer, Heidelberg (2013). https://doi.org/10.1007/978-3-642-33018-6_30
13. Lv, Y., Duan, Y., Kang, W., Li, Z., Wang, F.-Y.: Traffic flow prediction with big data: a deep learning approach. IEEE Trans. Intell. Transp. Syst. **16**, 1–9 (2014)
14. Farrokhmanesh, M., Hamzeh, A.: Music classification as a new approach for malware detection. J. Comput. Virol. Hacking Tech. **15**, 77–96 (2018)
15. Hashemi, H., Hamzeh, A.: Visual malware detection using local malicious pattern. J. Comput. Virol. Hacking Tech. **15**, 1–14 (2018)
16. Xiao, X., Zhang, S., Mercaldo, F., Hu, G., Sangaiah, A.K.: Android malware detection based on system call sequences and LSTM. Multimed. Tools Appl. **78**, 3979–3999 (2017)
17. Ye, Y., Chen, L., Hou, S., Hardy, W., Li, X.: DeepAM: a heterogeneous deep learning framework for intelligent malware detection. Knowl. Inf. Syst. **54**(2), 265–285 (2017)
18. Huang, W., Song, G., Hong, H., Xie, K.: Deep architecture for traffic flow prediction: deep belief networks with multitask learning. IEEE Trans. Intell. Transp. Syst. **15**(5), 2191–2201 (2014)
19. Gibert, D., Mateu, C., Planes, J., Vicens, R.: Using convolutional neural networks for classification of malware represented as images. J. Comput. Virol. Hacking Tech. **15**, 15–28 (2018)
20. Cui, Z., Xue, F., Cai, X., Cao, Y., Wang, G., Chen, J.: Detection of malicious code variants based on deep learning. IEEE Trans. Industr. Inf. **14**(7), 3187–3196 (2018)
21. LeCun, Y., Bengio, Y.: Convolutional networks for images, speech, and time series. Handb. Brain Theor. Neural Netw. **3361**(10), 1995 (1995)
22. Mikolov, T., Sutskever, I., Chen, K., Corrado, G.S., Dean, J.: Distributed representations of words and phrases and their compositionality. In: Advances in Neural Information Processing Systems, pp. 3111–3119 (2013)
23. Arp, D., Spreitzenbarth, M., Huebner, M., Gascon, H., Rieck, K.: Drebin: efficient and explainable detection of android malware in your pocket. In: 21th Annual Network and Distributed System Security Symposium (NDSS), February 2014

24. Spreitzenbarth, M., Echtler, F., Schreck, T., Freling, F.C., Hoffmann, J.: MobileSandbox: looking deeper into android applications. In: 28th International ACM Symposium on Applied Computing (SAC), March 2013
25. Brash, D.: Recent additions to the ARMv7-A architecture. In: 2010 IEEE International Conference on Computer Design (2010)
26. Abadi, M.C.A.D.: TensorFlow: learning functions at scale. In: Proceedings of the 21st ACM SIGPLAN International Conference on Functional Programming - ICFP 2016 (2016)
27. Powers, D.M.: Evaluation: from precision, recall and F-measure to ROC, informedness, markedness and correlation (2011)

Prediction of Neurological Deterioration of Patients with Mild Traumatic Brain Injury Using Machine Learning

Gem Ralph Caracol[1(✉)], Jin-gyu Choi[2], Jae-Sung Park[3],
Byung-chul Son[3], Sin-soo Jeon[3], Kwan-Sung Lee[3], Yong Sam Shin[3],
and Dae-joon Hwang[4]

[1] College of Information and Communication Engineering,
Sungkyunkwan University, Natural Sciences Campus, Seoul, South Korea
gemralph@gmail.com
[2] Yeouido St. Mary's Hospital, College of Medicine, The Catholic University
of Korea, Seoul, South Korea
[3] Seoul St. Mary's Hospital, College of Medicine, The Catholic University
of Korea, Seoul, South Korea
[4] College of Software, Sungkyunkwan University, Natural Sciences Campus,
Seoul, South Korea

Abstract. Possible Neurological Deterioration (ND) of patients with Traumatic Brain Injury (TBI) is difficult to identify especially the mild and moderate injuries. When ND happens, death or lifelong disability is prevalent. Early prediction of possible ND would allow medical and healthcare institutions to provide the needed medical treatment. This paper presents the results that show Machine Learning (ML) can be used to create predicative models with high prediction rates even with a small set of patient records (219 patient records with 54 variables). From the patient records, 20 randomized data sets with preconditions on the testing and training data were created and fed to selected Artificial Neural Network (ANN) and Classification Algorithms. Preconditions on testing and training data can affect the prediction models created by the different algorithms. The best prediction models created by the ANN algorithms (multilayer perceptron (MLP), recurrent neural network (RNN), and long short-term memory (LSTM)) and two classification algorithms (linear regression and logistic regression algorithms) are considered acceptable and could be applied as medical decision support to identify patients that may potentially have ND. Early prediction of a possible ND of a patient can now be easily carried out as soon as his or her records and medical test results are ready and match the 54 variables needed for prediction.

Keywords: Machine learning · Neurological deterioration prediction · Traumatic brain injury · Small data set

Abbreviations

AUC — Area Under the Curve
CT — Computed Tomography
DS — Dataset

© Springer Nature Singapore Pte Ltd. 2019
H. Nguyen (Ed.): RSSDS 2019, CCIS 1150, pp. 198–210, 2019.
https://doi.org/10.1007/978-981-15-1960-4_14

EDH   - Epidural Hematoma
FN   - False Negative
FP   - False Positive
GCS   - Glasgow Coma Scale
ICH   - Intra-cerebral Hematoma/ Contusion
IVH   - Intraventricular Hemorrhage
LSTM   - Long Term Short Term Memory
MLP   - Multilayer Perceptron
ND   - Neurological Deterioration
NN   - Neural Network
P   - Positive
RNN   - Recurrent Neural Network
ROC   - Receiver Operating Characteristic
SAH   - Subarachnoid Hemorrhage
SBP   - Systolic Blood Pressure
SDH   - Subdural Hematoma
TBI   - Traumatic Brain Injury
TN   - True Negative
TP   - True Positive
WBC   - White Blood Cell

1 Background

The severity of a traumatic brain injury (TBI) is often difficult to accurately classify and predicting the outcome in the first few hours of the trauma especially in mild and moderate injuries [2]. Errors of TBI management in emergency rooms (ER) and TBI preventable deaths are not common [19].

However clinical signs that MTBI patients will have neurological deterioration (ND) does not show during the first examination. MRI and CT scan results are often are normal. The medical staff will have to monitor these patients if they think (judgment call based on experience) ND may occur. What if ND occurs to patients that they think ND will not happen? This is a bad situation to a patient. Neurological deterioration is a clinical worsening of a brain, defined as a decrease of ≥ 2 GCS points ≥ 4 SIP score points, lasting longer than 8 h, requires surgical intervention, and or resulting in death [20].

Traumatic brain injury (TBI) continues to be a prevalent cause of death or lifelong disability worldwide [1–3]. TBI occurs with greater frequency than other neurological disorders [4]. Management errors of TBI in Emergency Rooms and TBI preventable deaths are also high [19].

Among TBI cases, 90% are categorized as mild TBI (MTBI) [5]. In general, MTBI is characterized by a 10% risk for intracranial abnormalities (epidural/subdural hematoma or contusion) and a 1% risk of intracranial hematoma with the need for immediate surgery [5]. Therefore, it is important to recognize the clinical signs and provide immediate medical treatment. However the clinical signs can only be seen after a new CT is taken after neurological deterioration (ND) takes place. The ND status is considered the main determinant in the treatment strategy for MTBI patients. To be able to predict that ND may occur, the medical staff can prepare in advance for any treatment strategy and/ or surgery operations. Currently there is no guideline or definite criteria to determine or predict possible ND of MTBI patients.

Furthermore, managing TBI is expensive and labor intensive [6]. TBI treatments are highly time sensitive, and the possibility of patient improvement and survival can be increased if optimal and prompt decisions are made during the course of treatment. Therefore, the accurate prediction of the outcome would allow medical and healthcare institutions to offer treatment that balances the expenses with the limited available resources.

As a branch of artificial intelligence, machine learning (ML) is widely used in medicine for classification and prediction, and there is currently an active ongoing study into existing predictive models based on ML methods [7]. Medical interest in ML has grown because of its advanced analytical capabilities [8]. Furthermore, ML is seen as a means of (i) improving management strategies, (ii) informing patients/family members accordingly about expectations, and (iii) facilitating the designs of future clinical trials [9, 10]. Therefore, ML predictive studies have been conducted for TBI.

One review reported that TBI outcome prediction was the most-studied topic in neurology [8]. Reference [11] used an artificial neural network (ANN) model to predict outcomes after mild severe head injury. Their research was focused on predicting the five categories of the Glasgow Outcome Score (death, persistent vegetative state, severe disability, moderate disability, and good recovery); the predictions of their model had an accuracy of 75.8%. Reference [12] compared ANN and logistic regression models in their ability to predict outcomes in head trauma on the basis of clinical data. Reference [6] used several ML techniques (e.g., logistic regression, decision trees, Bayesian networks, and neural networks) to accurately predict the outcome of a severe head injury. Their study demonstrated highly accurate predictions with logistic regression and decision trees; consequently, a hybrid predictive model was developed by combining both algorithms. Reference [13] demonstrated the importance of ranking variables according to their importance, and they created simple models to accurately predict the clinical outcomes for TBI patients. Their results can help clinicians improve management strategies and make better clinical decisions. Logistic regression models and Bayesian network analysis have been employed to assess the associations of key variables with the other remaining variables. Reference [14] developed a Bayesian ANN model that can forewarn of possible hypertensive events that could lead to neurological deterioration (ND) of TBI patients in the ICU. This model allows for more attentive monitoring and earlier clinical assessment to mitigate the onset of hypotension.

The aforementioned studies are evidence of the growing literature on the use of ML for TBI. However, a prediction model for Mild TBI isn't available. There is also no clear guideline in predicting neurological deterioration in MTBI. The physician in-charge will do his best to judge the situation given the patient data and based on experience. When ND happens, death or lifelong disability is prevalent. Therefore a clear guideline or a prediction model for predicting ND on MTBI patients is essential. The aim of this study is to identify a robust predictive ML model that could help medical practitioners make better clinical decisions for the early management of MTBI patients. This study also aims to present different ML algorithms that can be used in creating prediction models for predicting ND for patients with MTBI.

Now the question is: Can we use machine learning in predicting possible neuro-logical deterioration in MTBI? During the course of finding the answer to this question, that following problems or opportunities to work on were encountered: (1) Patient Data Privacy, (2) Small Data Set, and (3) What machine learning algorithms to use.

2   Methods

2.1   Data Collection and Preprocessing

Patient data in South Korea are protected by privacy laws and it is also difficult to use them even permission due to authorization rules. We were able to collect only 219 MTBI patient records from the hospital that gave us authorization. The records are composed of only the first examination data of each patient from January 2014 to May 2018. Even though we were authorized to use the patient records for our research, we decided to convert the data into binary format in case we need to allow other data scientists to test our findings. We also used "OneHot Encode" to make sure that there really is no identifying value in the binary data set except 0 and 1. Figure 1 shows the conversion process.

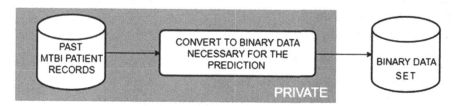

Fig. 1. Converting patient data to binary data.

In this process, we only allow the physician or neurosurgeon to identify the vari-ables that he considered can contribute in predicting ND of MTBI patients. He came up with 54 fields and the 55th field is the ND outcome as shown in Table 1.

Table 1.

No.	Fields	Criteria having a value of 1 otherwise 0
1	Age	≥ 65
2	Sex	Male
3	Hypertension	Patient has
4	Diabetes	Patient has
5	Blood diseases	Patient has
6	Chronic liver disease	Patient has
7	Chronic arthritic disease	Patient has
8	Solid Cancer	Patient has
9	Antiplatelet agent	Patient has
10	Anticoagulant	Patient has
11	GCS 13	GCS 13
12	GCS 14	GCS 14
13	GCS 15	GCS 15
14	SBP	<90
15	SBP = 90 to <140	= 90 to <140
16	SBP > 140	>140
17	Pulse normal	If normal
18	Pulse More than 100 pulse tachycardia	>100
19	Pulse rate < 60	<60
20	Respiratory Rate	>20
21	Body temperature	>37.5
22	WBC count (109/L)	>10
23	Hemoglobin	<12
24	Platelet count	>100
25	Prothrombin (INR)	≤ 1.2
26	Prothrombin (s)	>14
27	aPTT (s)	>40
28	CT time (h) after an accident < 3	<3 h
29	CT time (h) after an accident = 3 to 12	3 to 12 h
30	After CT crash hour (h) = 12 or more	12 h or more
31	Subdural hematoma	Patient has
32	Epidural hematoma	Patient has
33	Intracerebral hematoma/contusion	Patient has
34	SAH	Patient has
35	IVH	Patient has
36	Dominant lesion, type SDH	Patient has
37	Dominant lesion, type EDH	Patient has

(continued)

(*continued*)

No.	Fields	Criteria having a value of 1 otherwise 0
38	Dominant lesion, type ICH	Patient has
39	Dominant lesion, type SAH	Patient has
40	Dominant side Left (0)|Right (1)	Patient has
41	Supratentorial location	Patient has
42	Infratentorial location	Patient has
43	Interhemispheric location	Patient has
44	Maximal thickness EDH/SDH (mm) = 0	= 0
45	Maximal thickness EDH/SDH (mm) ≥ 0 to <5	> 0 to <5
46	Maximal thickness EDH/SDH (mm) > 5	5 and more
47	Size of ICH (cc) = 0	0
48	Size of ICH (cc) = ~ 10	>0 to <10
49	Size of ICH (cc) = 10	10 and more
50	Midline shift (mm) = 0	0
51	Midline shift (mm) = ~ 5	> 0 to ≤ 5
52	Midline shift (mm) > 5	> 5
53	Skull fracture	Patient has
54	Basal Cistern Compression	Patient has
55	Neurological deterioration	Patient has

2.2 Training and Validation Datasets

A total of 219 patient records were used in this study, with 25 cases having ND. Among the records, 80% (175 cases) were used as training data and 20% (44 cases) were used as validation data. Given that there were only few ND cases, we needed to create supervised datasets. We used the stratification process so that we can set conditions on the data sets. Each dataset that we generated had at least 5 (20%) and no more than 12 (50%) ND cases in the test data set. We created 20 datasets and fed into the different non-neural and neural network algorithms. Figure 2 shows the flow of producing the prediction results for validating this research. The past MTBI patient records and the conversion to binary data are in a shaded box to show that it is only accessible by physicians or any authorized hospital personnel.

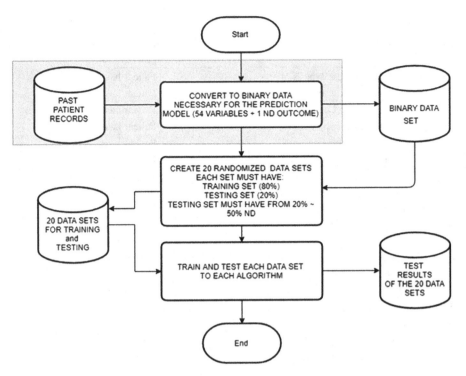

Fig. 2. The flowchart on testing and training the 20 data sets

2.3 Modeling Methods Using Neural Network and Non-neural Network Algorithms

The 20 datasets were fed into different non-neural and neural network algorithms which are considered commonly used binary classifiers to create predictive models. All ML algorithms we used were from KERAS and Scikit-Learn, both are conveniently accessible and easy to use. We chose three neural network classification algorithms from the Keras library [15], namely, multilayer perceptron (MLP), recurrent neural network (RNN), and long short-term memory (LSTM), and six classification algorithms from the scikit-learn library [16], namely, linear regression, decision tree classifier, support vector machine, k-nearest-neighbors classifier (KNeighborsClassifier), naive Bayes classifier (GaussianNB), and logistic regression. We also used scikit-learn metrics to generate the receiving operating characteristics (ROCs) for each prediction model which we consider as the best in each group of results.

We used the confusion matrix as guide in the statistical performance measures we used. Figure 3 shows the confusion matrix where P represents patients with neurological deterioration and N represents without neurological deterioration. TP represents the number of patients correctly predicted with neurological and TN represents the number of patients correctly predicted without neurological deterioration. FP represents the number of patients without neurological deterioration but incorrectly predicted with neurological deterioration. FN represents the number of patients with neurological deterioration but incorrectly predicted without neurological deterioration.

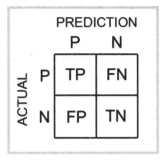

Fig. 3. The confusion matrix

The formula of the statistical performance measures that we that used in identifying prediction models that can be useful for predicting ND are as follows:

$$True\ Positive\ Rate\ =\ TP/(TP+FN) \tag{1}$$

$$True\ Negative\ Rate\ =\ TN/(TN+FP) \tag{2}$$

$$False\ Positive\ Rate\ =\ FP/(FP+TN) \tag{3}$$

$$Precision\ =\ TP/(TP+FP) \tag{4}$$

True Positive Rate is also known as sensitivity or recall. True Negative Rate is also known as Specificity. Sensitivity is the ability of a model to correctly identify the possibility of ND of MTBI patients. Specificity relates to the model's ability to correctly reject the possibility of having ND of MTBI patients. Precision is the accuracy of predicting ND and how often it is correct. False Positive Rate is the proportion of all negatives that still yield positive prediction outcomes. The specificity of the prediction is equal to 1 minus the false positive rate. AUC will also be created to show a model's ability to distinguish ND outcome. We consider this grading system as our guide to select the best model: more than 0.90 are considered excellent, 0.80 to 0.90 are good, within 0.70 and 0.80 are satisfactory, and less than 0.70 are poor [18].

3   Results

Twenty randomized data sets were created from MTBI patients (n = 219) and marked DS#1 to DS#20 respectively. Each data set is fed into 9 machine learning algorithms for ND predictions and obtained prediction results. Table 2 shows the summary of the algorithms' number of created models with sensitivity and/or precision is greater than 0.70.

Table 2. Number of models with either sensitivity or precision greater than 0.70.

Algorithms	Sensitivity	Precision	Average	Both	Highest Ave	Best DS
MLP	8	4	5	3	0.86	#3
RNN	11	4	5	3	0.86	#5
LSTM	7	3	5	1	0.77	#3
Linear regression	7	4	3	2	0.83	#2
Logistic regression	11	2	3	1	0.80	#5
SVM	1	0	0	0	0.60	–
Kneighbor	7	0	0	0	0.60	–
Decision tree	1	5	1	0	0.76	–
GausianNB	0	15	0	0	0.51	–

Tables 3, 4, 5, 6, 7 and 8 showing each algorithms' models with either sensitivity or precision greater than 0.70. Both sensitivity and precision that are satisfactory or better is the criteria for selecting and ranking a prediction model.

Table 3. MLP models with either sensitivity or precision greater than 0.70.

Multilevel perceptron (MLP)									
DS#	P	TP	TN	FP	FN	S sensitivity	Specificity	PR precision	Average (S, PR)	Rank
3	7	6	36	1	1	0.86	0.97	0.86	0.86	1st
5	5	4	37	1	2	0.67	0.97	0.80	0.73	–
8	7	3	37	4	0	1.00	0.90	0.43	0.71	–
9	5	4	38	1	1	0.80	0.97	0.80	0.80	2nd
19	6	4	37	2	1	0.80	0.95	0.67	0.73	–

Table 4. RNN models with either sensitivity or precision greater than 0.70.

Recurrent neural network (RNN)									
DS#	P	TP	TN	FP	FN	S sensitivity	Specificity	PR precision	Average (S, PR)	Rank
3	7	6	36	1	1	0.86	0.97	0.86	0.86	2nd
5	5	5	37	0	2	0.71	1.00	1.00	0.86	1st
8	7	3	37	4	0	1.00	0.90	0.43	0.71	–
9	5	4	38	1	1	0.80	0.97	0.80	0.80	3rd
19	6	4	38	2	0	1.00	0.95	0.67	0.83	–

Table 5. LSTM models with either sensitivity or precision greater than 0.70.

DS#	P	TP	TN	FP	FN	S sensitivity	Specificity	PR precision	Average (S, PR)	Rank
Long short term memory (LSTM)										
1	5	4	37	1	2	0.67	0.97	0.80	0.73	–
3	7	5	36	2	1	0.83	0.95	0.71	0.77	1st
5	5	4	37	1	2	0.67	0.97	0.80	0.73	–
8	7	3	37	4	0	1.00	0.90	0.43	0.71	–
19	6	4	38	2	0	1.00	0.95	0.67	0.83	–

Table 6. Linear Regression models with either sensitivity or precision greater than 0.70.

DS#	P	TP	TN	FP	FN	S sensitivity	Specificity	PR precision	Average (S,PR)	Rank
Linear Regression										
2	6	5	36	1	2	0.71	0.97	0.83	0.77	1st
3	7	5	35	2	2	0.71	0.95	0.71	0.71	2nd
6	4	4	38	2	0	1.00	0.95	0.67	0.83	–

Table 7. Logistic Regression models with either sensitivity or precision is greater than 0.70

DS#	P	TP	TN	FP	FN	S sensitivity	Specificity	PR precision	Average (S,PR)	Rank
Logistic Regression										
1	5	4	37	1	2	0.67	0.97	0.80	0.73	–
5	5	4	38	1	1	0.80	0.97	0.80	0.80	1st
6	4	3	38	3	0	1.00	0.93	0.50	0.75	–

Table 8. Decision Tree model with either sensitivity or precision greater than 0.70.

DS#	P	TP	TN	FP	FN	S sensitivity	Specificity	PR precision	Average (S,PR)	Rank
Decision Tree										
3	7	6	34	1	3	0.67	0.97	0.86	0.76	–

Table 9 shows the best models from each algorithm that was able to create prediction with either sensitivity or precision greater than 0.70.

Table 9. Best model in each algorithm.

Algorithm	P	TP	TN	FP	FN	S sensitivity	Specificity	PR precision	Average (S, PR)	Data Set	AUC	Rank
Best Models												
MLP	7	6	36	1	1	0.86	0.97	0.86	0.86	#3	0.950	2nd
RNN	5	5	37	0	2	0.71	1.00	1.00	0.86	#5	0.964	1st
LSTM	7	5	36	2	1	0.83	0.95	0.71	0.77	#3	0.913	5th
Linear Reg	6	5	36	1	2	0.71	0.97	0.83	0.77	#2	0.965	4th
Logistic Reg	5	4	38	1	1	0.80	0.97	0.80	0.80	#5	0.969	3rd

Figure 4 shows the ROC of each best model from the algorithms in Table 9.

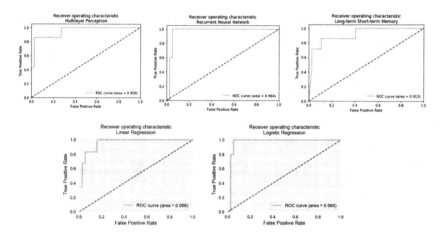

Fig. 4. ROC of the best prediction models in MLP, RNN, LSTM, linear regression and logical regression

4   Discussion

Early prediction of neurological deterioration of MTBI patients is considered very challenging. Management errors of TBI in emergency rooms and TBI preventable deaths are high [19]. Using machine learning for neurological deterioration in MTBI is also considered very challenging since data from patients are difficult to collect due to privacy laws and authorization rules from hospitals. In this study, we were able to only have 219 MTBI records. We decided to convert these patient data into binary format using "One Hot Encode" with many considerations in mind. Converting these data into binary format will allow us to use and even share it without leaking or providing patient details. Having 219 patient records and with only 25 with ND, conditions must be set in creating training and testing sets. This process is called stratification. We generated 20 stratified data sets and fed them to the 9 algorithms that are considered classifiers.

Five (5) algorithms namely: MLP, RNN, LSTM, Linear Regression and Linear Regression were able to produce relevant results. However four (4) algorithms namely: SVM, KNeighbors Classifier and Decision Tree didn't produce well. This could be the result of the type of data that we have – binary values.

From the 20 generated data sets or 180 models (9 algorithms x 20 data sets) created, RNN's DS #5 is considered the best model for predicting neurological deterioration in MTBI patients. RNN's DS#5 has a precision of 1.0 or 100% and a sensitivity of .71 or 71%.

Even though the sensitivity is only 71%, all patients with the possibility of having neurological deterioration are predicted because the precision is 100%. Those patients without ND but were predicted as having ND will just be given more monitoring which is considered good for the patient.

5   Conclusion

One of the aims of this study is to identify a robust predictive ML model that could help medical practitioners make better clinical decisions for the early management of MTBI patients. This study was able to create a usable prediction model in RNN DS#5.

The other aim of this study is identify ML algorithms that can create prediction models with small data sets. This study was also able to identify the MLP, RNN, LSTM, linear regression, and logistic regression algorithms can create predictive models that may be useful even with a small data set.

Furthermore this study was able to create prediction models with data values that are in binary value. Creating and sharing prediction models and even the actual data sets to other data scientists and or medical practitioners can be done without providing patient detail. It has a potential of creating a big binary data set if neurosurgeons and or hospitals can agree on the standard way in identification of variables to be used in prediction process.

Acknowledgements. We would like to acknowledge Seoul St. Mary's Hospital Catholic University of Korea for authorizing us to use the TBI patient records for this research.

Ethics Approval and Consent to Participate. This study was approved by the institutional review board in the authors' institution (KC17RESI0625). Since this study is a retrospective study based on medical records generated during the course of medical treatment, the patient's personal information was not disclosed and was exempted from the consent process by the Institutional review board.

Consent of Publication. This manuscript does not include identifying images or other personal or clinical details of participants.

Availability of Data. The binary dataset (reproduced dataset) is available for publication.

Competing Interests. The authors declare that they have no competing interests.

References

1. Ji, S.Y., Smith, R., Huynh, T., Najarian, K.: A comparative analysis of multi-level computer-assisted decision making systems for traumatic injuries. BMC Med. Inform. Decis. Mak. **9**, 2 (2009)
2. Lawrence, T.P., Pretorius, P.M., Ezra, M., Cadoux-Hudson, T., Voets, N.L.: Early detection of cerebral microbleeds following traumatic brain injury using MRI in the hyper-acute phase. Neurosci. Lett. **655**, 143–150 (2017). https://doi.org/10.1016/j.neulet.2017.06.046
3. Burke, J.F., Stulc, J.L., Skolarus, L.E., Sears, E.D., Zahuranec, D.B., Morgenstern, L.B.: Traumatic brain injury may be an independent risk factor for stroke. Neurology **81**(1), 33–39 (2013). https://doi.org/10.1212/WNL.0b013e318297eecf
4. Hirtz, D., Thurman, D.J., Gwinn-Hardy, K., Mohamed, M., Chadhuri, A.R., Zalutsky, R.: How common are the "common" neurologic disorders? Neurology **68**, 326–337 (2007)
5. Vos, P.E., et al.: Mild traumatic brain injury. Eur. J. Neurol. **19**, 191–198 (2012)

6. Pang, B.C.: Hybrid outcome prediction model for severe traumatic brain injury. J. Neurotrauma **24**, 136–146 (2017)

7. Alanazi, H.O., Abdullah, A.H., Qureshi, K.N.: A critical review for developing accurate and dynamic predictive models using machine learning methods in medicine and health care. J. Med. Syst. **41**, 69 (2017). https://doi.org/10.1007/s10916-017-0715-6

8. Celtikci, E.: A systematic review on machine learning in neurosurgery: the future of decision-making in patient care. Turk. Neurosurg. **28**, 167–173 (2018)

9. Murray, G.D., et al.: Multivariable prognostic analysis in traumatic brain injury: results from the IMPACT study. J. Neurotrauma **24**, 329–337 (2007)

10. MRC CRASH Trial Collaborators, Perel P., Arango, M., et al.: Predicting outcome after traumatic brain injury: practical prognostic models based on large cohort of international patients. BMJ **336**, 425–429 (2008)

11. Hsu, M.H., Li, Y.C., Chiu, W.T., Yen, J.C.: Outcome prediction after moderate and severe head injury using an artificial neural network. Stud. Health Technol. Inf. **116**, 241–245 (2005)

12. Eftekhar, B., Mohammad, K., Ardebili, H.E., Ghodsi, M., Ketabchi, E.: Comparison of artificial neural network and logistic regression models for prediction of mortality in head trauma based on initial clinical data. BMC Med. Inf. Decis. Making **5**, 3 (2005)

13. Zador, Z., Sperrin, M., King, A.T.: Predictors of outcome of traumatic brain injury: new insight using receiver operating curve indices and Bayesian network analysis. PLoS ONE **11**, e0158762 (2016)

14. Donald, R., Howells, T., Piper, I., et al.: Forewarning of hypotensive events using a Bayesian artificial neural network in neurocritical care. J. Clin. Monit. Comput. (2018). https://doi.org/10.1007/s10877-018-0139-y

15. Keras, C.F.: (2015). https://keras.io

16. Pedregosa, F., Varoquaux, G., Gramfort, A., et al.: Scikit-learn: machine learning in Python. J. Mach. Learn. Res. **12**, 2825–2830 (2011)

17. Ribeiro, M.T., Singh, S., Guestrin, C.: "Why Should I Trust You?" Explaining the Predictions of Any Classifier (2016). CoRR. http://arxiv.org/abs/1602.04938

18. Steyerberg, E.W., et al.: Assessing the performance of prediction models: a framework for traditional and novel measures. Epidemiology **21**, 128–138 (2010)

19. Kim, S.C., et al.: Preventable deaths in patients with traumatic brain injury. Clin. Exp. Emerg. Med. **2**(1), 51–58 (2015)

20. Ovesen, C., Christensen, A.F., Havsteen, I., et al.: Prediction and prognostication of neurological deterioration in patients with acute ICH: a hospital-based cohort study. BMJ Open **5**, e008563 (2015). https://doi.org/10.1136/bmjopen-2015-008563

Spherical Data Handling and Analysis with R package rcosmo

Daniel Fryer and Andriy Olenko(✉)

Department of Mathematics and Statistics, La Trobe University, Melbourne,
VIC 3086, Australia
{d.fryer,a.olenko}@latrobe.edu.au

Abstract. The R package **rcosmo** was developed for handling and analysing Hierarchical Equal Area isoLatitude Pixelation (HEALPix) and
Cosmic Microwave Background (CMB) radiation data. It has more than
100 functions. **rcosmo** was initially developed for CMB, but also can
be used for other spherical data. This paper discusses transformations
into **rcosmo** formats and handling of three types of non-CMB data:
continuous geographic, point pattern and star-shaped. For each type of
data we provide a brief description of the corresponding statistical model,
data example and ready-to-use R code. Some statistical functionality
of **rcosmo** is demonstrated for the example data converted into the
HEALPix format. The paper can serve as the first practical guideline to
transforming data into the HEALPix format and statistical analysis with
rcosmo for geo-statisticians, GIS and R users and researches dealing
with spherical data in non-HEALPix formats.

Keywords: Spherical data · Spatial statistics · rcosmo · HEALPix ·
Random field · Spatial point process · Directional · Star-shaped

1 Introduction

The package **rcosmo** was developed to offer the functionality needed to handle and analyse Hierarchical Equal Area isoLatitude Pixelation (HEALPix) and
Cosmic Microwave Background (CMB) radiation data. Comprehensive software
packages for working with HEALPix data are available in Python and MAT-
LAB, see, for example, [8,9] and [10]. The main aim of **rcosmo** was to provide
a convenient access to big CMB data and HEALPix functionality for R users.

The package has more than 100 functions. They can be broadly divided into
four groups:

- CMB and HEALPix data holding, subsetting and visualization,
- HEALPix structure operations,
- geometric methods,
- statistical methods.

This research was partially supported under the Australian Research Council's Discovery Project DP160101366.

H. Nguyen (Ed.): RSSDS 2019, CCIS 1150, pp. 211–225, 2019.
https://doi.org/10.1007/978-981-15-1960-4_15

The detailed summary of **rcosmo** structure, HEALPix, geometric and statistical functionality is provided in [5]. Technical description and examples of the core **rcosmo** functions can be found in the package documentation on CRAN [6]. This paper addresses a different important problem of using **rcosmo** for spherical-type data in non-HEALPix formats.

rcosmo was initially developed for CMB data that are HEALPix indexed and can be represented as GIS (geographic information system) raster images and modelled as random fields. However, there are other spherical coordinate systems and statistical models. Spherical data are of main interest for geosciences, environmetric and biological studies, but most of researches in these fields are not aware about advantages of the HEALPix data structure or do not have ready-to-use R code to transform their data into HEALPix formats. The aim of this publication is to demonstrate how to deal with three different types of non-CMB/non-HEALPix data. First, we consider continuous geo-referenced observations that are modelled by random fields. The second type of data are discrete spherical data that can be given as realisations of spatial point processes. Finally, the analysis of irregularly star-shaped geometric bodies is presented. Real data examples and illustrations of basic **rcosmo** statistical functions are given for each of the three types.

To reproduce the results of this paper the current version of the package **rcosmo** can be installed from CRAN. A development release is available from GitHub (https://github.com/frycast/rcosmo). A reproducible version of the code and the data used in this paper are available in the folder "esearch materials" from the website https://sites.google.com/site/olenkoandriy/.

2   Coordinate Systems for Spherical Data Representation

The HEALPix representation is the key element for indexing spherical data in **rcosmo**. This section recalls the main spherical coordinate systems and introduces basics of HEALPix. The following sections will demonstrate conversion from different data representations into the HEALPix format.

To index locations of observations the vast majority of spatial applications dealing with spherical data use one of the following coordinate system: Cartesian, geographic, spherical or HEALPix. The Cartesian and spherical coordinate systems are often appear in mathematical description of models. The geographic coordinates are the main indexing tools in GIS, Earth and planetary sciences, while HEALPix has become very popular in recent cosmological research dealing with CMB data.

For simplicity, this section considers the unit sphere with radius 1. Using the *Cartesian coordinate system* of the three pairwise orthogonal basis vectors denoted by $(1,0,0), (0,1,0)$ and $(0,0,1)$, a location on the sphere is specified by a triplet $(x,y,z) = x(1,0,0) + y(0,1,0) + z(0,0,1)$, where $x,y,z \in \mathbb{R}$ and $||(x,y,z)|| = \sqrt{x^2 + y^2 + z^2} = 1$.

The *spherical coordinates* (θ, φ) of a point are obtained from (x, y, z) by inverting the three equations

$$x = \sin(\theta)\cos(\varphi), \quad y = \sin(\theta)\sin(\varphi), \quad z = \cos(\theta),$$

where $\theta \in [0, \pi]$ and $\varphi \in [0, 2\pi)$.

For a point with the spherical coordinates (θ, φ) its geographic coordinates will be written as (θ_G, φ_G). *Geographic coordinates* are obtained from spherical coordinates by setting

$$\varphi_G = \begin{cases} \varphi, & \text{for } \varphi \in [0, \pi], \\ \varphi - 2\pi, & \text{for } \varphi \in (\pi, 2\pi), \end{cases} \quad \text{and} \quad \theta_G = \frac{\pi}{2} - \theta.$$

Thus $\varphi_G \in (-\pi, \pi]$ and $\theta_G \in [-\pi/2, \pi/2]$. When representing Earth's surface in any of the above coordinate systems we align the x-axis with the Earth's Prime Meridian, and have the z-axis pointing north. Commonly φ_G is referred to as longitude and θ_G is referred to as latitude and the both are often measured in degrees instead of radians.

HEALPix is a *Hierarchical Equal Area Isolatitude Pixelation* of the sphere. Detailed derivations of the HEALPix coordinates can be found in [7]. First, the unit sphere is divided into 12 equatorial base pixels. A planar projection of the base pixels is given in Fig. 1. The base pixelation divides the sphere into one equatorial and two polar regions. Referring to the indices shown in Fig. 1, pixels $5, 6, 7$ and 8 are "equatorial"; pixels $1, 2, 3$ and 4 are "north polar"; and pixels $9, 10, 11$ and 12 are "south polar."

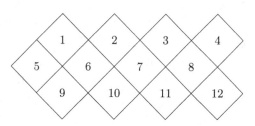

Fig. 1. HEALPix base pixel planar projection as 12 squares.

The base pixelation is defined to have the resolution parameter $j = 0$. For resolution $j = 1$, each base pixel is subdivided into 4 equiareal child pixels. This process is repeated for higher resolutions with each pixel at resolution $j = k$ being one of 4 child pixels from the subdivision of its parent pixel in resolution $j = k - 1$. At any resolution j, the number N_s of pixels per base pixel edge is $N_s = 2^j$ and the total number of pixels is $T = 12N_s^2$.

During this subdivision, pixel boundary and centre locations are chosen in such a way that all pixel centres lie on $4N_s - 1$ rings of constant latitude, making it easy to implement various mathematical methods, in particular Fourier transforms with spherical harmonics. Pixel indices are then assigned to child pixels in one of two ways, known as the "ring" and "nested" *ordering schemes*. In the ring ordering scheme, indices are assigned in the increasing order from east to west along isolatitude rings, and then increasing north to south, as in the example shown in Fig. 2. In the nested ordering scheme the children of base pixel $b \in \{1, 2, \ldots, 12\}$ are labelled with $T/12$ consecutive labels as shown in Fig. 3. This nested ordering scheme allows efficient implementation of local operations such as nearest-neighbour searches.

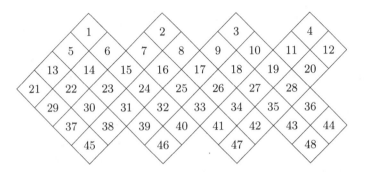

Fig. 2. HEALPix pixelation at resolution $j = 1$ in ring ordering scheme.

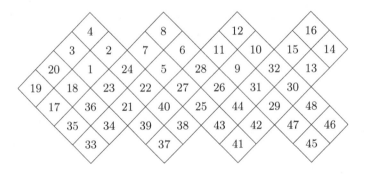

Fig. 3. HEALPix pixelation at resolution $j = 1$ in nested ordering scheme.

3   Continuous Geographic Data

In this section we demonstrate how **rcosmo** can be applied to handle continuous geo-references observations. Such observations are usually collected over dense geographic grids or obtained as results of spatial interpolation or smoothing. Continuous geographic data are common in meteorology, for example, maps with temperature, precipitation, wind direction, or atmospheric pressure values. Other examples of continuous data are land elevations, heights above mean sea level, and ground-level ozone measurements. There are also other numerous environmetrics examples. These data are usually represented and visualised as topographic/contour maps or GIS raster images. In geographic applications it is often assumed that the Earth has a spherical shape with a radius about 6378 km, but with an elevation that departs from this sphere in a very irregular manner. Therefore, most applied methods for the above data are based on spherical statistical models.

Traditionally theoretical models that are used for the continuous type of data are called *random fields* in statistics or *spatially dependent variables* in geostatistics. Below we introduce basic notations and background by reviewing some results about spherical random fields, see more details in [13] and [15].

We will denote a 3d sphere with radius 1 by

$$\mathbb{S}^2 = \left\{ \mathbf{x} \in \mathbb{R}^3 : \|\mathbf{x}\| = 1 \right\}.$$

A spherical random field on a probability space $(\Omega, \mathcal{F}, \mathbf{P})$, denoted by

$$T = \left\{ T(\theta, \varphi) = T_\omega(\theta, \varphi) : 0 \le \theta \le \pi, \quad 0 \le \varphi \le 2\pi, \ \omega \in \Omega \right\},$$

or $\widetilde{T} = \{\widetilde{T}(\mathbf{x}), \mathbf{x} \in \mathbb{S}^2\}$, is a stochastic function defined on the sphere $\mathbb{S}^2$.

The field $\widetilde{T}(\mathbf{x})$ is called isotropic (in the weak sense) on the sphere $\mathbb{S}^2$ if $E\widetilde{T}(\mathbf{x})^2 < \infty$ and its first and second-order moments are invariant with respect to the group $SO(3)$ of rotations in $\mathbb{R}^3$, i.e.

$$\mathbf{E}\widetilde{T}(\mathbf{x}) = \mathbf{E}\widetilde{T}(g\mathbf{x}), \quad \mathbf{E}\widetilde{T}(\mathbf{x})\widetilde{T}(\mathbf{y}) = \mathbf{E}\widetilde{T}(g\mathbf{x})\widetilde{T}(g\mathbf{y}),$$

for every $g \in SO(3)$ and $\mathbf{x}, \mathbf{y} \in \mathbb{S}^2$. This means that the mean $\mathbf{E}T(\theta, \varphi) = constant$ and that the covariance function $\mathbf{E}T(\theta, \varphi)T(\theta', \varphi')$ depends only on the angular distance $\Theta = \Theta_{PQ}$ between the points $P = (\theta, \varphi)$ and $Q = (\theta', \varphi')$ on $\mathbb{S}^2$. A wide class of non-isotropic random field models can be obtained by adding a deterministic component $T_{det}(\theta, \varphi)$ to the field T.

As an example of a spherical random field we use the total column ozone data from the Nimbus-7 polar orbiting satellite, see more details in [3]. This data set provides measurements of the total amount of atmospheric ozone in a given column of $1°$ latitude be $1.25°$ longitude grid. The CSV file available from the website https://hpc.niasra.uow.edu.au/ckan/dataset/tco contains 173405 rows with the measurements recorded on the 1st of October, 1988. We will be using the fields *lon: longitude, lat: latitude* and *ozone: TCO level 2 data*.

First we demonstrate how to use **rcosmo** to transform the geographic referenced ozone data to the HEALPix representation. The R code below loads the ozone data from the file *toms881001.csv* into R. Then geographic coordinates are transformed to spherical ones in radians. Finally, the function *HPDataFrame* creates an **rcosmo** object at the resolution 2048, i.e. on the 50,331,648 nodes grid.

```
> library(rcosmo)
> library(celestial)
> totalozone <- read.csv("toms881001.csv")
```

```
> sph <- geo2sph(data.frame(lon = pi/180*totalozone$lon,  lat =
  pi/180*totalozone$lat))
> df1 <- data.frame(phi = sph$phi, theta = sph$theta,
  I = totalozone$ozone)
> hp <- HPDataFrame(df1, auto.spix = TRUE, delete.duplicates
  = TRUE, nside = 2048)
```

Now we transform the result to an object of the CMBDataFrame class, which is the main class for statistical and geometric analysis in **rcosmo**.

```
cmb <- as.CMBDataFrame(hp)
> str(cmb)
Classes 'CMBDataFrame': 173251 obs. of 3 variables:
 $ I    : num   222 216 202 197 200 ...
 $ theta: num   2.75 2.78 2.81 2.83 2.85 ...
 $ phi  : num   4.35 4.34 4.34 4.33 4.33 ...
 - attr(*, "coords")= chr "spherical"
 - attr(*, "ordering")= chr "nested"
 - attr(*, "nside")= int 2048
 ...
```

To visualise the data we first centre them by subtracting the mean and then rescale to use the **rcosmo** colour scheme.

```
> cmb$I1 <- (cmb$I-mean(cmb$I))/100000
> plot(cmb, intensities = "I1", back.col = "white", size = 10)
```

Now we add the coastline of Australia to the obtained 3d plot. We use the R package *map* to extract longitude and latitude coordinates of the Australian boarder. Then, similarly to the above code we transform the border coordinates to a CMBDataFrame object and plot it on the ozone map.

```
> library(maps)
> library(mapdata)
> aus<-map("worldHires", "Australia", mar=c(0,0,0,0), plot =FALSE)
> aus1 <- data.frame(aus$x,aus$y)
> aus1 <- aus1[complete.cases(aus1),]
> sph1 <- geo2sph(data.frame(lon = pi/180*(aus1[,1]+180),
  lat = pi/180*(aus1[,2])))
> df2 <- data.frame(phi = sph1$phi,theta = sph1$theta, I = 1)
> hp1 <- HPDataFrame(df2, auto.spix = TRUE, delete.duplicates
  = TRUE, nside = 2048)
> cmb1 <- as.CMBDataFrame(hp1)
> plot(cmb1,  size = 10, col = "black", add=TRUE)
```

Setting the country to China

```
> chi <- map("worldHires", "China", mar=c(0,0,0,0), plot =FALSE)
```

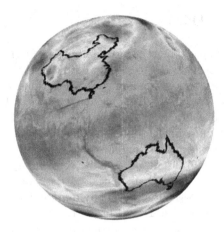

Fig. 4. Total column ozone map with Australia and China boundaries.

and repeating the above commands will add the boundaries of China to the plot. The result is shown in Fig. 4.

Now the data are in the HEALPix format and **rcosmo** functions can be used to analyse them. For example, the following code first computes the sample mean *alpha* of the total column ozone data. Then **rcosmo** commands *exprob* and *extrCMB* estimate the exceedance probability above the level *alpha* and get three largest ozone values and their locations within the spherical window *win1*.

```
> alpha <- mean(cmb[,"I", drop = TRUE])
> alpha
[1] 298.4333
> win1 <- CMBWindow(theta = c(0,pi/2,pi/2), phi = c(0,0,pi/2))
> exprob(cmb, win1, alpha,intensities = "I")
[1] 0.3557902
> extrCMB(cmb, win1, 3, intensities = "I")
A CMBDataFrame
# A tibble: 3 $\times$ 4
     I theta   phi       I1
  <dbl> <dbl> <dbl>    <dbl>
1  179.  3.07  3.32 -0.00119
2  180.  2.96  4.21 -0.00119
3  180.  2.94  4.27 -0.00118
```

To compute the estimated entropy for the ozone measurements within the region *win1* one can use

```
> entropyCMB(cmb, win1)
[1] 2.214391
```

4   Point Pattern Data

In this section we demonstrate **rcosmo** handling of geographic point pattern data. Comparing to continuous geographic data these points are not densely regularly spaced and often have random spatial locations. Some classical examples include studies of settlement distributions, locations of trees, seismological events, data aggregated over a set of zones to specific "central" locations. Spatial point data are also common in geographical epidemiology studies that deal with disease mapping, clustering and finding locations of possible sources. These data are usually represented and visualised as GIS vector images.

In statistical applications random spatial patterns of points are often modelled by spatial point processes. Points usually represent locations of objects and the associated marks are used to record properties of these objects.

A *spatial point process* X is a random countable subset of the sphere $\mathbb{S}^2$. This process is a measurable mapping defined on the probability space $(\Omega, \mathcal{F}, \mathbf{P})$ and taking values in finite/countable sets of points from $\mathbb{S}^2$. For every Borel subset $A \subset \mathbb{S}^2$ the corresponding random variable $N(A)$ denotes the number of points in this subset. For simplicity we restrict our consideration to simple point processes that have realizations with no coincident points.

The distribution of a point process X is defined by the joint distributions of the numbers of points $(N(A_1), ..., N(A_k))$ in the subsets $A_1, ..., A_k \subset \mathbb{S}^2, k \in \mathbb{N}$. A point process on $\mathbb{S}^2$ is isotropic if its distribution is invariant under the group $SO(3)$. The mean measure of a point process X assigns to every subset $A \subset \mathbb{S}^2$ the expected number of points in this subset.

The most popular point processes in applications are Poisson and Cox point processes. More details on the theory and applications of spatial point processes can be found in [1, 4] and the references therein.

As an example we use the Integrated Global Radiosonde Archive (IGRA) measurements. They were collected from radiosondes and pilot balloons at over 2700 stations, [11]. Observations include locations of stations, temperature, pressure, wind direction and speed, etc. For the following analysis we used latitudes and longitudes of stations and their elevations above sea level. The TXT file *igra2-station-list.txt* with IGRA stations data was downloaded from the website https://www1.ncdc.noaa.gov/pub/data/igra/ and saved as a CSV file.

First, the records with missing information were removed. As some missing locations were coded by "−9999" they were also deleted. Then the longitude was recorded using the range $[0, 360]$.

```
> x <- read.csv("igra2-station-list.csv", header=FALSE)
> x1 <- x[,c("V2","V3","V4")]
> x1 <- x1[complete.cases(x1),]
> x1 <- x1[x1$V3>-300,]
> x1$V3 <- x1$V3 + 180
```

Similar to Sect. 3, data were transformed to the CMBDataFrame class with the variable "I" denoting stations' elevation above sea level.

```
> sph <- geo2sph(data.frame(lon = pi/180*x1$V3, lat =
  pi/180*x1$V2))
> df1 <- data.frame(phi = sph$phi, theta = sph$theta, I = x1$V4)
> cmb <- as.CMBDataFrame(hp)
> str(cmb)
Classes 'CMBDataFrame' : 2688 obs. of  3 variables:
 $ I    : num   10 16 4 378 433 ...
 $ theta: num   1.272 1.145 1.13 0.93 0.931 ...
 $ phi  : num   2.06 4.1 4.11 4.31 4.34 ...
 - attr(*, "coords")= chr "spherical"
 - attr(*, "ordering")= chr "nested"
 - attr(*, "nside")= int 2048
...
```

After a rescaling the locations of IGRA stations were plotted with darker colours corresponding to higher elevated stations.

```
> cmb$I1 <- (cmb$I-mean(cmb$I))/1000000
> plot(cmb,intensities = "I1", size = 3, back.col = "white",
  add=TRUE)
```

Similar to Sect. 3 the boundaries of Australia and China were added to reference stations positions, see the resulting Fig. 5. It is obvious that the stations are not uniformly distributed over the globe and much higher elevated in China than in Australia.

Fig. 5. Locations of IGRA stations and Australia and China boundaries.

As the data are in the HEALPix format a few **rcosmo** functions were employed to analyse them. For example, the first Minkowski functional *fmf* can be used to estimate a relative area of HEALPix locations with the elevation above sea level.

```
> fmf(cmb, 0, intensities = "I")/(dim(cmb)[1]*pixelArea(cmb))
[1] 0.9869792
```

The minimum angular geodesic distance between IGRA stations was computed by

```
> minDist(cmb)
[1] 0.00049973
```

The marginal distribution plots in Fig. 6 show that elevation departs from the uniform distribution with respect to geographic coordinates.

```
> plotAngDis(cmb, intensities = "I")
```

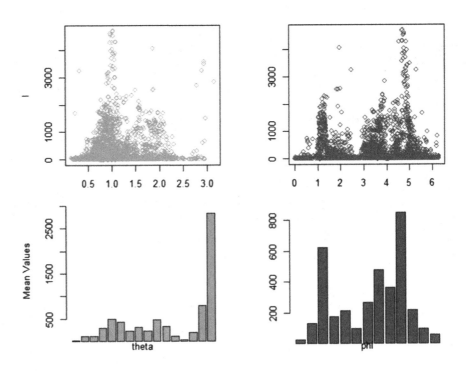

Fig. 6. Distribution of the elevation with respect to spherical angles.

5   Directional Data

This section demonstrates how to use **rcosmo** with directional and shape data. In this publication we restrict our consideration only to star-shaped data. More details on other models and methods in statistical directional and shape analysis can be found in [12] and [14].

In contrast to geographic data in Sects. 3 and 4, directional data are not necessarily located on a sphere, but rather are observed in radial directions from a common centre. However, they are usually indexed by points of the unit sphere. Directional and shape data are common in various fields. For example, in geo-sciences (direction of the Earth's magnetic pole, epicentres of earthquakes, directions of remnant rock magnetism), biology (movement directions of birds and fish, animal orientation), in physics (optical axes of crystals, molecular links, sources of cosmic rays), etc.

The main statistical tools to model and investigate directional data are *circular and spherical distributions and statistics,* see, for example, [12]. Probably the simplest spherical distribution is the uniform one with the constant density $1/(4\pi)$ for all $\mathbf{x} \in \mathbb{S}^2$. This is the only distribution that is invariant under both rotations and reflections. An important directional statistic is the sample mean direction, which is computed as the direction of the sum $\sum_{i=1}^{n} \mathbf{x}_i$ of the observed set of unit vectors $\mathbf{x}_i \in \mathbb{S}^2, i = 1, ..., n$.

Many of directional methods can be translated from spheres to *star-shaped surfaces,* with additional marks representing radial distances to observations. A body U in $\mathbb{R}^3$ is called star-shaped if there is a point $\mathbf{x}_0 \in \mathbb{R}^3$ such that for every $\mathbf{x} \in U$, the segment joining $\mathbf{x}_0$ and $\mathbf{x}$ belongs to U. The corresponding body's surface is also called star-shaped. The marks containing radial distances can be used to statistically investigate and compare various geometric properties of star-shaped data. For example, to estimate the mean directional asymmetry in a solid spherical angle ω one can use the excess from the overall mean distance

$$\frac{\sum_{x_i \in \omega} ||\mathbf{x}_i - \mathbf{x}_0||}{\#i : x_i \in \omega} \Big/ \frac{\sum_{x_i \in U} ||\mathbf{x}_i - \mathbf{x}_0||}{\#i : x_i \in U},$$

where $\#$ denotes a number of cases.

In this section we consider the shape data of the brain substructure amygdala studied in [2]. Structural abnormalities of amygdala are related to functional impairment in autism. The data consist of amygdala MRI measurements of 46 control and autistic persons and contain their group identifiers, age, left and right amygdala surface coordinates. The MATLAB file *chung.2010.NI.mat* available from the website http://pages.stat.wisc.edu/~mchung/research/amygdala/ includes 2562 surface points for each person.

First we load the full data set into R

```
> library(R.matlab)
> mat <- R.matlab::readMat("chung.2010.NI.mat")
```

Then we select two persons 10 and 13 (control and autistic) of the same age 17. Cartesian coordinates of left amygdala sampled points of person 10 were transformed by first centring them and then converting to spherical coordinates. The corresponding 3d plot is shown in the first Fig. 7.

```
> p1 <- data.frame(mat$left.surf[10,,])
> p1 <- apply(p1, 2, function(y) y - mean(y))
> library(rgl)
> plot3d(p1)
> library(sphereplot)
> p1s <- as.data.frame(car2sph(p1, deg = FALSE))
> names(p1s) <- c("theta", "phi", "I")
```

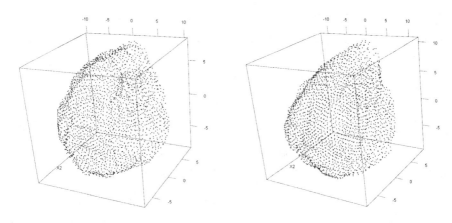

Fig. 7. Sampled points of left amygdala surfaces of persons 10 and 13.

In contrast to Sects. 3 and 4, now we let **rcosmo** to find an *nside* resolution that separates points so that each belongs to a unique pixel. Then we save the data as a CMBDataFrame and create a new variable I1 with rescaled distances $||\mathbf{x}_i - \mathbf{x}_0||$ to use the **rcosmo** colour scheme.

```
> hp1 <- HPDataFrame(p1s, auto.spix = TRUE)
> cmb1 <- as.CMBDataFrame(hp1)
> cmb1
A CMBDataFrame
# A tibble: 2,562 $\times$ 3
I theta      phi
<dbl> <dbl>   <dbl>
1  7.82 0.470 0.0289
2  8.24 0.474 3.05
3  8.87 2.65  0.00577
4  8.15 2.67  3.10
5  8.19 1.62  0.464
...
> pix(cmb1) <- pix(hp1)
> cmb1$I1 <- (cmb1$I-mean(cmb1$I))/1000
> plot(cmb1,intensities = "I1",back.col = "white", size = 3,
  xlab = '', ylab = '', zlab = '')
```

We repeat the same steps for the left amygdala of person 13. The second Fig. 7 shows sampled points of the left amygdala of this person. The spherical plots in Fig. 8 use colours to represent the values of $||\mathbf{x}_i - \mathbf{x}_0||$ in the corresponding directions for each person.

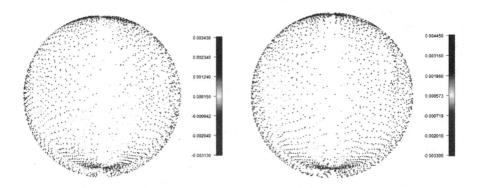

Fig. 8. Heat maps of $||\mathbf{x}_i - \mathbf{x}_0||$ for persons 10 and 13.

To analyse and compare shapes of the amygdalae we first use directional histograms. For example, Fig. 9 shows that directional distributions of sampled points with respect to θ are almost identical. Similar results were obtain for φ directions. Thus, directional sampling rates of amygdalae for persons 10 and 13 are almost identical.

```
> hist(cmb1$theta)
> hist(cmb2$theta)
```

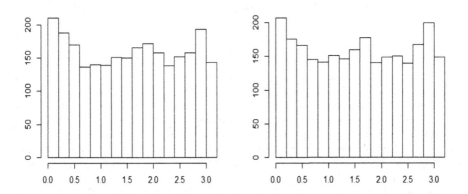

Fig. 9. Distributions of sampled points with respect to θ for persons 10 and 13.

However, basic statistical analysis of the variable I containing values of the sampled radial distances $||\mathbf{x}_i - \mathbf{x}_0||$ shows differences in the shapes of the control and autistic cases:

```
> mean(cmb1$I)
[1] 7.525838
> mean(cmb2$I)
[1] 8.396525
> fmf(cmb1, alpha=mean(cmb1$I))
[1] 0.0003348093
> fmf(cmb2, alpha=mean(cmb1$I))
[1] 0.0004266883
```

Person 13 has larger amygdala. The area where the observed values of $||\mathbf{x}_i - \mathbf{x}_0||$ exceed the mean value $mean(cmb1\$I)$ for person 13 is larger by more than 27% than for person 10.

To confirm that the difference between two subjects is not only in the amygdalae' sizes, but also in their shapes, one can study relative asymmetries. The **rcosmo** command *CMBWindow* was used to select a spherical angle. Then, means of 10 largest values of $||\mathbf{x}_i - \mathbf{x}_0||$ in this angle were computed by the function *extrCMB* and normalised by the overall mean values.

```
> win1 <- CMBWindow(theta = c(pi/2,pi,pi/2), phi = c(0,0,pi/2))
> mean(extrCMB(cmb1, win1, 10)$I)/mean(cmb1$I)
[1] 0.6875167
> mean(extrCMB(cmb2, win1, 10)$I)/mean(cmb2$I)
[1] 0.75863
```

The results demonstrate that not only absolute but also relative asymmetries of amygdalae for the control and autistic persons are different.

Acknowledgements. We would like to thank V.V. Anh, P. Broadbridge, N. Leonenko, M. Li, I. Sloan, and Y. Wang for their discussions of CMB and spherical statistical methods, and J. Ryan for developing and extending the **mmap** package.

References

1. Baddeley, A., Rubak, E., Turner, R.: Spatial Point Patterns. Methodology and Applications with R. Chapman and Hall/CRC, New York (2015)
2. Chung, M.K., Worsley, K.J., Nacewicz, B.M., Dalton, K.M., Davidson, R.J.: General multivariate linear modeling of surface shapes using SurfStat. NeuroImage **53**, 491–505 (2010). https://doi.org/10.1016/j.neuroimage.2010.06.032
3. Cressie, N., Johannesson, G.: Fixed rank kriging for very large spatial data sets. J. Roy. Stat. Soc.: Ser. B (Stat. Methodol.) **70**(1), 209–226 (2008). https://doi.org/10.1111/j.1467-9868.2007.00633.x
4. Diggle, P.J.: Statistical Analysis of Spatial and Spatio-Temporal Point Patterns. Chapman and Hall/CRC, New York (2013)
5. Fryer, D., Olenko, A., Li, M.: rcosmo: R Package for Analysis of Spherical, HEALPix and Cosmological Data (2019, submitted)
6. Fryer, D., Olenko, A., Li, M., Wang, Y.: rcosmo: Cosmic Microwave Background Data Analysis. R package version 1.1.0. (2019). https://CRAN.R-project.org/package=rcosmo

7. Gorski, K.M., et al.: HEALPix: a framework for high-resolution discretization and fast analysis of data distributed on the sphere. Astrophys. J. **622**(2), 759–771 (2005). https://doi.org/10.1086/427976

8. HEALPix: Data Analysis, Simulations and Visualization on the Sphere. https://healpix.sourceforge.io/. Accessed 30 May 2019

9. Healpy documentation homepage. https://healpy.readthedocs.io/. Accessed 30 May 2019

10. HEALPix Library for MATLAB. http://sufoo.c.ooco.jp/program/healpix.html. Accessed 30 May 2019

11. Integrated Global Radiosonde Archive homepage. https://www.ncdc.noaa.gov/data-access/weather-balloon/integrated-global-radiosonde-archive. Accessed 30 May 2019

12. Ley, C., Verdebout, T.: Modern Directional Statistics. CRC Press, Boca Raton (2017)

13. Marinucci, D., Peccati, G.: Random Fields on the Sphere. Representation, Limit Theorems and Cosmological Applications. Cambridge University Press, Cambridge (2011)

14. Srivastava, A., Klassen, E.P.: Functional and Shape Data Analysis. Springer, New York (2016)

15. Yadrenko, M.I.: Spectral Theory of Random Fields. Optimization Software Inc., New York (1983)

On the Parameter Estimation in the Schwartz-Smith's Two-Factor Model

Karol Binkowski[ID], Peilun He$^{(\boxtimes)}$[ID], Nino Kordzakhia[ID], and Pavel Shevchenko[ID]

Macquarie University, North Ryde, NSW 2109, Australia
{karol.binkowski,nino.kordzakhia,pavel.shevchenko}@mq.edu.au,
peilun.he@students.mq.edu.au

Abstract. The two unobservable state variables representing the short and long term factors introduced by Schwartz and Smith in [16] for risk-neutral pricing of futures contracts are modelled as two correlated Ornstein-Uhlenbeck processes. The Kalman Filter (KF) method has been implemented to estimate the "short" and "long" term factors jointly with unknown model parameters. The parameter identification problem arising within the likelihood function in the KF has been addressed by introducing an additional constraint. The obtained model parameter estimates are the Maximum Likelihood Estimators (MLEs) evaluated within the KF. Consistency of the MLEs is studied. The methodology has been tested on simulated data.

Keywords: Kalman Filter · Parameter estimation · Partially observed linear system

1 Introduction

Over more than four decades stochastic processes have been used for modelling of commodity futures prices. In early studies, the commodity prices were modelled using a geometric Brownian motion, [3]. For pricing of commodity derivatives, the mean-reverting processes were used for the first time in [8]; they are also known as Ornstein-Uhlenbeck (O-U) processes, primarily introduced in [14] for modelling of velocity process of Brownian particle under friction.

This work is based on the paper by Schwartz and Smith in [16], where the O-U two-factor model was used for modelling of short and long equilibrium commodity spot price levels. A commodity spot price S_t is modelled as the sum of two unobservable factors χ_t and ξ_t. Both processes χ_t and ξ_t are represented as the mean-reverting processes. In the mean-reverting model, when the commodity price is higher than the equilibrium price level, some new suppliers will enter the market and create downward pressure on the prices. Conversely, when the price is lower than the equilibrium price level, some high-cost suppliers will exit the market and put upward pressure on the prices. In the short term, due to these

H. Nguyen (Ed.): RSSDS 2019, CCIS 1150, pp. 226–237, 2019.
https://doi.org/10.1007/978-981-15-1960-4_16

movements, the price fluctuates temporarily, and it will eventually converge to its equilibrium level over the long term.

Kalman's filtering technique for estimation of the state variables using historical futures prices remains popular for its ability to reproduce realistic commodity futures term structure, see e.g. [1]. In [5], the authors improved the performance of the Kalman Filter by deriving the commodity spot prices from futures prices which have had incorporated an analyst's forecasts of spot prices. In [4], the Kalman Filter is used to study the effect of stochastic volatility and interest rates on the commodity spot prices using the market prices of long-dated futures and options. The Kalman technique was used in [15] for calibration and filtering of partially observable processes using particle Markov chain Monte Carlo approach. The authors of [6] developed the extended Kalman Filter for estimation of the state variables in the two-factor model for the commodity spot price and its yield developed in [17].

Our motivation is driven by the fact that the parameter estimation problem in the linear system using the Kalman Filter cannot be overlooked whilst the estimation of the state variables remains the priority. In the different setup the parameter estimation problem for bivariate O-U process using Kalman Filter has been studied in [7,13].

In this paper, we conduct the simulation study where the Kalman Filter is deployed for estimation of the model parameters jointly with the components of the logarithm of spot price process, which is used in the pricing formula for the futures contracts. In Sect. 2, we provide the notation and analytical formulae used for the formulation of linear partially observable system specific for commodity futures pricing given in [16]. In Sect. 3, we present the details of our implementation of KF algorithm designed for estimation of the parameters of the multi-dimensional partially observable linear system jointly with the estimation of unobserved state variables χ_t and ξ_t. The results of the simulation study are presented in Sect. 4.

2   Two-Factor Model

In this section, we provide a brief description of Schwartz-Smith's model from [16]. We discuss the risk-neutral setup used for futures pricing and present the formulae for futures prices, incorporating the short and long term dynamic factors.

2.1   A Commodity Spot Price Modelling

We model the logarithm of the spot price S_t using the additive model,

$$\log(S_t) = \chi_t + \xi_t,$$

where χ_t and ξ_t are the short and long dynamic factors, respectively. We assume that changes in χ_t are temporary and it converges to 0 in a long term, following an O-U process

$$d\chi_t = -\kappa\chi_t dt + \sigma_\chi dZ_t^\chi,$$

with the mean 0. The changes in the equilibrium level of ξ_t are expected to persist and the process itself is assumed to be mean-reverting

$$d\xi_t = (\mu_\xi - \gamma\xi_t)dt + \sigma_\xi dZ_t^\xi.$$

The processes Z_t^χ and Z_t^ξ are correlated standard Brownian motions with $dZ_t^\chi dZ_t^\xi = \rho_{\chi\xi}dt$, and $\rho = \rho_{\chi\xi}$. Given the initial values χ_0 and ξ_0, χ_t and ξ_t are jointly normally distributed with expected value

$$E[(\chi_t, \xi_t)] = (e^{-\kappa t}\chi_0, \frac{\mu_\xi}{\gamma}(1 - e^{-\gamma t}) + e^{-\gamma t}\xi_0), t \geq 0 \qquad (1)$$

and covariance matrix

$$Cov[(\chi_t, \xi_t)] = \begin{pmatrix} \frac{1-e^{-2\kappa t}}{2\kappa}\sigma_\chi^2 & \frac{1-e^{-(\kappa+\gamma)t}}{\kappa+\gamma}\sigma_\chi\sigma_\xi\rho_{\chi\xi} \\ \frac{1-e^{-(\kappa+\gamma)t}}{\kappa+\gamma}\sigma_\chi\sigma_\xi\rho_{\chi\xi} & \frac{1-e^{-2\gamma t}}{2\gamma}\sigma_\xi^2 \end{pmatrix}. \qquad (2)$$

Derivations of (1) and (2) are given in Appendix A. Therefore, the logarithm of the spot price is normally distributed with mean

$$E[\log(S_t)] = e^{-\kappa t}\chi_0 + \frac{\mu_\xi}{\gamma}(1 - e^{-\gamma t}) + e^{-\gamma t}\xi_0$$

and variance

$$Var[\log(S_t)] = \frac{1 - e^{-2\kappa t}}{2\kappa}\sigma_\chi^2 + \frac{1 - e^{-2\gamma t}}{2\gamma}\sigma_\xi^2 + 2\frac{1 - e^{-(\kappa+\gamma)t}}{\kappa + \gamma}\sigma_\chi\sigma_\xi\rho_{\chi\xi}.$$

Hence S_t, the commodity spot price, is log-normally distributed and

$$E(S_t) = \exp(E[\log(S_t)] + \frac{1}{2}Var[\log(S_t)]),$$

or

$$\log[E(S_t)] = e^{-\kappa t}\chi_0 + \frac{\mu_\xi}{\gamma}(1 - e^{-\gamma t}) + e^{-\gamma t}\xi_0$$

$$+ \frac{1}{2}\left(\frac{1 - e^{-2\kappa t}}{2\kappa}\sigma_\chi^2 + \frac{1 - e^{-2\gamma t}}{2\gamma}\sigma_\xi^2 + 2\frac{1 - e^{-(\kappa+\gamma)t}}{\kappa + \gamma}\sigma_\chi\sigma_\xi\rho_{\chi\xi}\right), \qquad (3)$$

where $\frac{\mu_\xi}{\gamma}$ is the mean parameter, σ_χ and σ_ξ are the volatilities, γ and κ are so-called the speed of mean-reversion parameters of χ and ξ processes, respectively.

2.2   Risk-Neutral Approach to Spot Price Modelling

In this section we introduce two additional parameters which can be interpreted as the market price of commodity spot price risk. The approach stems from the risk-neutral pricing theory for futures, developed in [3]. Hence, the Schwartz-Smith's model with additional parameters can be rewritten as follows

$$d\chi_t = (-\kappa\chi_t - \lambda_\chi)dt + \sigma_\chi dZ_t^{\chi^*},$$

$$d\xi_t = (\mu_\xi - \gamma\xi_t - \lambda_\xi)dt + \sigma_\xi dZ_t^{\xi^*},$$

where the parameters λ_χ and λ_ξ appear as the risk-neutral mean correction terms. Under the risk-neutral measure, χ_t and ξ_t are also jointly normally distributed with mean

$$E^*[(\chi_t, \xi_t)] = (e^{-\kappa t}\chi_0 - \frac{\lambda_\chi}{\kappa}(1 - e^{-\kappa t}), \frac{\mu_\xi - \lambda_\xi}{\gamma}(1 - e^{-\gamma t}) + e^{-\gamma t}\xi_0)$$

and covariance matrix

$$Cov[(\chi_t, \xi_t)]^* = Cov[(\chi_t, \xi_t)].$$

The logarithm of commodity spot price is normally distributed with mean

$$E^*[\log(S_t)] = e^{-\kappa t}\chi_0 - \frac{\lambda_\chi}{\kappa}(1 - e^{-\kappa t}) + \frac{\mu_\xi - \lambda_\xi}{\gamma}(1 - e^{-\gamma t}) + e^{-\gamma t}\xi_0$$

and variance

$$Var^*[\log(S_t)] = Var[\log(S_t)].$$

The spot price is log-normally distributed with

$$\log[E^*(S_t)] = E^*[\log(S_t)] + \frac{1}{2}Var^*[\log(S_t)] = e^{-\kappa t}\chi_0 + e^{-\gamma t}\xi_0 + A(t), \quad (4)$$

where

$$A(t) = -\frac{\lambda_\chi}{\kappa}(1 - e^{-\kappa t}) + \frac{\mu_\xi - \lambda_\xi}{\gamma}(1 - e^{-\gamma t})$$

$$+\frac{1}{2}\left(\frac{1 - e^{-2\kappa t}}{2\kappa}\sigma_\chi^2 + \frac{1 - e^{-2\gamma t}}{2\gamma}\sigma_\xi^2 + 2\frac{1 - e^{-(\kappa+\gamma)t}}{\kappa + \gamma}\sigma_\chi\sigma_\xi\rho_{\chi\xi}\right).$$

In (4) the parameters λ_χ and λ_ξ appear due to the adjustment made in (3).

2.3   Risk-Neutral Approach to Pricing of Futures

A futures contract is defined as an agreement to trade or own an asset in the future, [12]. We are interested to know what is the price of such contract at present. Let $F_{0,T}$ be the current market price of the futures contract with maturity T. For elimination of arbitrage, colloquially known as a "free-lunch", the futures price must be equal to the expected commodity spot price at its delivery time T. Hence, under the risk-neutral measure from Sect. 2.2, assuming zero interest rate, we obtain

$$\log(F_{0,T}) = \log[E^*(S_T)] = e^{-\kappa T}\chi_0 + e^{-\gamma T}\xi_0 + A(T).$$

We denote

$$x_t = \begin{pmatrix} \chi_t \\ \xi_t \end{pmatrix}; \quad c = \begin{pmatrix} 0 \\ \frac{\mu_\xi}{\gamma}(1 - e^{-\gamma\Delta t}) \end{pmatrix}, \quad G = \begin{pmatrix} e^{-\kappa\Delta t} & 0 \\ 0 & e^{-\gamma\Delta t} \end{pmatrix},$$

Δt is the discretization width.

Let w_t be a column-vector of normally distributed random variables,

$$E(w_t) = 0$$

and

$$W = Cov(w_t) = Cov[(\chi_{\Delta t}, \xi_{\Delta t})].$$

In discrete time, we will obtain the following AR(1) dynamics for the bivariate state variable x_t

$$x_t = c + Gx_{t-1} + w_t. \tag{5}$$

The relationship between x_t and the observed futures prices is given by

$$y_t = d_t + F_t' x_t + v_t, \tag{6}$$

where

$$y_t' = (\log(F_{T_1}), \log(F_{T_2}), \ldots, \log(F_{T_n})),$$

$$d_t' = (A(T_1), A(T_2), \ldots, A(T_n)),$$

$$F_t = \begin{pmatrix} e^{-\kappa(T_1-t)}, e^{-\kappa(T_2-t)}, \ldots, e^{-\kappa(T_n-t)} \\ e^{-\gamma(T_1-t)}, e^{-\gamma(T_2-t)}, \ldots, e^{-\gamma(T_n-t)} \end{pmatrix},$$

and v_t is a $n \times 1$ vector of independent, normally distributed random variables $E(v_t) = 0$ and $Cov(v_t) = V$ and $T_1, T_2, ..., T_n$ are the futures maturity times. We assume that V is a diagonal matrix with the vector $v = (s_1^2, s_2^2, \ldots, s_n^2)$ of non-zero elements on the diagonal. The number of the futures contracts is n. Let $\mathcal{F}_t$ be a σ-algebra generated by the futures contracts up to time t and $\theta = (\kappa, \gamma, \mu_\xi, \sigma_\chi, \sigma_\xi, \rho, \lambda_\chi, \lambda_\xi, v)$ be a vector of unknown parameters. The model (5)–(6) is similar to the model discussed in Chapter 3, [10].

The conditional log-likelihood function of $y = (y_1, y_2, \ldots, y_{n_T})$ is

$$l(\theta; y) = \sum_{t=1}^{n_T} p(y_t | \mathcal{F}_{t-1}),$$

where $p(y_t | \mathcal{F}_{t-1})$ is the probability density of y_t given the information available until $t - 1 = t - \Delta$ and n_T is the number of time-points. We assume that the prediction errors $e_t = y_t - E(y_t | \mathcal{F}_{t-1})$ have a multivariate normal distribution, then the log-likelihood function is

$$l(\theta; y) = -\frac{n n_T \log(2\pi)}{2} - \frac{1}{2} \sum_{t=1}^{n_T} [\log(\det(L_{t|t-1})) + e_t' L_{t|t-1}^{-1} e_t] \tag{7}$$

where $L_{t|t-1} = Cov(e_t | \mathcal{F}_{t-1})$. The vector of unknown parameters θ will be estimated by maximising the log-likelihood function from (7). However, the maximisation of $l(\theta; y)$ is inhibited by the parameter identification problem. This fact can be proved analyticaly by mathematical induction since the prediction error e_t and covariance matrix $L_{t|t-1}$ are invariant to label switching of coordinates of x_t.

3   Kalman Filter

In this section, we are using Kalman Filter to estimate the unobservable vector of state variables $x_t = (\chi_t, \xi_t)'$ using simulated y_t. We recall the Eqs. (5) and (6) for x_t and y_t, respectively

$$x_t = c + G x_{t-1} + w_t,$$

$$y_t = d_t + F_t' x_t + v_t.$$

For initialisation of the Kalman Filter we use the expectation and covariance matrix, suggested in [2]

$$a_0 = E(x_0) = \left(0, \frac{\mu_\xi}{\gamma}\right)'$$

and

$$P_0 = Cov(x_0) = \begin{pmatrix} \frac{\sigma_\chi^2}{2\kappa} & \frac{\sigma_\chi \sigma_\xi \rho_{\chi\xi}}{\kappa + \gamma} \\ \frac{\sigma_\chi \sigma_\xi \rho_{\chi\xi}}{\kappa + \gamma} & \frac{\sigma_\xi^2}{2\gamma} \end{pmatrix}.$$

The flowchart for Kalman Filter is given in Fig. 1. The recursive process is constructed by starting at $x_0 \sim N(a_0, P_0)$.

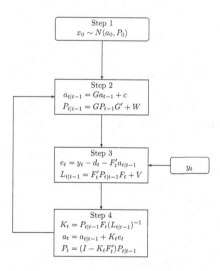

Fig. 1. Flowchart for Kalman Filter.

Next, we will evaluate the conditional expected value $a_{t|t-1}$ and the conditional covariance matrix $P_{t|t-1}$ of the state vector x_t. At Step 3 the new observation y_t is entered and we calculate the prediction error e_t and covariance matrix $L_{t|t-1}$.

Next, we update a_t and P_t through the Kalman's gain matrix K. Finally, we calculate the log-likelihood function at time t

$$l_t = -\log(\det(L_{t|t-1})) - e_t' L_{t|t-1}^{-1} e_t.$$

We complete the recursive process from $t = 1$ to n_T by summing up all l_t's to obtain the log-likelihood function

$$l(\theta; y) = \sum_{t=1}^{n_T} l_t.$$

Then we maximise $l(\theta; y)$ for obtaining the conditional maximum likelihood estimate (MLE) of θ.

The overview of R packages for Kalman Filter is given in [19]. The R packages, which were adapted to this study DSE [9], KFAS [11] and ASTSA [18], were mostly sensitive to the choice of the initial values of the model parameters.

4   Simulation Study

In this section, we present the results of the simulation study conducted for purposes of validating the use of Kalman Filter for estimation of the state vector x_t jointly with the model parameters θ. The simulation study has been programmed as follows.

1. Set $\theta = (\kappa, \gamma, \mu, \sigma_\chi, \sigma_\xi, \rho, v)$ as the vector of true values.
2. Simulate x_t and y_t using the true values of parameters set in θ.
3. Set the intervals for searching of unknown parameters. Locate the grid over the Cartesian product of these intervals.
4. Do grid-search for finding the "best" initial vector θ_0.
5. Maximise the log-likelihood function $l(\theta; y)$ using the "best" initial vector θ_0. For circumventing the parameter identification problem in x vector, we added the constraint $\kappa \geq \gamma$. In Schwartz-Smith's model, the speed of mean-reversion parameter γ of the long term factor ξ_t is naturally dominated by κ, the speed of mean-reversion of the short term factor χ_t.
   Obtain $\hat{\theta}$, the MLE of θ.
6. At $\hat{\theta}$ obtain the estimates of the state variables $\hat{\chi}$ and $\hat{\xi}$.

The grid search for the "best" initial set of the parameters' values allowed to overcome the problem of sensitivity to the initial values.

Further, for simplicity we assume $\lambda_\chi = \lambda_\xi = 0$ and $s_1^2 = s_2^2 = ... = s_n^2 = s^2$.

The model parameter estimates obtained by using the above procedure (1–6) are presented in Table 1.

For some sample sizes n from the range $(500, 8000)$, the "best" initial values are given in Table 2. These initial values were used for obtaining the corresponding optimal model parameter estimates in Table 1.

Table 1. $\hat{\theta}$ for $n \in [500, 8000]$; NLL stands for $-l(\theta; y)$.

n	κ	γ	μ	σ_χ	σ_ξ	ρ	s	NLL
500	1.2775	0.0350	−0.1037	1.3910	0.1811	−0.9517	0.0301	−12805
1000	1.4973	0.9896	−2.0078	1.1990	0.5409	−0.3623	0.0299	−25612
2000	1.5327	0.9834	−1.9932	1.2579	0.4351	−0.3604	0.0297	−51376
4000	1.4895	1.0139	−2.0361	1.3376	0.4736	−0.5880	0.0300	−102549
6000	1.4711	0.9913	−1.9931	1.3526	0.4261	−0.6439	0.0300	−153904
8000	1.4938	0.9960	−2.0008	1.3198	0.3936	−0.6078	0.0300	−205120
True (θ_0)	1.50	1.00	−2.00	1.30	0.30	−0.70	0.03	

Table 2. "Best" initial values for $\hat{\theta}$.

n	κ	γ	μ	σ_χ	σ_ξ	ρ	s
500	2.2525	0.7575	1.7500	1.5025	1.0050	−0.5000	0.5000
1000	1.5050	0.7575	−0.5000	1.0050	1.5025	−0.5000	0.7500
2000	0.7575	0.7575	−2.7500	1.5025	1.5025	0.5000	0.5000
4000	2.2525	0.7575	1.7500	1.0050	1.5025	0.5000	0.2500
6000	1.5050	1.5050	−0.5000	0.5075	1.5025	0.5000	0.2500
8000	2.2525	0.7575	−0.5000	0.5075	1.0050	−0.5000	0.2500

The convergence of the parameter estimates can be seen in Fig. 2, where the estimation errors $\hat{\theta}_i - \theta_i, i = 1, 2, ..., 7$ are plotted versus the sample size n, θ is the vector of true parameter values.

The paths of the estimated state variables $\hat{\chi}$ and $\hat{\xi}$ were obtained through Kalman Filter along with the simulated trajectories of χ and ξ and their 95%-Confidence Intervals (CIs) based on the true values of the model parameters are presented in Fig. 3.

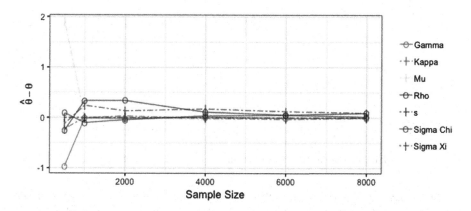

Fig. 2. Componentwise estimation error plots for θ versus n.

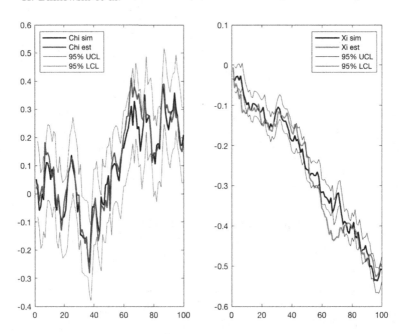

Fig. 3. Estimated $(n = 8000), \hat{\chi}_t, \hat{\xi}_t$ and simulated χ_t, ξ_t paths with their 95% CI.

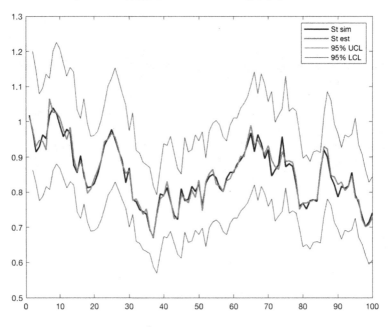

Fig. 4. Estimated $\hat{S}_t$ and simulated S_t with its 95% CI.

The plots of the paths of the estimated $\hat{S}_t = \exp(\hat{\chi}_t + \hat{\xi}_t)$ and the spot prices $S_t = \exp(\chi_t + \xi_t)$ computed using the simulated paths of χ and ξ along with 95% CI based on the true model parameter values θ are presented in Fig. 4.

In the AWS Australian Sydney computing center, for computations in Matlab we used c5.18xlarge instances (CPU 36) taking $10\,h$ for $n = 8000$.

5   Conclusions

In this paper, the parameter estimation problem has been studied in the linear partially observable system, which is specific for commodity futures prices developed in the Schwartz-Smith's two-factor model in the risk-neutral setting. In the simulation study, the Kalman Filter algorithm has been implemented and tested for estimation of the parameters of the bivariate O-U process x jointly with the estimation of its unobserved components χ and ξ. In this study, we suggested the remedy for rectifying the parameter identification problem arising within MLE procedure in the Kalman Filter. The simulation study illustrated the robustness of the grid-search and consistency of the estimates of the model parameters and state vector x.

A   Derivations of (1) and (2)

In Sect. 2.1, we define the components of a bivariate Ornstein-Uhlenbeck process as

$$d\chi_t = -\kappa\chi_t dt + \sigma_\chi dZ_t^\chi \tag{8}$$

$$d\xi_t = (\mu_\xi - \gamma\xi_t)dt + \sigma_\xi dZ_t^\xi, \tag{9}$$

where $dZ_t^\chi, dZ_t^\xi \sim N(0, \sqrt{\Delta t})$ are correlated standard Brownian motions. Here we show how to derive (1) and (2). Firstly, from (8),

$$\Delta\chi_t = -\kappa\chi_t\Delta t + \sigma_\chi\sqrt{\Delta t}\epsilon_\chi.$$

Therefore,

$$\chi_{t+1} = (1 - \kappa\Delta t)\chi_t + \sigma_\chi\sqrt{\Delta t}\epsilon_\chi. \tag{10}$$

Similarly, from (9), we get

$$\xi_{t+1} = (1 - \gamma\Delta t)\xi_t + \mu_\xi\Delta t + \sigma_\xi\sqrt{\Delta t}\epsilon_\xi, \tag{11}$$

where $\epsilon_\chi, \epsilon_\xi \sim N(0, 1)$. Let $Corr(\epsilon_\chi, \epsilon_\xi) = \rho$ and $w = \begin{pmatrix} \sigma_\chi\sqrt{\Delta t}\epsilon_\chi \\ \sigma_\xi\sqrt{\Delta t}\epsilon_\xi \end{pmatrix}$, then

$$W = Var(w) = \begin{pmatrix} \sigma_\chi^2\Delta t & \rho\sigma_\chi\sigma_\xi\Delta t \\ \rho\sigma_\chi\sigma_\xi\Delta t & \sigma_\xi^2\Delta t \end{pmatrix}. \tag{12}$$

Let $X_t = \begin{pmatrix} \chi_t \\ \xi_t \end{pmatrix}$, $c = \begin{pmatrix} 0 \\ \mu_\xi\Delta t \end{pmatrix}$ and $G = \begin{pmatrix} 1 - \kappa\Delta t & 0 \\ 0 & 1 - \gamma\Delta t \end{pmatrix}$. Then from (10) and (11) we get

$$X_{t+1} = c + GX_t + w_{t+1}.$$

Let $\phi = 1 - \kappa\Delta t$, $\psi = 1 - \gamma\Delta t$. Then

$$E(X_t) = \begin{pmatrix} (1 - \kappa\Delta t)\chi_{t-1} \\ (1 - \gamma\Delta t)\xi_{t-1} + \mu_\xi\Delta t \end{pmatrix}$$

$$= \begin{pmatrix} (1 - \kappa\Delta t)^n\chi_0 \\ (1 - \gamma\Delta t)^n\xi_0 + (1 - \gamma\Delta t)^{n-1}\mu_\xi\Delta t + ... + (1 - \gamma\Delta t)^0\mu_\xi\Delta t \end{pmatrix}$$

$$= \begin{pmatrix} \phi^n\chi_0 \\ \psi^n\xi_0 + \mu_\xi\Delta t\frac{1-(1-\gamma\Delta t)^n}{\gamma\Delta t} \end{pmatrix}$$

$$= \begin{pmatrix} \phi^n\chi_0 \\ \psi^n\xi_0 + \frac{\mu_\xi}{\gamma}(1 - \psi^n) \end{pmatrix}, \tag{13}$$

and

$$Var(X_t) = G \cdot Var(X_{t-1}) \cdot G' + W = G^n Var(X_0)(G')^n + G^{n-1}W(G')^{n-1} + ... + G^0 W(G')^0.$$

If we assume $Var(X_0) = 0$, we can get

$$Var(X_t) = G^{n-1}W(G')^{n-1} + ... + G^0W(G')^0$$

$$= \begin{pmatrix} \sigma_\chi^2\Delta t\sum_{i=0}^{n-1}\phi^{2i} & \rho\sigma_\chi\sigma_\xi\Delta t\sum_{i=0}^{n-1}(\phi\psi)^i \\ \rho\sigma_\chi\sigma_\xi\Delta t\sum_{i=0}^{n-1}(\phi\psi)^i & \sigma_\xi^2\Delta t\sum_{i=0}^{n-1}\psi^{2i} \end{pmatrix}$$

$$= \begin{pmatrix} \sigma_\chi^2\Delta t\frac{1-\phi^{2n}}{1-\phi^2} & \rho\sigma_\chi\sigma_\xi\Delta t\frac{1-(\phi\psi)^n}{1-\phi\psi} \\ \rho\sigma_\chi\sigma_\xi\Delta t\frac{1-(\phi\psi)^n}{1-\phi\psi} & \sigma_\xi^2\Delta t\frac{1-\psi^{2n}}{1-\psi^2} \end{pmatrix}. \tag{14}$$

When $n \to \infty$, $\Delta t = t/n \to 0$, $\Delta t^2 = 0$, then

$$\phi^n = (1 - \frac{\kappa t}{n})^n \to e^{-\kappa t},$$

$$\psi^n = (1 - \frac{\gamma t}{n})^n \to e^{-\gamma t},$$

$$(\phi\psi)^n = (1 - (\kappa + \gamma)t/n)^n \to e^{-(\kappa+\gamma)t},$$

$$1 - \phi^2 = 2\kappa\Delta t, 1 - \psi^2 = 2\gamma\Delta t, 1 - \phi\psi = (\kappa + \gamma)\Delta t.$$

From Eqs. (13) and (14), we have

$$E(X_t) = \begin{pmatrix} e^{-\kappa t}\chi_0 \\ e^{-\gamma t}\xi_0 + \frac{\mu_\xi}{\gamma}(1 - e^{-\gamma t}) \end{pmatrix}$$

and

$$Var(X_t) = \begin{pmatrix} \frac{\sigma_\chi^2}{2\kappa}(1 - e^{-2\kappa t}) & \frac{\rho\sigma_\chi\sigma_\xi}{\kappa+\gamma}(1 - e^{-(\kappa+\gamma)t}) \\ \frac{\rho\sigma_\chi\sigma_\xi}{\kappa+\gamma}(1 - e^{-(\kappa+\gamma)t}) & \frac{\sigma_\xi^2}{2\gamma}(1 - e^{-2\gamma t}) \end{pmatrix},$$

in the linearised form $Var(X_{\Delta t}) \approx W$ from (12).

References

1. Ames, M., Bagnarosa, G., Matsui, T., Peters, G., Shevchenko, P.V.: Which risk-factors drive oil futures price curves? Available at SSRN 2840730 (2016)
2. Binkowski, K., Shevchenko, P., Kordzakhia, N.: Modelling of commodity prices. CSIRO technical report (2009)
3. Black, F.: The pricing of commodity contracts. J. Financ. Econ. **3**, 167–179 (1976)
4. Cheng, B., Nikitopoulos, C.S., Schlögl, E.: Pricing of long-dated commodity derivatives: do stochastic interest rates matter? J. Bank. Finance **95**, 148–166 (2018)
5. Cortazar, G., Millard, C., Ortega, H., Schwartz, E.S.: Commodity price forecasts, futures prices and pricing models (2016). http://www.nber.org/papers/w22991.pdf
6. Ewald, C.O., Zhang, A., Zong, Z.: On the calibration of the Schwartz two-factor model to WTI crude oil options and the extended Kalman filter. Ann. Oper. Res. **282**, 1–12 (2018)
7. Favetto, B., Samson, A.: Parameter estimation for a bidimensional partially observed Ornstein-Uhlenbeck process with biological application. Scand. J. Stat. **37**(2), 200–220 (2010)
8. Gibson, R., Schwartz, E.S.: Stochastic convenience yield and the pricing of oil contingent claims. J. Finance **45**, 959–976 (1990)
9. Gilbert, P.: Brief user's guide: dynamic systems estimation (DSE). Available in the file doc/DSE-guide. pdf distributed together with the R bundle DSE (2005). http://cran.r-project.org
10. Harvey, A.C.: Forecasting, Structural Time Series Models and the Kalman Filter. Cambridge University Press, Cambridge (1990)
11. Helske, J.: KFAS: exponential family state space models in R. arXiv preprint arXiv:1612.01907 (2016)
12. Hull, J.C.: Options, Futures, and Other Derivatives, 8th edn. Prentice Hall (2012)
13. Kutoyants, Y.A.: On parameter estimation of the hidden Ornstein-Uhlenbeck process. J. Multivar. Anal. **169**, 248–263 (2019)
14. Ornstein, L.S., Uhlenbeck, G.E.: On the theory of the Brownian motion. Phys. Rev. **36**(5), 823 (1930)
15. Peters, G., Briers, M., Shevchenko, P., Doucet, A.: Calibration and filtering for multi factor commodity models with seasonality: incorporating panel data from futures contracts. Methodol. Comput. Appl. Probab. **15**, 841–874 (2013)
16. Schwartz, E., Smith, J.E.: Short-term variations and long-term dynamics in commodity prices. Manag. Sci. **46**(7), 893–911 (2000)
17. Schwartz, E.S.: The stochastic behavior of commodity prices: implications for valuation and hedging. J. Finance **52**(3), 923–973 (1997)
18. Shumway, R.H., Stoffer, D.S.: Time Series Analysis and Its Applications: With R Examples. Springer, Heidelberg (2017). https://doi.org/10.1007/978-3-319-52452-8
19. Tusell, F.: Kalman filtering in R. J. Stat. Softw. **39**, 1–27 (2011)

Interval Estimators for Inequality Measures Using Grouped Data

Dilanka S. Dedduwakumara$^{(\boxtimes)}$ and Luke A. Prendergast

Department of Mathematics and Statistics, La Trobe University,
Melbourne, Australia
18748354@students.latrobe.edu.au

Abstract. Income inequality measures are often used as an indication of economic health. How to obtain reliable confidence intervals for these measures based on sampled data has been studied extensively in recent years. To preserve confidentiality, income data is often made available in summary form only (i.e. histograms, frequencies between quintiles, etc.). In this paper, we show that good coverage can be achieved for bootstrap and Wald-type intervals for quantile-based measures when only grouped (binned) data are available. These coverages are typically superior to those that we have been able to achieve for intervals for popular measures such as the Gini index in this grouped data setting. To facilitate the bootstrapping, we use the Generalized Lambda Distribution and also a linear interpolation approximation method to approximate the underlying density. The latter is possible when groups means are available. We also apply our methods to real data sets.

Keywords: Histograms · Inequality measures · Bootstrap confidence intervals · Generalized Lambda Distribution

1 Introduction

Income data are generally made available in binned formats by governing bodies to preserve the confidentiality of the individual participants. Obtaining inferences from such summary information has been recently discussed by Deduwakumara and Prendergast (2018), in the context of obtaining confidence intervals for quantiles using estimates of the underlying distribution using grouped data. As we will show in what follows, we can obtain reliable confidence intervals for some inequality measures using bootstrap and Wald-type approaches.

Motivated by these findings, we compare the interval estimators for inequality measures when the data are available in grouped form only. For comparison, we use the well-known Gini, Theil and Atkinson indices and the newly proposed quantile ratio index (Prendergast and Staudte 2018). We begin by introducing these measures before discussing some distribution estimation strategies in Sect. 3. In Sect. 4, we report findings of simulations for interval estimators of the inequality measures. Two real data examples are presented in Sect. 5, followed by a brief discussion in Sect. 6.

H. Nguyen (Ed.): RSSDS 2019, CCIS 1150, pp. 238–252, 2019.
https://doi.org/10.1007/978-981-15-1960-4_17

2   Some Inequality Measures

Let f, F and Q denote the density, distribution and quantile functions respectively for the population of interest. For $p \in [0,1]$, let $x_p = Q(p) = F^{-1}(p)$ denote the p-th quantile. We find it convenient to consider continuous probability distributions to model incomes while acknowledging that, in practice, a population of incomes has a finite number, N, of individuals. Let $x_1, \ldots, x_n$ denote a simple random sample of incomes from the population and let $\widehat{x}_p$ be the estimated p-th quantile.

2.1   Gini Index

Suppose $X \sim F$ where X represents a randomly chosen income from the population and let $\mu = E(X)$ denote mean income. Easily the most commonly used inequality measure is the Gini index (Gini 1914), which measures the deviation of the income distribution from perfect equality. It can be defined as,

$$G = 1 - \frac{1}{\mu} \int_0^\infty [1 - F(x)]^2 \, dx$$

with $G \in [0,1]$. Here, $G = 1$ indicates that one individual holds all wealth (e.g. one individual with income greater than zero) and $G = 0$ represents the equality of incomes for all. The Gini index can be estimated for a simple random sample of size n, with the ordered values of $x_1, \ldots, x_n$ by,

$$\hat{G} = \frac{2 \sum_i i x_i}{n \sum_i x_i} - \frac{n+1}{n}.$$

For more details on the Gini index and estimation see, for example, Dixon *et al.* (1988) and Damgaard and Weiner (2000).

2.2   Theil Index

Based on information theory, Theil (1967) proposed an entropy-based measure which is defined to be

$$T = \int_0^\infty \left(\frac{x}{\mu}\right) \log \left(\frac{x}{\mu}\right) f(x) \, dx$$

where $T \in [0, \infty)$. In practice where a population consists of finite number of N incomes, the upper bound is $\ln(N)$. The Theil index can be estimated by

$$\widehat{T} = \frac{1}{n} \sum_i \frac{x_i}{\bar{x}} \ln \left(\frac{x_i}{\bar{x}}\right)$$

where $\bar{x}$ is the sample mean and where $\widehat{T} \in [0, \ln(n)]$. Further properties of the Theil index can be found in Theil (1967), Allison (1978) and Shorrocks (1980).

2.3   Atkinson Index

The Atkinson index was initially introduced by Atkinson (1970). This measure depends on the sensitivity parameter, ϵ $(0 < \epsilon < \infty)$, which represents the level of inequality aversion. As this parameter increases, more weight is shifted to the distribution at the lower end and vice versa. It is defined as

$$A = 1 - \left[\int_0^\infty \left(\frac{x}{\mu} \right)^{1-\epsilon} f(x) \, dx \right]^{\frac{1}{1-\epsilon}}.$$

where $A \in [0, 1]$.

Atkinson values represent the proportion of total income that would be needed to achieve an equal level of social welfare if incomes were perfectly distributed. Depending on the value of ϵ, the sample estimate is

$$\hat{A} = \begin{cases} 1 - \dfrac{1}{\bar{x}} \left(\dfrac{1}{n} \sum_i x_i^{1-\epsilon} \right)^{\frac{1}{1-\epsilon}}, & \text{for } 0 \leq \epsilon < 1 \\[2ex] 1 - \dfrac{1}{\bar{x}} \left(\prod_i x_i \right)^{\frac{1}{n}}, & \text{for } \epsilon = 1 \end{cases}$$

We use the value of $\epsilon = 0.5$ for our analysis which is the default value used in the package ineq (Zeileis 2014) in R software (R Core Team 2017). More details for the Atkinson index can be found in Atkinson (1970), Biewe and Jenkins (2006) and Shorrocks (1980).

2.4   Quantile Ratio Index

Prendergas and Staude (2018, 2019) introduced the quantile ratio index (QRI) which uses the ratio of symmetric quantiles and which is simpler than similarly defined inequality measures given by Prendergast and Staudte (2016b). The QRI is denoted as

$$I = 1 - \int_0^1 \frac{x_{p/2}}{x_{1-p/2}} \, dp = 1 - \int_0^1 R(p) \, dp$$

where $I \in [0, 1]$. Note that $R(p)$ is the ratio of symmetric quantiles so that I can be seen to be based on the average ratio of incomes chosen symmetrically from the poorer and richer halves of the incomes respectively. For a suitably large J, I is estimated as $J^{-1} \sum_j \left[1 - \widehat{R}(p_j) \right]$ where $p_j = (j - 1/2)/J$ and $\widehat{R}(p_j)$ is the ratio of the estimated $(p_j/2)$-th and $(1 - p_j/2)$-th quantiles. Prendergast and Staudte (2018) show that $J = 100$ is large enough to obtain good estimates of I and so this will be our choice in what follows.

3 Density Estimation Methods

We now consider two methods for estimating the density from grouped data. The first requires bins and frequencies, and the second also requires the bin means. The methods were used by Dedduwakumara and Prendergast (2018) to obtain intervals for quantiles from histograms.

3.1 GLD Estimation Method

Due to flexibility in approximating a wide range of distributions, the Generalized Lambda Distribution (GLD) is commonly used and particularly favoured in fields such as economics and finance. Defined in terms of its quantile function, several parameterizations for the GLD exist. Following is the FKML parameterization for the GLD given by Freimer *et al.* (1988) which is often favoured since it is defined for all parameter choices, with the only restriction being that the scale parameter must be greater than zero. The GLD quantile function is

$$Q(p) = \lambda + \frac{1}{\eta} \left[\frac{(p^\alpha - 1)}{\alpha} - \frac{(1-p)^\beta - 1}{\beta} \right]. \tag{1}$$

The GLD has been used in different contexts to obtain various interval estimators (e.g. Su 2009; Prendergast and Staudte 2016a) when the full data set is available. However, using the percentile matching methods presented by Karian and Dudewicz (1999) and Tarsitano (2005), the GLD parameters can still be estimated when data is in grouped format with frequencies and bins. This method is available in the **bda** package (Wang 2015).

3.2 Linear Interpolation Method

The linear interpolation method was proposed by Lyon *et al.* (2016) as a method of estimating the underlying distribution of binned data when the group (bin) means are also available. Within each bin, a linear density is estimated using the lower and upper bounds of the bin and the associated mean, and the final bin is fitted with an unbounded exponential tail. The slope of the linear density is determined by the mean in relation to the bin midpoint. Closed form solutions for the density and the quantile functions are extensively provided by Lyon *et al.* (2016) and following is a summary of the density results.

Assume there are J intervals in the grouped data bounded by $[a_{j-1}, a_j), j = 1, \ldots, J$ where $a_0 > -\infty$ and $a_J = \infty$. Let the midpoint, mean and relative frequency of the jth bin be denoted by x_j^c, $\bar{x}_j$ and $\widehat{f}_j$. The linear density for the jth bin is

$$h_j(x) = \alpha_j + \beta_j x, \qquad x \in [a_{j-1}, a_j) \tag{2}$$

where the estimates of α_j, β_j are given by,

$$\widehat{\beta}_j = \widehat{f}_j \frac{12(\bar{x}_j - x_j^c)}{(a_j - a_{j-1})^3}, \quad \widehat{\alpha}_j = \frac{\widehat{f}_j}{a_j - a_{j-1}} - \widehat{\beta}_j x_j^c. \tag{3}$$

The density estimate for the final unbounded interval using an exponential tail is provided by,

$$h_J(x) = \frac{\eta}{\lambda} \exp\left\{ -\frac{(x - a_{J-1})}{\lambda} \right\} \tag{4}$$

where $\widehat{\eta} = \widehat{f}_J$ and $\widehat{\lambda} = \bar{x}_J - a_{J-1}$.

4   Interval Estimators Using Grouped Data

In this section, we propose and describe our bootstrap and Wald-type methods to produce intervals for inequality measures using grouped information. The variance of the QRI estimator depends on the underlying income distribution density function applied to income quantiles (Prendergast and Staudte 2018). Therefore, provided we can obtain good estimates of the density from grouped data, then the QRI is well-suited to obtaining Wald-type intervals in this setting. Aside from bootstrapping, to obtain the variance of, for example, the Gini index, it is common to use the jackknife approach or other methods that require the full data set. Consequently, obtaining an approximation to the variances for the Gini, Thiel and Atkinson measure estimators from grouped data is not straightforward and therefore an area for further research.

For the bootstrapping procedure, we obtain the bootstrap samples from the estimated quantile function arising from the estimated GLD or linear interpolation densities. We then use the percentile bootstrap interval described below. While there are other bootstrap methods available that often have improved performance over the percentile method, they require the full data set and it is not immediately clear on how to use them when data is only available in grouped format; e.g. the bootstrap t interval requires the variance of the estimator, the BCa method (Efron 1987) and Efron's ABC method (Diciccio and Efron 1992) requires the full sample data to calculate the acceleration parameter. However, we did try a variation of the bootstrap t interval whereby the α parameter was estimated as usual, but where the estimate and its standard error were also approximated from the bootstrap samples given the lack of the full data set. Coverages were usually no better, and often worse than those for the percentile approach so we do not present them in what follows for brevity. Further variations of bootstrap methods to accommodate the lack of the full data set may result in improved results and this is an area for future research.

Bootstrap Confidence Intervals. In the following algorithm, we describe the estimation of percentile bootstrap confidence intervals in detail.

Step 1: Estimate the GLD and linear interpolation densities using available summary information of bin points and frequencies (and bin means for the linear interpolation approach).

Step 2: Take 500 bootstrap samples of size n using the estimated quantile functions from the two estimation methods using the inverse transform sampling method. That is, randomly generate n numbers, $y_1, \ldots, y_n$ in $[0,1]$ from the uniform distribution and then the ith observation for the jth bootstrap is $y_{ji} = \widehat{Q}(y_i)$ where $\widehat{Q}$ is the estimated quantile function.

Step 3: Construct the percentile bootstrap 95% confidence intervals by taking the 2.5% and 97.5% quantiles of the 500 bootstrapped estimates of the inequality measures.

For the GLD method, we consider the available bin points as the empirical percentiles in the percentile matching method, providing the estimated parameters for the GLD. By using the GLD quantile function (Sect. 3.1) and the estimated parameters, we can easily take the bootstrap samples using the inverse transform sampling method as in Step 2. For the linear interpolation approach, we use the following two quantile functions to generate data depending on the value of p (Lyon *et al.* 2016). For the bounded interval of $[a_{j-1}, a_j)$, the following quantile function is used for $p \in [0,1)$ is,

$$\widehat{x}_p = \frac{-\widehat{\alpha}_j + \sqrt{2\widehat{\beta}_j p + \widehat{C}_j}}{\widehat{\beta}_j} \tag{5}$$

where, $\widehat{C}_j = [\widehat{\alpha}_j^2 - 2\widehat{\beta}_j \widehat{F}_{j-1} + 2\widehat{\beta}_j \widehat{\alpha}_j a_{j-1} + \widehat{\beta}_j^2 (a_{j-1})^2]$, $\widehat{\beta}_j$ and $\widehat{\alpha}_j$ as in (3).

Further the fitted exponential tail yields the following quantile function when the cumulative relative frequency up to final (Jth) interval is denoted by $\widehat{F}_J$,

$$\widehat{x}_p = a_{J-1} - \widehat{\lambda} \ln \left(1 - \frac{p - \widehat{F}_{J-1}}{\widehat{\eta}}\right). \tag{6}$$

Wald-Type Confidence Intervals for the QRI. Obtaining confidence intervals for the QRI from full data sets is studied by Prendergast and Staudte (2018). The variance of the estimator depends on the density function and quantiles. Therefore, given a good estimation of the density which in turn would be expected to give good estimates to quantiles, QRI intervals from grouped data are possible.

The $(1 - \alpha) \times 100$ confidence interval for I is given by $\hat{I} \pm z_{1-\alpha/2}\sqrt{\mathrm{Var}(\hat{I})}$, where $\mathrm{Var}(\hat{I})$ is adopted from Prendergast and Staudte (2018) where we use $J = 100$. Here, $z_{1-\alpha/2}$ is the $1 - \alpha/2$ percentile from the standard normal distribution. $\mathrm{Var}(\hat{I})$ consists of the variances and co-variances terms of ratios of symmetrically chosen quantiles (see Prendergast and Staudte 2018). We then require estimates for population quantiles and density function. As described earlier, first we estimate the underlying density and quantile functions using the GLD and linear interpolation methods. Then those estimated quantile functions can be used to estimate the symmetrically chosen quantiles.

5   Simulations and Examples

We begin by reporting our findings for simulation studies conducted with a variety of distributions before considering real data examples.

5.1   Simulations

To assess coverage, we consider the lognormal distribution with $\mu = 0$ and $\sigma = 1$ and the Singh-Maddala distribution with parameter values $a = 1.6971$, $b = 87.6981$ and $q = 8.3679$ where these parameters were from fitted US family incomes reported by McDonald (1984). We also consider the Dagum distribution with the parameter choices of $a = 4.273$ $b = 14.28$ and $p = 0.36$ which were used in Kleiber (2008) and were estimated from fitted US family incomes in 1969. The χ_2^2, Pareto type II distribution with scale one and shape equal to two and the exponential distribution with rate one were also considered. Table 1 provides the population inequality values of each measure.

Table 1. True values of inequality measures for each distribution.

F	Gini	Theil	Atkinson	I
Lognormal	0.520	0.500	0.221	0.664
Singh-Maddala	0.355	0.206	0.106	0.579
Dagum	0.335	0.191	0.097	0.548
χ_2^2	0.500	0.423	0.215	0.702
Pareto (2)	0.667	1.000	0.383	0.740
Exponential (1)	0.500	0.423	0.215	0.702
Weibull (10)	0.067	0.007	0.004	0.167

From Table 2 for quintile-grouped data and using the linear interpolation method, intervals for I produces coverage probabilities close to the nominal level of 0.95 together with narrow mean width for all settings and with both bootstrap and the Wald-type intervals. Given that the computation of the interval is much more efficient for the Wald-type interval, there does not appear to be an advantage for using the bootstrap. However, for the Gini, Theil and Atkinson

Table 2. Empirical coverage probabilities and average widths (in brackets) of Bootstrapped interval estimates of inequality measures from quintiles estimated using linear interpolation method at nominal level 95%, each based on 1000 replications and 500 bootstrap repetitions.

F	n	Bootstrap				Wald-type
		Gini	Theil	Atkinson	I	I
Lognormal	50	0.788 (0.164)	0.734 (0.327)	0.785 (0.129)	0.947 (0.162)	0.968 (0.163)
	100	0.813 (0.119)	0.761 (0.250)	0.804 (0.097)	0.960 (0.112)	0.965 (0.112)
	250	0.837 (0.075)	0.720 (0.161)	0.813 (0.062)	0.967 (0.069)	0.962 (0.070)
	500	0.840 (0.054)	0.650 (0.115)	0.798 (0.045)	0.955 (0.048)	0.956 (0.049)
Singh-Maddala	50	0.909 (0.128)	0.921 (0.151)	0.911 (0.072)	0.948 (0.165)	0.949 (0.164)
	100	0.925 (0.091)	0.927 (0.108)	0.914 (0.052)	0.933 (0.114)	0.959 (0.116)
	250	0.940 (0.058)	0.948 (0.069)	0.938 (0.034)	0.933 (0.072)	0.947 (0.072)
	500	0.946 (0.041)	0.952 (0.049)	0.946 (0.024)	0.941 (0.050)	0.948 (0.051)
Dagum	50	0.902 (0.128)	0.886 (0.143)	0.869 (0.069)	0.939 (0.169)	0.946 (0.168)
	100	0.914 (0.093)	0.902 (0.105)	0.904 (0.051)	0.952 (0.117)	0.951 (0.118)
	250	0.902 (0.059)	0.878 (0.067)	0.893 (0.033)	0.940 (0.073)	0.948 (0.074)
	500	0.925 (0.042)	0.891 (0.048)	0.918 (0.024)	0.943 (0.052)	0.954 (0.052)
χ^2_2	50	0.930 (0.158)	0.939 (0.285)	0.931 (0.126)	0.954 (0.170)	0.964 (0.170)
	100	0.930 (0.111)	0.933 (0.204)	0.930 (0.090)	0.955 (0.117)	0.952 (0.118)
	250	0.938 (0.071)	0.939 (0.131)	0.939 (0.058)	0.951 (0.072)	0.952 (0.073)
	500	0.948 (0.050)	0.950 (0.093)	0.946 (0.041)	0.945 (0.051)	0.960 (0.051)
Pareto (2)	50	0.633 (0.172)	0.391 (0.490)	0.603 (0.177)	0.968 (0.163)	0.969 (0.162)
	100	0.637 (0.121)	0.351 (0.373)	0.590 (0.131)	0.970 (0.112)	0.971 (0.113)
	250	0.571 (0.077)	0.172 (0.242)	0.484 (0.084)	0.949 (0.069)	0.959 (0.070)
	500	0.500 (0.054)	0.083 (0.173)	0.362 (0.060)	0.973 (0.048)	0.961 (0.049)
Exponential (1)	50	0.916 (0.158)	0.934 (0.288)	0.921 (0.126)	0.939 (0.169)	0.965 (0.170)
	100	0.929 (0.111)	0.938 (0.204)	0.924 (0.090)	0.952 (0.116)	0.966 (0.118)
	250	0.936 (0.071)	0.949 (0.131)	0.935 (0.058)	0.929 (0.072)	0.962 (0.073)
	500	0.943 (0.050)	0.945 (0.093)	0.947 (0.041)	0.961 (0.050)	0.963 (0.051)

measures, the coverages are comparatively weaker but improves as the sample size increases for most of the distributions.

Table 3 shows that the intervals based on the GLD and quintiles for the Gini, Theil and Atkinson measures have poor coverage. Coverages are typically very good for the QRI intervals, albeit more conservative than those using the linear interpolation method. However, coverages become low for the lognormal suggesting that quintiles do not provide enough information to get a good approximation using the GLD.

Table 3. Empirical coverage probabilities and average widths (in brackets) of Boot-strapped interval estimates of inequality measures from quintiles estimated using GLD method at nominal level 95% for, each based on 1000 replications and 500 bootstrap repetitions.

F	n	Bootstrap				Wald-type
		Gini	Theil	Atkinson	I	I
Lognormal	50	0.495 (0.168)	0.406 (0.387)	0.598 (0.150)	0.967 (0.173)	0.974 (0.172)
	100	0.446 (0.117)	0.366 (0.260)	0.510 (0.101)	0.975 (0.126)	0.971 (0.105)
	250	0.373 (0.071)	0.269 (0.141)	0.453 (0.059)	0.899 (0.085)	0.924 (0.065)
	500	0.271 (0.049)	0.165 (0.090)	0.359 (0.039)	0.661 (0.063)	0.713 (0.046)
Singh-Maddala	50	0.862 (0.134)	0.937 (0.151)	0.953 (0.090)	0.979 (0.168)	0.989 (0.180)
	100	0.783 (0.094)	0.920 (0.107)	0.930 (0.063)	0.984 (0.119)	0.973 (0.125)
	250	0.735 (0.060)	0.918 (0.068)	0.911 (0.040)	0.974 (0.075)	0.955 (0.078)
	500	0.646 (0.042)	0.887 (0.048)	0.803 (0.028)	0.965 (0.054)	0.925 (0.056)
Dagum	50	0.844 (0.133)	0.955 (0.140)	0.988 (0.085)	0.990 (0.174)	0.988 (0.192)
	100	0.759 (0.094)	0.909 (0.099)	0.991 (0.060)	0.991 (0.123)	0.981 (0.132)
	250	0.561 (0.060)	0.799 (0.063)	0.982 (0.038)	0.982 (0.079))	0.959 (0.083)
	500	0.299 (0.042)	0.575 (0.045)	0.981 (0.027)	0.967 (0.057)	0.941 (0.059)
χ_2^2	50	0.652 (0.169)	0.544 (0.359)	0.749 (0.158)	0.980 (0.170)	0.989 (0.172)
	100	0.583 (0.121)	0.488 (0.269)	0.663 (0.111)	0.971 (0.117)	0.978 (0.118)
	250	0.605 (0.073)	0.512 (0.147)	0.666 (0.065)	0.970 (0.073)	0.979 (0.073)
	500	0.568 (0.051)	0.467 (0.096)	0.624 (0.044)	0.974 (0.051)	0.969 (0.051)
Pareto (2)	50	0.558 (0.237)	0.508 (1.029)	0.609 (0.289)	0.973 (0.161)	0.989 (0.161)
	100	0.579 (0.197)	0.549 (1.056)	0.607 (0.251)	0.971 (0.111)	0.977 (0.111)
	250	0.626 (0.152)	0.647 (0.982)	0.663 (0.201)	0.968 (0.069)	0.972 (0.069)
	500	0.650 (0.123)	0.697 (0.903)	0.687 (0.169)	0.976 (0.048)	0.977 (0.049)
Exponential (1)	50	0.653 (0.172)	0.559 (0.388)	0.722 (0.163)	0.973 (0.169)	0.980 (0.171)
	100	0.589 (0.119)	0.513 (0.259)	0.667 (0.110)	0.970 (0.117)	0.983 (0.118)
	250	0.578 (0.074)	0.483 (0.151)	0.651 (0.066)	0.982 (0.073)	0.973 (0.073)
	500	0.561 (0.051)	0.470 (0.095)	0.615 (0.044)	0.973 (0.051)	0.969 (0.051)

When the data is summarised in deciles rather than quintiles (i.e. more bins and more information), Table 4 shows improved coverage is achieved with the GLD method. However, coverage is still poor for the Gini, Theil and Atkinson measures when compared to the good coverages achieved for the QRI. Again, the similar coverages for the bootstrap and Wald-type intervals suggest that the Wald-type is a good choice since it is simple and quick to compute.

Table 4. Empirical coverage probabilities and average widths (in brackets) of Bootstrapped interval estimates of inequality measures from deciles estimated using GLD method at nominal level 95% for, each based on 1000 replications and 500 bootstrap repetitions.

F	n	Bootstrap				Wald-type
		Gini	Theil	Atkinson	I	I
Lognormal	50	0.754 (0.262)	0.733 (0.963)	0.762 (0.273)	0.926 (0.156)	0.948 (0.156)
	100	0.789 (0.209)	0.787 (0.892)	0.781 (0.227)	0.943 (0.108)	0.953 (0.109)
	250	0.760 (0.152)	0.761 (0.749)	0.756 (0.173)	0.938 (0.068)	0.943 (0.068)
	500	0.740 (0.113)	0.744 (0.585)	0.730 (0.130)	0.927 (0.048)	0.920 (0.048)
Singh-Maddala	50	0.791 (0.148)	0.760 (0.248)	0.769 (0.103)	0.912 (0.160)	0.958 (0.161)
	100	0.781 (0.102)	0.756 (0.167)	0.747 (0.068)	0.922 (0.111)	0.965 (0.113)
	250	0.786 (0.060)	0.748 (0.083)	0.715 (0.037)	0.941 (0.070)	0.954 (0.071)
	500	0.756 (0.041)	0.706 (0.052)	0.660 (0.025)	0.945 (0.050)	0.955 (0.050)
Dagum	50	0.735 (0.146)	0.631 (0.222)	0.740 (0.101)	0.898 (0.163)	0.937 (0.167)
	100	0.744 (0.099)	0.632 (0.138)	0.733 (0.067)	0.941 (0.115)	0.956 (0.118)
	250	0.709 (0.060)	0.564 (0.074)	0.685 (0.039)	0.957 (0.073)	0.960 (0.074)
	500	0.710 (0.042)	0.499 (0.047)	0.681 (0.027)	0.957 (0.052)	0.949 (0.052)
χ_2^2	50	0.807 (0.202)	0.783 (0.551)	0.845 (0.196)	0.941 (0.165)	0.958 (0.166)
	100	0.775 (0.141)	0.736 (0.392)	0.803 (0.134)	0.954 (0.115)	0.952 (0.116)
	250	0.799 (0.084)	0.763 (0.216)	0.779 (0.077)	0.969 (0.071)	0.959 (0.072)
	500	0.753 (0.057)	0.714 (0.136)	0.742 (0.050)	0.970 (0.050)	0.957 (0.051)
Pareto (2)	50	0.747 (0.283)	0.682 (1.374)	0.775 (0.355)	0.930 (0.159)	0.948 (0.160)
	100	0.787 (0.236)	0.745 (1.414)	0.800 (0.312)	0.945 (0.110)	0.939 (0.111)
	250	0.815 (0.185)	0.817 (1.370)	0.839 (0.258)	0.935 (0.068)	0.911 (0.069)
	500	0.812 (0.149)	0.856 (1.244)	0.845 (0.214)	0.905 (0.048)	0.928 (0.048)
Exponential (1)	50	0.802 (0.200)	0.762 (0.537)	0.822 (0.192)	0.920 (0.165)	0.953 (0.167)
	100	0.830 (0.142)	0.780 (0.395)	0.826 (0.135)	0.943 (0.115)	0.959 (0.116)
	250	0.785 (0.087)	0.743 (0.232)	0.781 (0.080)	0.968 (0.071)	0.957 (0.072)
	500	0.756 (0.057)	0.720 (0.139)	0.748 (0.051)	0.972 (0.050)	0.953 (0.051)

In Fig. 1 we look at what happens to estimates using the linear interpolation method for each measure (e.g. an estimate based on a bootstrap sample) as skew increases. In this case, we use the lognormal distribution while increasing the σ parameter from 0.5 to 2. The estimates are centered according to the true value so a value of zero indicates a perfect estimate. We exclude the Theil index from the analysis since its upper bound is unrestricted. As the distribution becomes more skewed, the Gini and Atkinson estimators have an increase in bias and variability whereas the quantile-based measure (I) indicates smaller variability

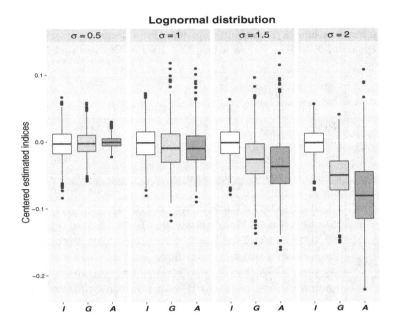

Fig. 1. Boxplots of 1000 centered (with respect to the true values) simulated estimates of inequality measures from quintiles, estimated using linear interpolation method from the Lognormal distribution with mean 0 and various standard deviation values where $n = 250$

and smaller bias throughout for all of the choices of σ. This helps to explain why the coverages are poor for the Gini and Atkinson measures.

6 Applications

6.1 Example 1: Household Income Reported with Group Means

In this example, we present household income data reported with group means by the Survey of Consumer Finances and Expenditures carried out by the Macquarie University and the University of Queensland which can be found in Podder (1972) and Kakwani and Podder (1976). The data is summarised in Table 5.

The confidence intervals produced by 500 bootstrapped samples using the linear interpolation (LI) and GLD methods are given in Table 6. As the final interval is unbounded, we arbitrarily set the upper limit of that bin to $500,000. As can be seen, the confidence intervals and the estimates generated by the two methods are similar.

Table 5. Australian household income data for 1967-68

Income	Number of households	Mean income
Below $1000	310	674.39
$1000–$2000	552	1426.10
$2000–$3000	1007	2545.79
$3000–$4000	1193	3469.35
$4000–$5000	884	4470.33
$5000–$6000	608	5446.60
$6000–$7000	314	6460.93
$7000–$8000	222	7459.14
$8000–$9000	128	8456.66
$9000–$11000	112	9788.38
$11000 and over	110	15617.69

Table 6. Interval and point estimates of the inequality measures generated using the linear interpolation (LI) and GLD methods for the data presented in Table 7.

Method	Bootstrap				Wald-type
	Gini	Theil	Atkinson	I	I
LI	0.319	0.178	0.088	0.509	0.510
	(0.311, 0.327)	(0.168, 0.188)	(0.084, 0.092)	(0.503, 0.517)	(0.502, 0.517)
GLD	0.329	0.177	0.104	0.519	0.521
	(0.321, 0.337)	(0.165, 0.190)	(0.098, 0.109)	(0.512, 0.528)	(0.513, 0.529)

6.2 Example 2: Comparison of Equalized Disposable Household Income Data

In this example, we compare two assumed-independent income distributions reported in deciles from ABS (2011) (see Table 7) to assess whether the income inequality measures of the two distributions are significantly different from one another. It is simple to adapt the previous intervals to the two-sample setting. For example, for the bootstrap approach we simply estimate the difference at each iteration and then form the interval by taking percentiles from the bootstrapped differences. For the Wald-type approach we can get the variance of the difference as a sum of the variances for each estimator of the QRI. For estimation purposes, the highest income has been considered as $5000 for both years.

From Table 8, it can be seen that all intervals for the difference in the measures do not include zero. These intervals then suggest that income inequality has change over the years. We can conclude that inequality of the equalized disposable household income in Western Australia has been significantly increased from 1996-97 to 2009-10.

Table 7. Equalized disposable household income at top of selected percentiles (\$) in Western Australia.

Percentile	1996-97	2009-10
10th	263	347
20th	311	454
30th	364	565
40th	434	663
50th	518	770
60th	586	882
70th	665	1071
80th	778	1296
90th	955	1652

Table 8. Point and interval estimates of inequality measures generated using GLD method for Equalized disposable household income in Western Australia presented in Table 7

Year		Bootstrap				Wald-type
		Gini	Theil	Atkinson	I	I
1996-97	Average Est.	0.262	0.107	0.053	0.488	0.489
	CI	(0.253, 0.271)	(0.099, 0.115)	(0.049, 0.057)	(0.473, 0.503)	(0.483, 0.496)
2009-10	Average Est.	0.326	0.174	0.083	0.538	0.538
	CI	(0.318, 0.334)	(0.163, 0.185)	(0.079, 0.088)	(0.528, 0.548)	(0.531, 0.545)
Difference	Average Est.	0.064	0.067	0.030	0.050	0.049
	CI	(0.051, 0.077)	(0.054, 0.08)	(0.025, 0.037)	(0.032, 0.07)	(0.040, 0.058)

7   Discussion

To preserve confidentiality, it is common for income data to be summarised in grouped format. We therefore considered interval estimators for several measures, including the popular Gini index and a newly proposed quantile-based measure, the QRI. Since grouped data contains bin boundaries and frequencies (and therefore quantile estimates of the data), the QRI is naturally suited to this setting. We showed that bootstrap intervals and a Wald-type interval, both using estimated densities form the grouped data, had typically excellent coverage (i.e. close to nominal). The other measures, however, often had intervals with poor coverage. Further research could include consideration of how to get good approximations to the variances of the Gini, Theil and Atkinson estimators when dealing with grouped data. This was possible for the QRI since the variance of the estimator can be approximated using the estimated density function. For the other measures it is not so straightforward. In summary, when faced with grouped data, if confidence intervals are needed then the QRI is a good option for measuring inequality.

References

ABS: Household income and income distribution, Australian bureau of statistics report 6523.0, 2009–10 (2011)

Allison, P.D.: Measures of inequality. Am. Sociol. Rev. 865–880 (1978)

Atkinson, A.B.: On the measurement of inequality. J. Econ. Theory **2**(3), 244–263 (1970)

Biewen, M., Jenkins, S.P.: Variance estimation for generalized entropy and atkinson inequality indices: the complex survey data case. Oxf. Bull. Econ. Stat. **68**(3), 371–383 (2006)

Damgaard, C., Weiner, J.: Describing inequality in plant size or fecundity. Ecology **81**(4), 1139–1142 (2000)

Dedduwakumara, D.S., Prendergast, L.A.: Confidence intervals for quantiles from histograms and other grouped data. Commun. Stat. Simul. Comput. 1–14 (2018)

Diciccio, T., Efron, B.: More accurate confidence intervals in exponential families. Biometrika **79**(2), 231–245 (1992)

Dixon, P.M., Weiner, J., Mitchell-Olds, T., Woodley, R.: Erratum to 'bootstrapping the gini coefficient of inequality'. Ecology **69**(4), 1307 (1988)

Efron, B.: Better bootstrap confidence intervals. J. Am. Stat. Assoc. **82**(397), 171–185 (1987)

Freimer, M., Kollia, G., Mudholkar, G.S., Lin, C.T.: A study of the generalized tukey lambda family. Commun. Stat. Theory Methods **17**(10), 3547–3567 (1988)

Gini, C.: Sulla misura della concentrazione e della variabilità dei caratteri (1914)

Kakwani, N.C., Podder, N.: Efficient estimation of the Lorenz curve and associated inequality measures from grouped observations. Econometrica: J. Econ. Soc. 137–148 (1976)

Karian, Z.A., Dudewicz, E.J.: Fitting the generalized lambda distribution to data: a method based on percentiles. Commun. Stat. Simul. Comput. **28**(3), 793–819 (1999)

Kleiber, C.: A guide to the Dagum distributions. In: Chotikapanich, D. (ed.) Modeling Income Distributions and Lorenz Curves. Economic Studies in Equality, Social Exclusion and Well-Being, vol. 5, pp. 97–117. Springer, New York (2008). https://doi.org/10.1007/978-0-387-72796-7_6

Lyon, M., Cheung, L.C., Gastwirth, J.L.: The advantages of using group means in estimating the Lorenz curve and Gini index from grouped data. Am. Stat. **70**(1), 25–32 (2016)

McDonald, J.B.: Some generalized functions for the size distribution of income. Econometrica, 647–663 (1984)

Podder, N.: Distribution of household income in Australia. Econ. Rec. **48**(2), 181–200 (1972)

Prendergast, L.A., Staudte, R.G.: Exploiting the quantile optimality ratio in finding confidence intervals for quantiles. STAT **5**(1), 70–81 (2016a)

Prendergast, L.A., Staudte, R.G.: Quantile versions of the Lorenz curve. Electron. J. Stat. **10**(2), 1896–1926 (2016b)

Prendergast, L.A., Staudte, R.G.: A simple and effective inequality measure. Am. Stat. **72**(4), 328–343 (2018)

Prendergast, L.A., Staudte, R.G.: Decomposing the quantile ratio index with applications to australian income and wealth data. Eur. J. Pure Appl. Math. (2019, to appear 8-June-2019)

R Core Team: R: a language and environment for statistical computing. R Foundation for Statistical Computing, Vienna, Austria (2017)

Shorrocks, A.F.: The class of additively decomposable inequality measures. Econometrica, 613–625 (1980)

Su, S.: Confidence intervals for quantiles using generalized lambda distributions. Comput. Stat. Data Anal. **53**(9), 3324–3333 (2009)

Tarsitano, A.: Estimation of the generalized lambda distribution parameters for grouped data. Commun. Stat. Theory Methods **34**(8), 1689–1709 (2005)

Theil, H.: Economics and information theory. Technical report (1967)

Wang, B.: bda: Density estimation for grouped data. R package version 5.1.6 (2015)

Zeileis, A.:. ineq: measuring inequality, concentration, and poverty. R package version 0.2-13 (2014)

Exact Model Averaged Tail Area Confidence Intervals

Paul Kabaila[1] and Rheanna Mainzer[2]([⊠])

[1] Department of Mathematics and Statistics, La Trobe University,
Bundoora, VIC 3086, Australia
`P.Kabaila@latrobe.edu.au`
[2] School of Mathematics and Statistics, The University of Melbourne,
Melbourne, VIC 3010, Australia
`rheanna.mainzer@unimelb.edu.au`

Abstract. Properties of the model averaged tail area (MATA) confidence interval proposed by Turek and Fletcher (CSDA 2012) depend critically on the data-based weights assigned to each tail area equation. By restricting attention to weights based on exponentiating minus AIC/2 and other similar weights it is not possible to find a MATA confidence interval with the desired minimum coverage probability. In the simple scenario that there are two nested normal linear regression models over which we average, a weight function is proposed that results in a MATA interval with correct minimum coverage for many combinations of the known quantity that the coverage depends on. This weight function is shown to outperform current popular choices of weight functions for the MATA interval.

Keywords: Optimized weight · Coverage probability · Expected length

1 Introduction

Model selection can be a difficult process when there are many suitable candidate models for a given data set. Once one has decided on an appropriate model, inference is usually made assuming that this model had been chosen a priori. Thus the uncertainty arising from model selection is not adequately accounted for. In response to this problem, Turek and Fletcher (2012) have proposed the model averaged tail area (MATA) confidence interval. The endpoints of this MATA interval are obtained by solving a weighted average of the tail area equations that define the confidence interval endpoints for each model. Fletcher (2018) illustrates the application of the MATA interval to some ecological data sets. The coverage probability and expected length of this MATA interval depend critically on the data-based weights assigned to each tail area equation. Following Buckland et al. (1997), Turek and Fletcher (2012) consider weights proportional to the exponential of minus AIC/2, $\text{AIC}_c/2$ and BIC/2, where AIC_c is

© Springer Nature Singapore Pte Ltd. 2019
H. Nguyen (Ed.): RSSDS 2019, CCIS 1150, pp. 253–262, 2019.
https://doi.org/10.1007/978-981-15-1960-4_18

AIC corrected for small samples. By restricting attention to weights based on exponentiating minus AIC/2 and other similar weights it is not possible to find a MATA confidence interval with the desired minimum coverage probability.

In this paper we propose a new weight function for the MATA confidence interval for the simple scenario that there are two nested linear regression models with normally distributed random errors over which we average: the full model and a simpler model. The simpler model is obtained from the full model by setting τ, a linear combination of the regression parameters, to 0. The parameter of interest θ is a distinct specified linear combination of the regression parameters. Let $\widehat{\theta}$ and $\widehat{\tau}$ denote the least squares estimators of θ and τ, respectively. Let $\gamma = \tau/(\mathrm{var}(\widehat{\tau}))^{1/2}$ and $\widehat{\gamma} = \widehat{\tau}/(\mathrm{var}(\widehat{\tau}))^{1/2}$. Large values of $\widehat{\gamma}^2$ provide evidence against the simpler model. Thus, any reasonable model weight should be a continuous decreasing function of $\widehat{\gamma}^2$. In this simple scenario, the weight functions based on exponentiating minus AIC/2, $\mathrm{AIC}_c/2$ and BIC/2 are of the form

$$w(\widehat{\gamma}^2; d, n, m) = \frac{1}{1 + (1 + \widehat{\gamma}^2/m)^{n/2} \exp(-d/2)},$$

where n is the dimension of the response vector, p is the dimension of the regression parameter vector, $m = n - p$ and d is equal to 2 for AIC, $2n/(m - 1)$ for AIC_c and $\ln(n)$ for BIC.

Kabaila, Welsh and Mainzer (2016) introduce a new class of weight functions, based on exponentiating criteria related to Mallows' C_P, that have the form

$$w_{\mathrm{M}}(\widehat{\gamma}^2; d, m) = \frac{1}{1 + (1 + \widehat{\gamma}^2/m)^{m/2} \exp(-d/2)}.$$

Note that $w_{\mathrm{M}}(\widehat{\gamma}^2; d, m)$ differs from the weight functions based on exponentiating minus AIC/2, $\mathrm{AIC}_c/2$ and BIC/2 only in the power $m/2$ in the denominator. Kabaila, Welsh and Mainzer (2016) conclude that, in the simple scenario under consideration, the MATA confidence interval with weight function $w_{\mathrm{M}}(\widehat{\gamma}^2; d = 0, m)$ outperforms the MATA confidence interval with weight function based on AIC or BIC in terms of coverage probability and scaled expected length, where the scaling is with respect to the standard interval based on the full model with the same minimum coverage probability as the MATA interval. Therefore, we use the MATA interval with weight function $w_{\mathrm{M}}(\widehat{\gamma}^2; d = 0, m)$ as the standard against which we compare MATA intervals based on new weight functions.

It follows from Kabaila, Welsh and Mainzer (2016) that the weight function $w_{\mathrm{M}}(\widehat{\gamma}^2; d = 0, m)$, and other weight functions in the class of weight functions that depend only on m and $\widehat{\gamma}^2$, result in a MATA interval with coverage probability and scaled expected length that are functions of m, the known correlation $\rho = \mathrm{corr}(\widehat{\theta}, \widehat{\tau})$ and the unknown parameter γ. As illustrated by Fig. 1, the MATA confidence interval obtained using the weight $w_{\mathrm{M}}(\widehat{\gamma}^2; d = 0, m)$ performs poorly for some values of ρ. The main innovation in the present paper is that we optimize the weight function within the class of weight functions that depend only on m and $\widehat{\gamma}^2$ to find a MATA interval with the correct minimum coverage probability (so that it is an 'exact' confidence interval) for all ρ satisfying $0 \leq |\rho| \leq 0.9$.

The paper is structured as follows. The MATA confidence interval for the case of two nested linear regression models is described in detail in Sect. 2. The new optimized weight function is introduced in Sect. 3, and an empirical example to illustrate the properties of the MATA confidence interval is given in Sect. 4. The remainder of the paper describes the computational method used to find the parameters of the new optimized weight function (Sect. 5) and considerations for improving the optimized weight function further (Sect. 6).

2   Description of the MATA Confidence Interval

We may describe the MATA confidence interval generally as follows. Suppose that we have K models, indexed by k. Also suppose that θ is the scalar parameter of interest which has an interpretation that is the same for all of these models. Let $\widehat{\theta}_k$ denote the usual estimator of θ, under the model k. Also let se_k denote the standard error of $\widehat{\theta}_k$, under the model k. Finally let $w_1, \ldots, w_K$ denote data-based weights. Suppose that G_k, the exact or approximate cdf of $(\widehat{\theta}_k - \theta)/\mathrm{se}_k$, is known for $m = 1, \ldots, K$. The MATA confidence interval for θ is obtained by supposing that

$$\sum_{k=1}^{K} w_k\, G_k \left(\frac{\widehat{\theta}_k - \theta}{\mathrm{se}_k} \right) \tag{1}$$

has approximately a $U(0,1)$ distribution. The upper and lower endpoints of the MATA confidence interval with nominal coverage probability $1 - \alpha$ are given by the solutions for θ of the equations $(1) = \alpha/2$ and $(1) = 1 - \alpha/2$, respectively.

In this paper we consider the case that there are two nested linear regression models over which we average. Let Y be a n-vector of responses, X be a known $n \times p$ model matrix with $p < n$ linearly independent columns, β be an unknown parameter p-vector and $\varepsilon \sim N(\mathbf{0}, \sigma^2 I_n)$, where σ^2 is an unknown positive parameter. The full model is given by

$$Y = X\beta + \varepsilon.$$

Let a and c be specified nonzero p-vectors, and let t be a specified number. The reduced model is the full model with $\tau = c^\top \beta - t = 0$. Suppose that the parameter of interest is $\theta = a^\top \beta$. Define $\widehat{\beta}$ to be the least squares estimator of β. Then $\widehat{\theta} = a^\top \widehat{\beta}$ and $\widehat{\tau} = c^\top \widehat{\beta} - t$ are the least squares estimators of θ and τ, respectively. Also let $\widehat{\sigma}^2$ be the usual unbiased estimator of σ^2, i.e. $\widehat{\sigma}^2 = (Y - X\widehat{\beta})^\top (Y - X\widehat{\beta})/(n - p)$. Define $v_\theta = \mathrm{Var}(\widehat{\theta})/\sigma^2 = a^\top (X^\top X)^{-1} a$ and $v_\tau = \mathrm{Var}(\widehat{\tau})/\sigma^2 = c^\top (X^\top X)^{-1} c$. The known correlation between $\widehat{\theta}$ and $\widehat{\tau}$ is $\rho = a^\top (X^\top X)^{-1} c/(v_\theta v_\tau)^{1/2}$. Let $\gamma = \tau/(\sigma v_\tau^{1/2})$ and denote the estimator of γ by $\widehat{\gamma} = \widehat{\tau}/(\widehat{\sigma}\, v_\tau^{1/2})$.

Suppose that $w : [0, \infty) \to [0, 1]$ is a decreasing continuous function, such that $w(z)$ approaches 0 as $z \to \infty$. Let F_ν denote the distribution function of

Student's t-distribution with ν degrees of freedom. For the scenario considered here, (1) becomes

$$w(\widehat{\gamma}^2)\, F_{m+1}\left(\left(\frac{m+1}{\widehat{\gamma}^2+m}\right)^{1/2} \frac{\delta - \rho\widehat{\gamma}}{(1-\rho^2)^{1/2}}\right) + \left(1 - w(\widehat{\gamma}^2)\right) F_m(\delta), \qquad (2)$$

where $\delta = (\widehat{\theta} - \theta)/(\widehat{\sigma} v_\theta^{1/2})$. Define δ_u to be the solution in δ of the equation (2) $= u$. The MATA confidence interval for θ with nominal coverage probability $1 - \alpha$ is given by

$$\left[\widehat{\theta} - v_\theta^{1/2}\, \widehat{\sigma}\, \delta_{1-\alpha/2},\ \widehat{\theta} + v_\theta^{1/2}\, \widehat{\sigma}\, \delta_{\alpha/2}\right].$$

We focus on two properties of the MATA confidence interval to assess its performance: the coverage probability and the scaled expected length, where the scaling is with respect to the standard confidence interval based on the full model, with the same minimum coverage probability as the MATA confidence interval. Exact expressions for the coverage probability and scaled expected length of this MATA confidence interval are given by Theorems 1 and 2 of Kabaila, Welsh and Abeysekera (2016). These exact expressions allow us to evaluate the coverage probability and scaled expected length of the MATA confidence interval without the use of simulations or large sample approximations. Because m and ρ are known, we denote the coverage probability and scaled expected length of the MATA interval by $CP(\gamma)$ and $SEL(\gamma)$, respectively.

3   The New Optimized Weight Function

We consider the new optimized weight function with the form

$$w_O(z) = \exp\left(-\left(b_0 + b_1 z^{1/2} + b_2 z + b_3 z^{3/2}\right)\right),$$

where $z = \widehat{\gamma}^2$. This form of weight function is the result of a wide-ranging search for a form of weight function that, when optimized, leads to a MATA confidence interval with excellent coverage and scaled expected length properties. The parameters b_0, b_1, b_2 and b_3 of $w_O(z)$ are chosen by minimizing $SEL(\gamma = 0)$, subject to constraints on $CP(\gamma)$, $SEL(\gamma)$ and $w_O(z)$. That is, we constrain $CP(\gamma) \geq 1 - \alpha$ for all γ, we constrain $SEL(\gamma)$ to be less than or equal to a specified upper bound, denoted by $SEL_{\max}$, for all γ and to be close to 1 at a specified γ value, and we constrain $w_O(z)$ to be a decreasing continuous function that approaches 0 as $z \to \infty$. A description of the computational method used to find the parameters b_0, b_1, b_2 and b_3 of $w_O(z)$ is given in Sect. 5. This weight function has the advantage that all of it's derivatives are continuous and that it depends on relatively few parameters. Note that the weight function $w_O(\widehat{\gamma}^2)$ depends on m through the optimized values of b_0, b_1, b_2 and b_3. In other words, like $w_M(\widehat{\gamma}^2; d = 0, m)$, this weight function depends only on m and $\widehat{\gamma}^2$.

3.1 Performance of the Optimized Weight Function

We evaluate the performance of the MATA confidence interval with the new optimized weight function $w_O(\hat{\gamma}^2)$ by comparing the coverage probability and scaled expected length of this confidence interval against the coverage probability and scaled expected length of the MATA confidence interval with weight function $w_M(\hat{\gamma}^2; d = 0; m)$. We have chosen to compare $w_O(\hat{\gamma}^2)$ with $w_M(\hat{\gamma}^2; d = 0; m)$ for the following reasons. Kabaila (2018) provides an upper bound on the minimum coverage probability of the MATA confidence interval for weights proportional to $\exp(-\text{BIC}/2)$ and $\exp(-\text{AIC}/2)$. Based on this upper bound, a model weight based on the BIC is not recommended. Kabaila, Welsh and Mainzer (2017) suggest that, for appropriate values of d, properties of the MATA confidence interval for the weight function $w_M(\hat{\gamma}^2; d; m)$ are comparable to properties of the MATA confidence interval for weight functions based on AIC and BIC. These authors also observe that using the weight function $w_M(\hat{\gamma}^2; d; m)$, as opposed to $w(\hat{\gamma}^2; d; n; m)$, reduces the number of known quantities that determine the coverage probability and scaled expected length of the MATA confidence interval from 4 to 3, thereby providing a considerable gain in simplicity of analysis. Note that $w_O(\hat{\gamma}^2)$ and $w_M(\hat{\gamma}^2; d = 0; m)$ belong to the same class of weight functions, i.e. weight functions that depend only on m and $\hat{\gamma}^2$.

For $\rho \in \{0.3, 0.6, 0.9\}$ and $m = 23$, Fig. 1 presents graphs of $CP(\gamma)$, $SEL(\gamma)$ and the weight function of the 95% MATA confidence interval, for both weight functions $w_M(z; d = 0, m)$ (left side of figure) and $w_O(z)$ (right side of figure). For $m = 23$, we obtain the parameters of $w_O(z)$ to be $b_1 = 0.9387$, $b_2 = 0.8942$, $b_3 \approx 0$ and $b_4 \approx 0$. Note that if $SEL(\gamma) < 1$ then the MATA interval is, on average, shorter than the standard interval (based on the full model) with the same minimum coverage probability for this given γ. Figure 1 illustrates the following properties. The optimized weight function $w_O(z)$ takes a lower value when $z = 0$ and initially decays with increasing z much more rapidly than the weight function $w_M(z; d = 0, m)$. The MATA confidence interval with optimized weight function $w_O(z)$ has minimum coverage probability very close to 0.95, whereas the MATA confidence interval with weight function $w_M(z; d = 0, m)$ has minimum coverage probability substantially less than 0.95. The MATA confidence interval with optimized weight function $w_O(z)$ has scaled expected length less than 1 when $\gamma = 0$, $|\max_\gamma SEL(\gamma) - 1|$ not too much larger than $|SEL(\gamma = 0) - 1|$, and scaled expected length that approaches 1 as γ becomes large. By comparison, the MATA confidence interval with weight function $w_M(z; d = 0, m)$ has $|\max_\gamma SEL(\gamma) - 1|$ much larger than $|SEL(\gamma = 0) - 1|$ for $\rho \in \{0.6, 0.9\}$, and does not approach 1 as γ increases for any value of $\rho \in \{0.3, 0.6, 0.9\}$. Therefore, the MATA confidence interval with optimized weight function $w_O(z)$ performs much better than the MATA confidence interval with weight function $w_M(z, d = 0, m)$.

4 Empirical Example

In this section we consider the supervisor performance data set found in Chatterjee and Hadi (p. 58–59, 2012). The response Y is the employee's overall rating of the job being done by the supervisor. The covariates are: X_1: handles employee complaints, X_2: does not allow special privileges, X_3: opportunity to learn new things, X_4: raises based on performance, X_5: too critical of poor performance, X_6: rate of advancing to better jobs. For further information on this data set, we refer the reader to pages 58–59 of Chatterjee and Hadi (2012). For this data set, X has $n = 30$ rows and $p = 7$ columns (6 variables and an intercept). Thus $m = n - p = 23$. We are interested in a confidence interval for the expected value of the response of individual 25. Therefore, $a = (1, 54, 42, 28, 66, 75, 33)$. Let $c = (0, 0, 0, 0, 1, 0, 0)$ and $t = 0$ so that the simpler model is the full model without X_4. The known correlation between $\widehat{\theta}$ and $\widehat{\tau}$ is $\rho = 0.7196$.

Figure 2 presents graphs of $CP(\gamma)$ and $SEL(\gamma)$ where $m = 23$, $\rho = 0.7196$ and $1 - \alpha = 0.95$, for both $w_M(z; d = 0, m)$ (plots on the left hand side of this figure) and $w_O(z)$ (plots on the right hand side of this figure). Similarly to Fig. 1, $\min_\gamma CP(\gamma)$ is less than 0.95 for the weight function $w_M(z; d = 0, m)$ and very close to 0.95 for the weight function $w_O(z)$. Also, $|\min_\gamma SEL(\gamma) - 1|$ is much smaller than $|\max_\gamma SEL(\gamma) - 1|$ for the weight function $w_M(z; d = 0, m)$. However, these quantities are comparable for the weight function $w_O(z)$.

5   Computational Method Used to Find the Parameters of the New Optimized Weight Function

We obtain the parameters b_0, b_1, b_2 and b_3 of $w_O(z)$ in MATLAB using fmincon, which carries out nonlinear constrained optimization. The objective function which is to be minimized is $SEL(\gamma = 0)$.

We use the following algorithm to ensure that $CP(\gamma) \geq 1 - \alpha$ for all γ, $SEL(\gamma) \leq SEL_{\max}$ for all γ and $w_O(z)$ is a decreasing continuous function that approaches 0 as $z \to \infty$. We specify the following constraints on $w_O(z)$ and the parameters of $w_O(z)$: $b_0 \geq 0$, if $b_3 = 0$ then $b_1 \geq 0$ and $b_2 \geq 0$, otherwise $b_2^2 \leq b_1 b_3$ and $b_3 \leq 0$. We also initialize the algorithm as follows. We specify the sets $\Gamma_{CP} = \{0, 1.6, 1.8, 2, 4, 6\}$ and $\Gamma_{SEL} = \{2, 3, 4, 5, 6\}$. These sets were chosen after an initial exploration of the behaviour of the functions $CP(\gamma)$ and $SEL(\gamma)$.

1. Run fmincon to compute the parameters that describe the weight function such that $SEL(\gamma = 0)$ is minimized, subject to the constraints on w_O described above and the constraints that (a) $CP(\gamma) \geq 1 - \alpha$ for all $\gamma \in \Gamma_{CP}$, (b) $SEL(\gamma) \leq SEL_{\max}$ for all $\gamma \in \Gamma_{SEL}$ and (c) $SEL(10) \leq 1.005$.
2. Compute the coverage probability, scaled expected length and weight function for an evenly spaced fine grid of values between 0 and $\gamma_{\max} = 10$. If all constraints are satisfied then we have found the parameters of w_O and we stop. If, on the other hand, any of the constraints are not satisfied then go to the next step.
3. If $CP(\gamma) < 1 - \alpha$ for any $\gamma \in [0, \gamma_{\max}]$ then add $\widetilde{\gamma}$ to the set Γ_{CP}, where $\widetilde{\gamma}$ is the value of γ that minimizes $CP(\gamma)$. If $SEL(\gamma) > SEL_{\max}$ for any $\gamma \in [0, \gamma_{\max}]$ then add $\overline{\gamma}$ to the set Γ_{SEL}, where $\overline{\gamma}$ is the value of γ that maximizes $SEL(\gamma)$. Go to step 1.

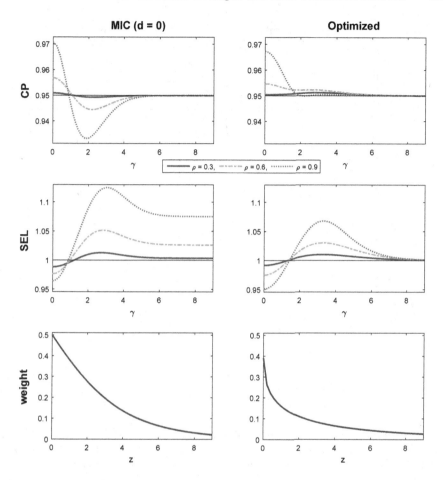

Fig. 1. Graphs of the coverage probability, scaled expected length and weight function of the MATA interval, for weight functions $w_M(z; d = 0, m)$ (left side of figure) and $w_O(z)$ (right side of figure), for $\rho \in \{0.3, 0.6, 0.9\}$ and $m = 23$.

Although the known correlation ρ does not appear in the expression for $w_O(z)$, the optimization method described above (to choose the parameters b_0, b_1, b_2 and b_3) results in this weight function implicitly depending on ρ. Interestingly, we found that if the coverage probability satisfies the constraints on $CP(\gamma)$ and $SEL(\gamma)$ for $\rho = \widetilde{\rho}$, where $\widetilde{\rho} \in [0, 1)$, then it also satisfies these constraints for all ρ satisfying $|\rho| \leq |\widetilde{\rho}|$. By obtaining the parameters of $w_O(z)$ for $\rho = 0.9$ we obtain a new class of optimized weight functions that, for $0 \leq |\rho| \leq 0.9$, results in a MATA confidence interval with correct minimum coverage probability, maximum and minimum scaled expected lengths which are comparable distances from 1 and scaled expected length approximately 1 when the data provide evidence against the simpler model.

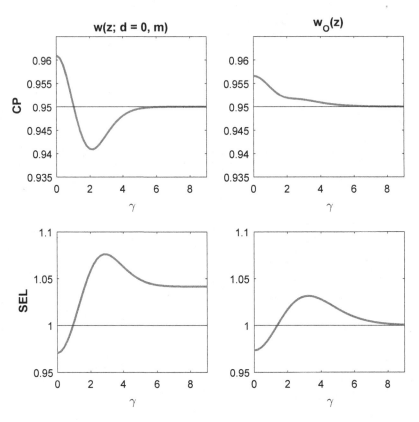

Fig. 2. Graphs of the coverage probability and scaled expected length of the MATA interval with weight functions $w(z; d = 0, m)$ (left side of figure) and the new optimzed weight function $w_O(z)$ (right side of figure), where $\rho = 0.71964$ and $m = 23$.

6 Can We Do Better if We Optimize the Weight Function for both m and ρ?

In this section we explore how much better the MATA confidence interval can perform when the weight function is allowed to also depend on the value of ρ. We consider again the supervisor performance data set described in Sect. 4. For a fair comparison, we specify that the maximum scaled expected length of the MATA interval with weight function that has been optimized for both ρ and m and the maximum scaled expected length of the MATA interval with weight function $w_O(z)$ should be the same. Thus, we set $SEL_{\max} = 1.05$ in the algorithm described in Sect. 4. The improvement can then be seen by comparing the minimum scaled expected lengths of the MATA interval for the two different optimized weight functions. Figure 3 compares the performance of $w_O(z)$ (plots on the left hand side of this figure) with the weight function which has been optimized for both m and ρ (plots on the right hand side of this figure). The maximum and minimum scaled expected lengths of the MATA interval with weight function $w_O(z)$ are 1.0316 and 0.9734, respectively. The maximum and minimum scaled expected length of the MATA interval with weight function optimized over ρ and γ are 1.0500 and 0.9459, respectively. Figure 3 illustrates that, for this example, there is not a great deal to gain by optimizing the weight function over both ρ and m.

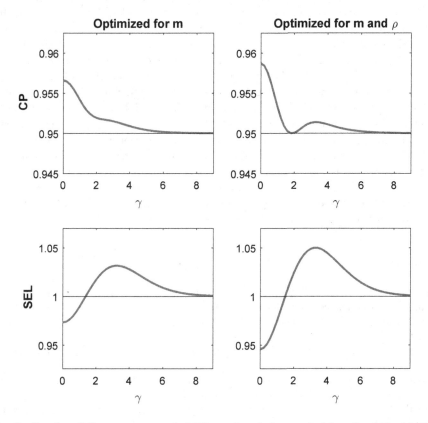

Fig. 3. Graphs of the coverage probability and scaled expected length of the MATA confidence interval with weight function $w_O(z)$ (left side of figure) and weight function which has been optimized over both m and ρ (right side of figure), where $\rho = 0.71964$ and $m = 23$.

7   Conclusion

Properties of the MATA confidence interval depend critically on the data-based weights assigned to each tail area equation. In this paper we consider the case where there are two nested normal linear regression models over which we average. We have introduced a new class of optimized weight functions that, for

$0 \leq |\rho| \leq 0.9$, result in a MATA confidence interval with correct minimum coverage probability, maximum and minimum scaled expected lengths which are comparable distances from 1 and scaled expected length approximately 1 when the data strongly contradicts the simpler model. This weight function is shown to perform much better, in terms of coverage probability and scaled expected length of the resulting MATA interval, than the best weight function found by Kabaila, Welsh and Mainzer (2017). This improvement in performance easily outweighs the increased time required for the computation of the MATA interval using the new optimized weight function.

In the simple scenario of two nested linear regression models, Kabaila, Welsh and Mainzer (2017) proved that the MATA confidence interval with weight functions of the kind considered in this paper belong to a subclass of the class of confidence intervals defined by Kabaila and Giri (2009). Since the MATA confidence interval is not optimal in the class of confidence intervals given by Kabaila and Giri (2009), it is not the best confidence interval to use in this scenario. However, the MATA confidence interval has the attractive property that it is easily computed in more complicated scenarios, such as when there are more than two nested linear regression models. We hope that the guidance given here on how to choose the weight function for the MATA confidence interval in the simple scenario of two nested linear regression models can aid the choice of the weight function for the MATA confidence interval in more complicated scenarios.

References

Buckland, S.T., Burnham, K.P., Augustin, N.H.: Model selection: an integral part of inference. Biometrics **53**, 603–618 (1997)

Chatterjee, S., Hadi, A.S.: Regression Analysis by Example, 5th edn. Wiley, Hoboken (2012)

Fletcher, D.: Model Averaging. SS. Springer, Heidelberg (2018). https://doi.org/10.1007/978-3-662-58541-2

Kabaila, P.: On the minimum coverage probability of model averaged tail area confidence intervals. Can. J. Stat. **46**, 279–297 (2018)

Kabaila, P., Welsh, A.H., Abeysekera, W.: Model-averaged confidence intervals. Scand. J. Stat. **43**, 35–48 (2016)

Kabaila, P., Giri, K.: Confidence intervals in regression utilizing prior information. J. Stat. Plan. Inference **139**, 3419–3429 (2009)

Kabaila, P., Welsh, A.H., Mainzer, R.: The performance of model averaged tail area confidence intervals. Commun. Stat.-Theory Methods **46**, 10718–10732 (2017)

Turek, D., Fletcher, D.: Model-averaged Wald intervals. Comput. Stat. Data Anal. **56**, 2809–2815 (2012)

Author Index

Printed in the United States
By Bookmasters